Transactions on Computer Systems and Networks

Transactions on Computer Systems and Networks is a unique series that aims to capture advances in evolution of computer hardware and software systems and progress in computer networks. Computing Systems in present world span from miniature IoT nodes and embedded computing systems to large-scale cloud infrastructures, which necessitates developing systems architecture, storage infrastructure and process management to work at various scales. Present day networking technologies provide pervasive global coverage on a scale and enable multitude of transformative technologies. The new landscape of computing comprises of self-aware autonomous systems, which are built upon a software-hardware collaborative framework. These systems are designed to execute critical and non-critical tasks involving a variety of processing resources like multi-core CPUs, reconfigurable hardware, GPUs and TPUs which are managed through virtualisation, real-time process management and fault-tolerance. While AI, Machine Learning and Deep Learning tasks are predominantly increasing in the application space the computing system research aim towards efficient means of data processing, memory management, real-time task scheduling, scalable, secured and energy aware computing. The paradigm of computer networks also extends it support to this evolving application scenario through various advanced protocols, architectures and services. This series aims to present leading works on advances in theory, design, behaviour and applications in computing systems and networks. The Series accepts research monographs, introductory and advanced textbooks, professional books, reference works, and select conference proceedings.

Dinesh K. Sharma · H. S. Hota ·
Aaron Rasheed Rababaah
Editors

Machine Learning for Real World Applications

Springer

Editors
Dinesh K. Sharma
Department of Business, Management and Accounting
University of Maryland Eastern Shore
Princess Anne, MD, USA

H. S. Hota
Department of Computer Science and Application
Atal Bihari Vajpayee Vishwavidyalaya
Bilaspur, Chhattisgarh, India

Aaron Rasheed Rababaah
College of Engineering and Applied Sciences
American University of Kuwait
Kuwait, Kuwait

ISSN 2730-7484 ISSN 2730-7492 (electronic)
Transactions on Computer Systems and Networks
ISBN 978-981-97-1899-3 ISBN 978-981-97-1900-6 (eBook)
https://doi.org/10.1007/978-981-97-1900-6

This Springer imprint is published by the registered company Springer Nature Singapore Pte Ltd.
The registered company address is: 152 Beach Road, #21-01/04 Gateway East, Singapore 189721, Singapore

Dedicated to
Our Beloved Mothers
Dinesh K. Sharma

H. S. Hota
Aaron Rasheed Rababaah

Foreword

It is both a privilege and a pleasure to introduce *Machine Learning for Real-World Applications*, a compelling edited book that delves into the fascinating world of machine learning and its transformative potential across diverse domains. In an age defined by rapid technological advancements, this collection of chapters showcases the remarkable impact of machine learning in solving real-world challenges and shaping the future of multiple sectors.

The book's foundation recognizes that machine learning, a significant field within artificial intelligence, has become a cornerstone technology influencing current and future trends in industry, healthcare, finance, security, and much more. Its growing prominence is evident in the attention it has received from the research and scientific communities, especially in the realm of deep learning, where significant advances have pushed the boundaries of intelligent models to new heights. This comprehensive volume covers an impressive array of topics, providing readers with a thorough journey through fundamental and advanced machine learning concepts. The chapters are thoughtfully crafted, addressing subjects within the book's scope from the ground up. This ensures that readers, whether they are undergraduate and postgraduate students, researchers, or industry professionals, will find a reliable source of theoretical and technical learning content.

One of the book's unique strengths is its emphasis on practical implementation. While the theoretical aspects are essential in building a solid foundation, it is in the application of machine learning models to real-world scenarios that their true potential is realized. Each chapter presents practical use cases, offering readers valuable insights into how these models can be implemented effectively to address complex challenges in healthcare, finance, agriculture, security, and beyond. The authors have demonstrated exceptional expertise in their respective fields, presenting cutting-edge research and innovative applications of machine learning algorithms. These applications carry significant weight, serving as rich experimental work supporting the presented methods' validity. Moreover, the detailed guidelines for implementing these models in similar and different scenarios ensure that readers can translate theoretical knowledge into practical solutions.

I commend the diversity of topics covered in this book. The chapters span various applications, from predicting financial trends and optimizing agricultural practices to detecting Parkinson's disease and enhancing network security. This diversity highlights the versatility of machine learning and its potential to revolutionize various industries and domains. Furthermore, the book is open to addressing challenges and opportunities that lie ahead for researchers in the field of machine learning. By shedding light on these areas, the authors provide valuable insights for future research directions, guiding those seeking to contribute to the continuous advancement of machine learning technologies.

This book is a valuable resource that bridges the gap between theory and practice in machine learning. It is a must-read for anyone interested in staying at the forefront of this dynamic field. I extend my heartfelt appreciation to the authors for their invaluable contributions, which have enriched the collective knowledge of the machine learning community. This edited book will inspire and motivate researchers, practitioners, and students alike to further explore the frontiers of machine learning. The book's applications in real-world scenarios offer a glimpse into the limitless possibilities of artificial intelligence and machine learning, encouraging us to imagine a future where these technologies will continue to shape and enhance our lives in remarkable ways.

Dr. Dong Wang
Professor
Fayetteville State University
Fayetteville, NC, USA

Preface

Machine learning (ML) is a major field in artificial intelligence (AI). ML is recognized as one of the key technologies that is shaping current and future trends in many sectors, including industry, government, healthcare, education, transportation, internet of things (IoT), smart cities, agriculture, machine vision, business, security, Natural Language Processing (NLP), finance, and many more. In recent years, ML, especially deep learning (DL), has gained significant advances and attention from the research and scientific communities that have radically changed the way we understand how intelligent models work and pushed the envelope to much higher limits. DL is the most suitable for image data processing. We recognize that these advances have influenced researchers to conduct exciting work exploiting these new concepts in ML, especially in deep learning, and to implement them in real-world applications.

This book presents a comprehensive coverage of ML, from fundamental to advanced concepts. The presented material addresses subjects within the book's scope from the ground up, giving the reader a reliable source of theoretical and technical learning content. The book emphasizes not only the theoretical aspects but also the practical and implementation aspects of all presented models in real-world applications. These applications carry significant weight as a means of providing rich experimental work that supports the validity of the presented methods and detailed guidelines on how to implement such models in similar and different scenarios and contexts. Furthermore, the book's chapters shed light on challenges and opportunities that researchers may utilize to steer their future research work. The book will be helpful to undergraduate and postgraduate students, researchers, and industry personal.

Rapid technological advancements have brought about significant changes in various aspects of our lives. From how we communicate and interact with others to how we navigate our daily routines, technology has become an integral part of our existence. In recent years, artificial intelligence (AI) and machine learning (ML) have garnered immense attention and revolutionized numerous industries.

ML has emerged as a powerful tool in data analysis, enabling us to extract valuable insights and make informed decisions. With advancements in technology and the availability of vast amounts of data, ML algorithms have proven instrumental in solving complex real-world problems across various domains. This edited book, "Advances in Machine Learning for Real-World Applications," brings together chapters that showcase machine learning techniques' application in diverse areas, ranging from healthcare and finance to cyber security and environmental management. This book consists of 15 chapters as follows:

Chapter 1 proposes an advanced ML approach to investigate the predictability of Metaverse Coins. A residual-driven, two-stage random forest model is created to achieve this goal.

Chapter 2 suggests an AI-based approach to combating driver fatigue and drowsiness and reducing the risk of fatigue-related traffic crashes.

Chapter 3 emphasizes the effective integration of clustering and classification or regression ML algorithms. By combining these techniques, this chapter explores how enhanced models can be developed to address real-world problems more effectively.

Chapter 4 delves into the design and development of a ML-based depression identification decision support system. In the complex and nuanced field of mental health, the identification and understanding of depression pose significant challenges. The cornerstone of this endeavor is the utilization of state-of-the-art machine learning (ML) techniques, incorporating advanced methodologies such as feature selection and extraction.

Chapter 5 focuses on ML algorithmic models for pairs trading in financial markets. This chapter demonstrates how ML can be leveraged to make informed investment decisions and improve trading strategies.

Chapter 6 evaluates ML techniques for optimized fertilizer recommendation. By leveraging advanced algorithms, this chapter showcases how ML can assist in improving agricultural practices and enhancing crop yield.

Chapter 7 explores the application of neural network-based techniques to network intrusion detection. By utilizing ML algorithms, this chapter showcases how network security can be enhanced through the identification and prevention of intrusions.

Chapter 8 introduces a smart machine learning algorithm designed to enhance the safety of elderly individuals by detecting falls in daily life. Focused on creating a fall-detection mobile app for paralyzed or leg-injured individuals, PFAST uses real-world training data to optimize system parameters. Utilizing sensors in Android devices, the app addresses the common issue of sudden falls, providing a comprehensive solution to improve the well-being of the elderly. The chapter covers the app's acronym, functions, necessity, potential issues, algorithm choice, sensor utilization, and user interaction through UML diagrams.

Chapter 9 focuses on the detection of Parkinson's disease using speech and handwriting analysis. In this scenario, ML algorithms are utilized to detect and diagnose Parkinson's disease, illustrating the value of advanced analysis approaches in healthcare.

Chapter 10 presents a CCTV AI-based system with custom object detection and video upscaling model. The chapter discusses the development of an intelligent surveillance system that utilizes ML algorithms for object detection and video enhancement.

Chapter 11 focuses on forecasting the end of COVID-19 in India using ML and population density clustering. By leveraging ML techniques, this chapter explores how accurate predictions can aid in public health planning and decision-making.

Chapter 12 examines the impact of the COVID-19 lockdown on internet addiction among children and adolescents. Using ML analysis, the authors explore the patterns and risk factors associated with internet addiction during this challenging period.

Chapter 13 discusses the classification of building architectural styles using convolutional neural network (CNN) models. This chapter highlights how ML can be applied to architectural analysis, providing insights into building styles and design trends.

Chapter 14 explores the synergistic integration of a genetic algorithm (GA) with an artificial neural network (ANN) for optimizing critical parameters of ANN in medical diagnosis. Specifically, the focus is on the automated determination of the optimal number of hidden layers, learning rate, and momentum for training the error back-propagation algorithm (EBPA). In the dynamic landscape of medical diagnosis, the intersection of artificial intelligence and computational optimization techniques has paved the way for innovative solutions to enhance diagnostic accuracy and efficiency.

Chapter 15 considers three major sources of energy used in the transportation sector of the USA. They are natural gas, petroleum, and fossil fuels. In this chapter, authors perform a predictive modeling exercise based on monthly data spanning over half a century (January 1973–August 2023). The authors use extreme gradient boosting, an ensemble ML approach, for this purpose. The results indicate that out of the sources, energy used from fossil fuels is the most predictable, followed by petroleum and natural gas. The findings of this study will be useful for practitioners and investors in the field of energy used in the transportation section of the USA.

In this edited book, the chapters present cutting-edge research and practical applications of machine learning algorithms in various domains. The authors have made significant contributions to the field, and their work serves as a valuable resource for researchers, practitioners, and students interested in the application of ML in real-world scenarios.

This edited book offers a comprehensive perspective on the applications of AI and ML, demonstrating their transformative potential in diverse domains. It is our hope that readers will find this collection of chapters both informative and inspiring, and that it will spark further research and innovation in the field. We extend our gratitude to the authors for their valuable contributions and hope that readers will find this book a valuable resource in exploring the frontier of AI and ML applications.

Princess Anne, MD, USA — Dinesh K. Sharma
Bilaspur, India — H. S. Hota
Kuwait City, Kuwait — Aaron Rasheed Rababaah

Acknowledgements

This book is a collection of the works of eminent researchers who have contributed to various chapters of the book. We are thankful to all the authors for their valuable contributions. We would like to express our gratitude to all reviewers for their help and professional support. We greatly appreciate the cordial support of Prof. Amlan Chakrabarti, Series Editor, to publish in the book series "Transactions on Computer Systems and Networks." We humbly acknowledge our gratitude to Professor Hari Sharma, Associate Dean and Chair, School of Business, Virginia State University, USA, for his unconditional support, valuable suggestions, and continuous encouragement.

We would like to thank the Department of Business, Management and Accounting at the University of Maryland Eastern Shore, Princess Anne, Maryland, USA; the Department of Computer Science and Application at Atal Bihari Vajpayee University, Bilaspur, India; and the College of Engineering and Applied Sciences at the American University of Kuwait, Kuwait. We would like to thank the team at Springer, for their help and support in publishing this book.

About This Book

This book is a comprehensive exploration of advanced machine learning (ML) techniques applied in real-world scenarios. Its primary objective is to offer valuable insights into the theoretical, practical, and implementation aspects of models within practical applications. Readers, including practitioners, educators, and academics, will find significant value in the wide array of covered domains. By delving into diverse applications such as predictive models, smart vehicles, pattern classification, decision support systems, business intelligence, smart agriculture systems, security applications, defense, and healthcare, this book presents comprehensive experimental work, establishing the validity of the proposed approaches. Moreover, it provides clear instructions on applying these models across comparable and distinct scenarios and contexts, making it a practical and useful resource. Beyond its applicative focus, the book also serves as a guide for researchers, presenting both challenges and opportunities to inspire future research efforts. Each chapter elucidates unique problems and solutions, enabling researchers to steer their investigations towards cutting-edge breakthroughs. Targeting senior undergraduate, graduate, and postgraduate students, academic researchers, and industry personnel, the book caters to a broad audience, spanning disciplines such as business, computer, industrial, and architectural engineering, computer science, information systems, and mathematics. Its inclusive scope makes it a valuable resource for various educational and professional settings, where readers can deepen their understanding of ML applications and drive innovation in their respective fields.

In conclusion, this book stands as a definitive reference for exploring advanced ML techniques and their successful integration into real-world applications. Its multidisciplinary approach, extensive domain coverage, and insightful guidance make it a valuable addition to the libraries of both academia and industry.

Contents

Editors and Contributors

About the Editors

Dinesh K. Sharma is a Professor of Quantitative Methods and Computer Applications in the Department of Business, Management, and Accounting at the University of Maryland Eastern Shore, USA. Professor Sharma has over 30 years of teaching experience, has served on numerous committees to supervise Ph.D. students, and is also an external Ph.D. thesis examiner for several universities in India. He has published over 260 refereed journal articles and conference proceedings, three edited books, with three more on the way, as well as 17 best paper awards. Professor Sharma is a member of Decision Sciences (USA), a life member of the Operational Research Society of India, and served as a program chair and coordinator of several international conferences in many countries. Dr. Sharma's research interests include mathematical programming, artificial intelligence, machine learning, inventory and supply chain management, healthcare management, portfolio management, and data analytics.

H. S. Hota is a Professor of Computer Science in the Department of Computer Science and Application and Dean of the Faculty of Science at Atal Bihari Vajpayee University in Bilaspur, India. His research interests include Artificial Intelligence and Machine Learning, Business Intelligence, Information Security, including Cloud Security, Data Mining, and Evolutionary Computing. He has published over 75 refereed journal articles in reputable journals and more than 60 articles in conference proceedings. He has also won five best paper awards at international conferences. He has also been to several countries and spoken at international conferences as a keynote speaker. He has three patents related to the application of machine learning in different domains to his credit, and he has published one book. Three edited books are forthcoming from internationally reputed publishers. He is also a member of the editorial boards of five international journals as well as a reviewer for many reputed journals. He is a member of many academic bodies, including the Computer Society of India and the Indian Science Congress.

Aaron Rasheed Rababaah is a Professor, College of Engineering and Applied Sciences at the American University of Kuwait (January 2016–Present). He held the following academic positions in the USA: Associate Professor and Assistant Professor at the University of Maryland Eastern Shore, Post-Doc Researcher and Instructor at Tennessee State University and Adjunct Instructor at Indiana University of South Bend. He finished a B.Sc. in Industrial Engineering from the University of Jordan, Jordan, a M.Sc. in Applied Computer Science from the University of Indiana, USA and a Ph.D. in Computer Information and Systems Engineering from Tennessee State University, USA. He has 5 years of experience in IE and 12 years in teaching at the college level. He has a research interest in Machine Intelligence and Computing Education He has published four books, 90 refereed journal and conference papers and 50 professional presentations. He was awarded several awards in education, research, and industry. He has supervised, advised, and referred senior projects, master theses, doctoral dissertations, and number of journals. He has been funded by USA Department of Defense, US Army.

Contributors

Talal Al-Failakawi American University of Kuwait, Salmiya, Kuwait

Shereef Abu Al-Maati American University of Kuwait, Salmiya, Kuwait

Hisham Al-Ruwaih American University of Kuwait, Salmiya, Kuwait

Yousef Al-Tabtabaei American University of Kuwait, Salmiya, Kuwait

Ali Alfeeli College of Engineering and Applied Sciences, American University of Kuwait, Salmiya, Kuwait

Ali Alghanemi College of Engineering and Applied Sciences, American University of Kuwait, Salmiya, Kuwait

Ali Alsarraf College of Engineering and Applied Sciences, American University of Kuwait, Salmiya, Kuwait

Aqeel Alsarraf College of Engineering and Applied Sciences, American University of Kuwait, Salmiya, Kuwait

Lubna Amin American University of Kuwait, Salmiya, Kuwait

Bidushi Chakraborty Institute of Engineering and Management, Kolkata, WB, India

Deepak Gupta Department of Computer Science and Engineering, Motilal Nehru National Institute of Technology Allahabad, Prayagraj, Uttar Pradesh, India

Yousef Hamadah American University of Kuwait, Salmiya, Kuwait

Barenya Bikash Hazarika Faculty of Computer Technology, Assam down town University, Panikhaiti, Guwahati, Assam, India

H. S. Hota Atal Bihari Vajpayee Vishwavidyalaya, Bilaspur, Chhattisgarh, India

Rishabh K. Hota BMS College of Engineering, Bengaluru, India

Rabin K. Jana Indian Institute of Management Raipur, Raipur, CG, India

Ruchir Kumar Kadwey Students of BMS College of Engineering, Bangalore, India

Bonomali Khuntia Department of Computer Science and Engineering, Centurion University, Bhubaneswar, India;
Department of Computer Science, Berhampur University, Berhampur, India

Yousef Landoulsi American University of Kuwait, Salmiya, Kuwait

B. Mokgweetsi Department of Statistics, University of Botswana, Gaborone, Botswana

Daniel Okunbor Fayetteville State University, Fayetteville, NC, USA

Ashalata Panigrahi Department of Computer Science & Engineering, NIST University, Institute Park, Pallur Hills, Berhampur, India

Manas Ranjan Patra Department of Computer Science and Engineering, Centurion University, Bhubaneswar, India;
Department of Computer Science & Engineering, NIST University, Berhampur, India

Aaron Rasheed Rababaah American University of Kuwait, Salmiya, Kuwait

Sediakgotla Department of Statistics, University of Botswana, Gaborone, Botswana

Anamika Shukla Sharma Govt. E. R. Rao PG Science College, Bilaspur, Chhattisgarh, India

Dinesh K. Sharma University of Maryland Eastern Shore, Princess Anne, USA

M. R. Shrisha BMS College of Engineering, Bangalore, India

Manish Kumar Shukla Department of Computer Science, Indira Gandhi National Tribal University Amarkantak (M.P), Amarkantak, India

Karam Ratan Singh Department of Basic and Applied Science, National Institute of Technology Arunachal Pradesh, Jote, India

Sauransh Singh Students of BMS College of Engineering, Bangalore, India

Shikhar Singh Students of BMS College of Engineering, Bangalore, India

Vikash Ku. Singh Department of Computer Science, Indira Gandhi National Tribal University Amarkantak (M.P), Amarkantak, India

R. Sivasamy Department of Statistics, University of Botswana, Gaborone, Botswana

Ritu Raj Sondhiya Department of Computer Science, Indira Gandhi National Tribal University Amarkantak (M.P), Amarkantak, India

Shrey Srivatsava Students of BMS College of Engineering, Bangalore, India

Khalid Sultan College of Engineering and Applied Sciences, American University of Kuwait, Salmiya, Kuwait

Satyananda Swain Department of Computer Science and Engineering, Centurion University, Bhubaneswar, India

Nilesh Verma University of Waikato, Hamilton, New Zealand

Abbreviation

ADF	Augmented Dickey-Fuller
AI	Artificial Intelligence
AMCNN	Attention-based Multiscale Convolutional Neural Network
ANN	Artificial Neural Network
ATR	Automatic Target Recognition
BEM	Boundary Element Method
CAD	Computer Aided Design
CCTV	Closed-Circuit Television
CDSS	Clinical Decision Support System
CIFAR	Canadian Institute for Advanced Research
CLIQUE	Clustering Quest
CNN	Convolutional Neural Network
CS	Compressed Sensing
CSPDarknet	Cross-Stage Partial Network Darknet
CUDA	Compute Unified Device Architecture
CV	Cross-Validation
DBSCAN	Density-Based Spatial Clustering of Applications with Noise
DENCLUE	Density Clustering
DGCN	Dynamical Graph Convolutional Networks
DIDSS	Depression Identification Decision Support System
DL	Deep Learning
DNN	Deep Neural Network
DT	Decision Tree
EBPA	Error Back Propagation Algorithm
ECM	Error Correction Model
EM	Electromagnetic Modelling
FAR	False Alarm Rate
FDTD	Finite Difference Time Domain
FFMPEG	Fast Forward Moving Picture Experts Group
FFN	Feed Forward Networks
FL	Fuzzy Logic

FOM	Frequency of Mouth
FRS	Functional Requirements Specification
FWM	Finite Element Method
GA	Genetic Algorithms
GAN	Generative Adversarial Network
GM	Geometric Mean
GMM	Gaussian Mixture Model
GPP	General-Purpose Possessor
GUI	Graphical User Interface
HR	High Resolution
HRP	High Reflectivity Points
IA	Index of Agreement
IAT	Internet Addiction Test
ID	Identifier
IDE	Integrated Development Environment
IDS	Intrusion Detection Systems
IMDB	Internet Movie Database
KKT	Karush-Kuhn-Tucker
KNN	K-Nearest Neighbor
LiDAR	Light Detection and Ranging
LR	Low Resolution
LSTM	Long Short-Term Memory
LVQ	Learning Vector Quantization
MCA	Model Classification Accuracies
MCS	Model Confidence Set
ML	Machine Learning
MLP	Multi-Layered Perceptron
MoM	Method of Moments
MPEG	Moving Picture Experts Group
MRI	Magnetic Resonance Imaging
NCCL	NVIDIA Collective Communications Library
NN	Neural Network
NP	Non-deterministic Polynomial-time
NSE	Nash-Sutcliffe Efficiency
OLS	Ordinary Least Squares
OPTICS	Ordering Points to Identify Clustering Structure
PANet	Path Aggregation Network
PET	Positron Emission Tomography
RBFN	Radial Basis Function Network
R-CNN	Region-Based Convolutional Neural Network
RF	Random Forest
RM	Reflectivity Map
RNN	Recurrent Neural Network
RT	Ray-Tracing
RTX	Ray Tracing Texel eXtreme

SAR	Synthetic Aperture Radar
SMO	Sequential Minimal Optimization
SOM	Self-Organizing Map
SPECT	Single Poton Emission Computed Tomography
SSRE	Salient Subspace Reflectivity Extraction
STING	Statistical Information Grid
SVM	Support Vector Machines
TLLH	Two-Level Learning Hierarchy
TOI	Target Objects of Interest
UAV	Unmanned Arial Vehicle
VESM	Virtual Environment Simulation Model
VGG	Visual Geometry Group
VRAM	Video Random Access Memory
WHO	World Health Organization
XGBoost	Extreme Gradient Boosting
YAML	YAML Ain't Markup Language
YOLO5	You Only Look Once Version 5

Chapter 1
Predictability of Metaverse Coins Using an Advanced Machine Learning Approach

Rabin K. Jana

Abstract Metaverse coins are a type of crypto asset used for making transactions on the Metaverse blockchain. Most of these cryptocurrencies have come into existence in the last 2–3 years and garnered investors' attention. This chapter examines the predictability of the top four Metaverse coins based on their market capitalization—Decentraland, The Sandbox, Stacks, and Theta Network. A residual-driven two-stage random forest model is proposed for this purpose. It is an extension of the traditional random forest model. The residual of the first stage is used as an explanatory feature for the second stage of forecasting. The feature importance study confirms the capability of residues to influence the future movements of the considered Metaverse coins. The predictive modeling exercise indicates that the future movements of all four coins can be estimated with high precision based on the values of four performance measures for the test, training, and full datasets. This study's findings will help investors select appropriate investment channels for diversification and risk mitigation.

Keywords Metaverse coin · Predictive modeling · Ensemble machine learning · Boruta

1.1 Introduction

Metaverse is a virtual platform built with the help of Web3 and blockchain technologies. It is one of the most fascinating internet revolutions of recent times. In this digital space, people can play games, virtually interact with each other, create and trade assets, forge social connections through their virtual "avatars", and do several other things (Barrera and Shah 2023). The currencies used for all such purposes are known as Metaverse tokens or Metaverse coins (Vidal-Tomás 2022). They are a specific type of cryptocurrency (Tiwari et al. 2018; Jana et al. 2019). Like the Metaverse platform,

R. K. Jana (✉)
Indian Institute of Management Raipur, Atal Nagar, Raipur, CG 493661, India
e-mail: rkjana@iimraipur.ac.in

D. K. Sharma et al. (eds.), *Machine Learning for Real World Applications*, Transactions on Computer Systems and Networks, https://doi.org/10.1007/978-981-97-1900-6_1

these digital currencies are expanding exponentially. The coin market is forecasted to exceed $800 billion by 2030 (Grand View Research 2030). Many of the present technological giants developed sophisticated hardware and software infrastructure by investing a few billion dollars to facilitate the inception, development, and growth of Metaverse coins (Q.ai 2022).

The literature on Metaverse is in a nascent stage. The possible transformation that the Metaverse could bring into the workplace has been explored (Purdy 2022). The author argued that the emergence of the new digital platform would be much different from the presently existing hybrid working majorly contributed by the COVID-19 pandemic. Another study examined the contributions of Metaverse in redesigning the architecture of society (Schumacher 2022). Authors have explored the prospects of Metaverse coins (Akkus et al. 2022). It was argued that there is a huge potential for these crypto assets as prospective investment opportunities for investors.

Like any asset, cryptocurrency price movements are impacted by different socio-economic factors, geopolitical factors, and several other events (Jana and Ghosh 2023). The COVID-19 pandemic, a recent black swan event, contributed immensely to creating a turmoil market situation (Jana and Das 2020; Jana et al. 2022a). Metaverse coins are no exception. These recently created crypto assets also encountered similar treatments. The study of temporal patterns and inherent dynamics is crucial for any asset (Ghosh et al. 2019, 2021). The other essential study is predictive modeling (Jana et al. 2020, 2022b; Ghosh and Jana 2023). Information obtained from these two types of studies greatly assists investors in choosing alternative investment avenues, especially during volatile market conditions. The literature reports only a few studies in this direction for Metaverse coins. One such study investigated the movements of a set of Metaverse coins (Vidal-Tomás 2022). It was reported that Metaverse coins do not co-move with conventional cryptocurrencies. The corresponding correlation with the market is also not significant.

The contributions of this study can be summarized as follows: The present research is the first to study the predictability of Metaverse coins and void a significant gap in the literature. For this purpose, we consider four major Metaverse coins and compare their predictability based on available data. We also study their temporal patterns. The second contribution is in terms of the methodology. We propose an ensemble machine learning approach and suggest improvements. Random forest (RF) was selected as the base algorithm (Breiman 2001). We obtained the initial forecasts by using the RF algorithm. Then, we calculated residuals for each observation. The residuals at any stage of the forecasting contain unexplained components. We incorporated residuals of the first stage forecasts by introducing a new corresponding feature and used them in the second stage forecasting. The suggested change in the baseline RF model helped immensely in obtaining future values with supreme accuracy for the considered Metaverse coins. From this point of view, the proposed methodology contributes to the predictive modeling literature.

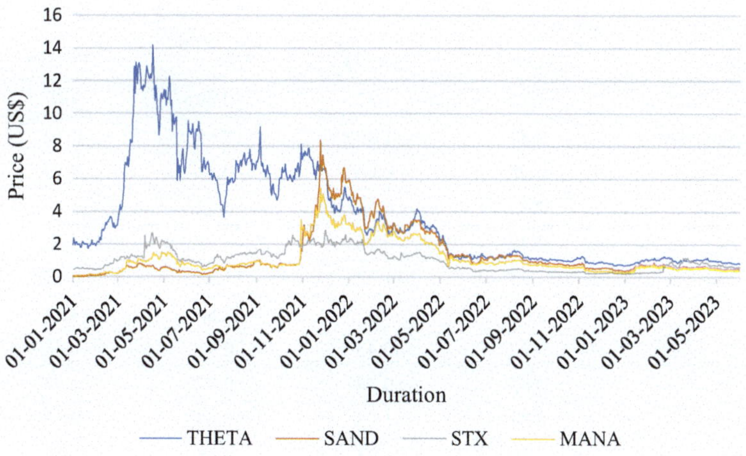

Fig. 1.1 Temporal patterns of THETA, SAND, STX, and MANA

1.2 Data

The data duration for this study is from January 01, 2021 to May 31, 2023. We collected daily observations of four Metaverse coins—Decentraland (MANA), The Sandbox (SAND), Stacks (STX), and Theta Network (THETA).[1] The temporal patterns of these time series are shown in Fig. 1.1. The visual inspection of temporal patterns indicates highly nonlinear behavior in all the series. We also find the descriptive statistics values to obtain more information about these time series and present them in Table 1.1.

The Shapiro–Wilk test statistic values (Table 1.1) indicate that all four time series are non-normal. The NN test of Teraesvirta confirms that all the series are highly nonlinear. The temporal patterns and the descriptive statistics support each other regarding the complicated nature of the past price movements of considered Metaverse coins.

1.3 Proposed Ensemble Predictive Model

The proposed predictive model consists of a residual-driven, two-stage RF model. Boruta feature selection algorithm is also a part of the methodology. The importance of features used for the proposed RF model is obtained using the Boruta algorithm. Both RF and Boruta are ensemble models. We briefly present the research methodology here.

[1] Data source: https://www.investing.com/.

Table 1.1 Descriptive statistics

Measure	MANA	SAND	STX	THETA
Mean	1.17	1.37	0.99	3.67
Standard deviation	0.98	1.49	0.64	3.12
Min	0.08	0.04	0.21	0.72
Q1	0.6	0.55	0.42	1.1
Median	0.78	0.75	0.88	2.68
Q3	1.23	1.32	1.39	6.07
Max	5.48	8.4	2.88	14.21
MAD	0.37	0.43	0.71	2.46
IQR	0.63	0.78	0.97	4.97
CV	0.84	1.08	0.64	0.85
Skewness	1.59	1.91	0.68	1.12
Kurtosis	1.79	3.01	−0.58	0.43
Shapiro–Wilk test	0.78***	0.72***	0.91***	0.84***
Teraesvirta NN test	16.90***	57.65***	29.84**	13.97***

NN: Neural Network, ***: significant at 1% level of significance

1.3.1 BORUTA Algorithm

Among several feature selection algorithms, Boruta is popular (Kursa et al. 2010). This ensemble algorithm is used to identify the importance of features used for predictive modeling. The literature reports several applications of Boruta in diverse domains (Ghosh et al. 2019; Jana and Ghosh 2022). The way Boruta works is similar to that of the classical RF algorithm. However, it accommodates enhanced randomness levels to select relevant features. The main steps of the Boruta algorithm can be summarized as follows:

Step 1: Generate replicas of the given features by expanding the information system. They are defined as shadow attributes.

Step 2: Trim down the correlation between the target variables and shadow attributes using an appropriate perturbation technique.

Step 3: Evaluate Z-scores for all attributes.

Step 4: Find the maximum of shadow attributes' Z-score. Denote this by S, say. Identify the attributes whose Z-scores are higher than S.

Step 5: Arrange the attributes for the two-tailed test for equality with S.

Step 6: Mark the attributes as important whose Z-scores are significantly greater than S. The remaining attributes are identified as less important features.

Step 7: Eliminate shadow attributes.

Step 8: Execute Step 1–Step 7 for a pre-defined number of runs.

1.3.2 The Residual-Driven RF Algorithm

We present the proposed residual-driven RF algorithm used for predicting the future movements of Metaverse coins. RF is a widespread ensemble ML algorithm that combines multiple decision trees to produce a single output (Breiman 2001). It is helpful for solving classification and regression problems. RF is an extension of bagging. Together with bagging, it introduces randomness in features and constructs decision trees that are uncorrelated to each other. The three hyperparameters of RF are the number of sampled features, node size, and number of trees. The algorithm has been well employed in predictive modeling of time series from various domains (Ghosh et al. 2019, 2020; Jana et al. 2021). The suggested residual-driven RF algorithm can be summarized as follows:

Step 1. From the training dataset, select bootstrapped samples.

Step 2. Construct a regression tree for each subset of samples drawn and perform the following:

2.1. From the feature set, select a subset of features randomly.
2.2. Choose the most suitable feature.
2.3. Continue 2.1–2.2 until the construction of the decision tree is complete.

Step 3. Produce the final forecast.

Step 4. Calculate the residuals (*Res*) for the full dataset.

Step 5. Include *Res* to the set of original features.

Step 6. Continue Step 1–Step 3.

1.4 Results and Analysis

This section presents the outcome of the proposed research methodology.

1.4.1 Findings of the Boruta Algorithm

We begin with the findings of the Boruta algorithm to identify the importance of features in predicting the future movements of Metaverse coins. We use the following technical indicators as individual features for predictive modeling: B5, B10, EMA5, EMA10, HP, LAG1, LAG2, LAG3, LAG4, LAG5, LOSS, LP, MA5, MA10, MTM5,

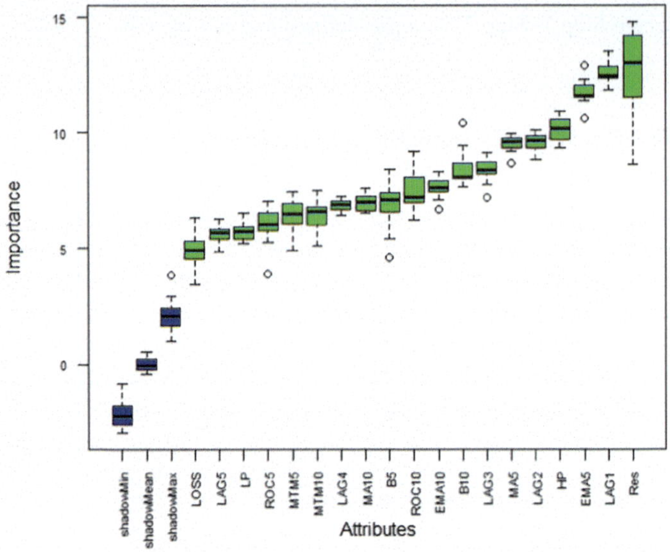

Fig. 1.2 Feature importance for MANA

MTM10, Res, ROC5, ROC10. More details about these technical indicators may be obtained from Jana et al. (2020). In this process, we also study the contributions of the top three features in this predicting exercise. Figures 1.2, 1.3, 1.4 and 1.5 represent the importance scores of the individual features through box plots generated by executing the Boruta algorithm for 200 runs. We observed no significant change in feature importance beyond 200 iterations. A box plot with a green color indicates that the feature is significant. Such a feature possesses better predictive ability than a shadow feature.

We considered 18 features initially for all the Metaverse coins. The *Res* is included as the 19th feature. The outcome of the Boruta algorithm confirms that all 19 features are significant for all the time series. Therefore, all 19 features have been used for predicting the future values of the considered Metaverse coins. The ranking of these features concerning the average importance (AvI) score is presented in Table 1.2.

Table 1.2 presents the ranking of features for all four Metaverse coins. It is apparent that *Res*, LAG1, and EMA5 are the three most influential features for all crypto assets. The only exception is that ROC10 occupies the third spot for SAND. For MANA and SAND, LAG1 is the most influential feature. For STX and THETA, the *Res* appears in the top spot. For MANA, *Res* occupies the second spot. The EMA5 is the third most important feature for MANA, STX, and THETA and the second most important for SAND. Therefore, the importance of *Res* together with LAG1 and EMA5 was proven to be significant for predicting future prices of the Metaverse coins.

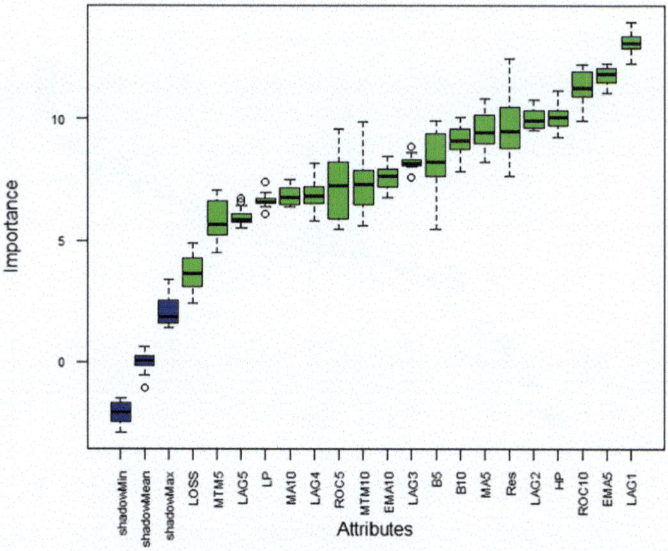

Fig. 1.3 Feature importance for SAND

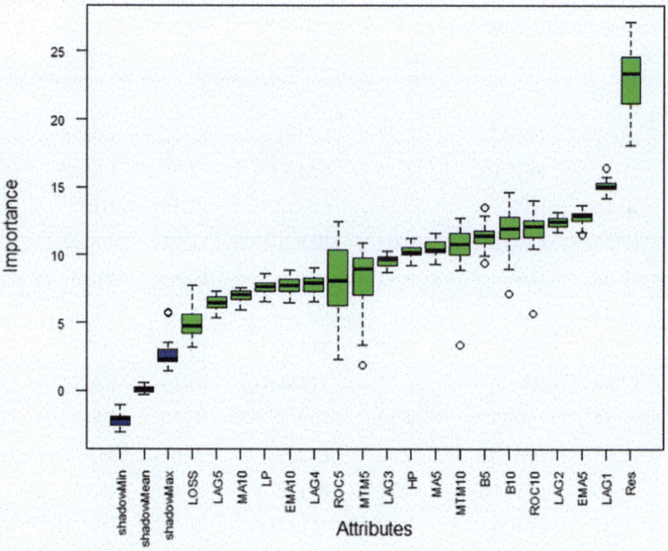

Fig. 1.4 Feature importance for STX

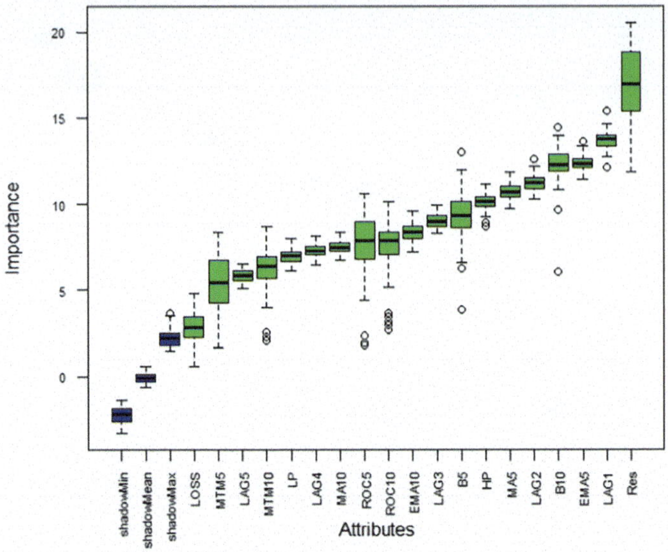

Fig. 1.5 Feature importance for THETA

Table 1.2 Feature ranking with AvI

MANA	AvI	SAND	AvI	STX	AvI	THETA	AvI
LAG1	12.59	LAG1	13.10	*Res*	23.36	*Res*	16.93
Res	12.57	EMA5	11.78	LAG1	15.12	LAG1	13.72
EMA5	11.74	ROC10	11.34	EMA5	12.78	EMA5	12.41
HP	10.16	HP	10.11	LAG2	12.42	B10	12.20
LAG2	9.56	LAG2	10.03	ROC10	11.94	LAG2	11.25
MA5	9.53	*Res*	9.71	B10	11.80	MA5	10.73
B10	8.48	MA5	9.56	B5	11.39	HP	10.11
LAG3	8.38	B10	9.15	MA5	10.56	B5	9.27
EMA10	7.64	B5	8.22	MTM10	10.49	LAG3	9.02
ROC10	7.54	LAG3	8.19	HP	10.30	EMA10	8.35
MA10	7.01	EMA10	7.59	LAG3	9.61	MA10	7.52
B5	6.89	MTM10	7.29	ROC5	8.17	ROC5	7.51
LAG4	6.87	ROC5	7.23	MTM5	8.16	ROC10	7.46
MTM10	6.45	LAG4	6.89	LAG4	7.86	LAG4	7.28
MTM5	6.44	MA10	6.83	EMA10	7.79	LP	6.98
ROC5	6.03	LP	6.61	LP	7.65	MTM10	6.15
LP	5.74	LAG5	5.96	MA10	7.01	LAG5	5.84
LAG5	5.67	MTM5	5.81	LAG5	6.39	MTM5	5.29
LOSS	4.89	LOSS	3.64	LOSS	4.95	LOSS	2.85

1.4.2 Findings of Predictive Modeling

We noticed that the behavioral patterns of the considered Metaverse coins are complicated. Therefore, it is challenging to predict their future movements with very high accuracy. We propose a two-stage residual-driven ensemble ML approach to achieve this tough goal and use 18 technical indicators and the *Res* as independent features. To verify the quality of prediction, we evaluate the following measures: Mean Absolute Error (MAE), Mean Square Error (MSE), Index and Agreement, and Nash–Sutcliffe Efficiency (NSE). Details of these measures may be obtained from Ghosh et al. (2019). It may be noted that for superior predictive performance, MAE and MSE values should be close to zero, and IA and NSE values should be close to one. We use the "rattle" package of R for implementing the proposed approach. The dataset is divided into training (70%), testing (15%), and validation (15%) segments. We report the predictive performance based on the training, testing, and full datasets in Tables 1.3, 1.4 and 1.5, respectively.

There was a total of 870 observations in the full dataset. Therefore, the training dataset contains 609 observations, and the test dataset contains 131 observations. From Tables 1.3, 1.4 and 1.5, we observed that the predictive model's performance

Table 1.3 Predictive performance (training dataset)

Series	MANA	SAND	STX	THETA
MAE	0.0278	0.0331	0.0189	0.0765
MSE	0.0061	0.0056	0.0023	0.0236
IA	0.9982	0.9992	0.9985	0.9994
NSE	0.9929	0.9970	0.9943	0.9978

Table 1.4 Predictive performance (testing dataset)

Series	MANA	SAND	STX	THETA
MAE	0.0494	0.0636	0.0364	0.1387
MSE	0.0097	0.0182	0.0060	0.1640
IA	0.9978	0.9983	0.9961	0.9947
NSE	0.9913	0.9931	0.9850	0.9796

Table 1.5 Predictive performance (full dataset)

Series	MANA	SAND	STX	THETA
MAE	0.0340	0.0424	0.0236	0.0943
MSE	0.0065	0.0104	0.0026	0.0508
IA	0.9983	0.9988	0.9984	0.9987
NSE	0.9932	0.9953	0.9938	0.9948

is the best for the training dataset, followed by the full and test datasets. Also, the MAE and MSE values for all the series are considerably low, and the IA and NSE values are close to one. Based on the IA and NSE values, we may conclude that the forecast quality of the proposed predictive model is highly accurate for all four Metaverse coins. It isn't easy to differentiate the predictability of a particular crypto asset compared to the remaining ones. Table 1.4 shows that the IA and NSE values for THETA are marginally inferior compared to MANA, SAND, and STX. To identify the Metaverse coin that is most predictable, we further perform the model confidence set (MCS) test (Hansen et al. 2011). The outcome of the test is furnished in Table 1.6. Therefore, we find that THETA is the least predictable. The IA and NSE values also indicated this. STX is the most predictable, followed by MANA and SAND.

Next, based on the test dataset, we compare the observed and forecasted values for all the Metaverse coins in Figs. 1.6, 1.7, 1.8 and 1.9. MANA_F in Fig. 1.6, SAND_F in Fig. 1.7, STX in Fig. 1.8, and THETA_F in Fig. 1.9 denote the forecasted values of the respective time series. These figures support the analytical findings for all the cryptocurrencies.

Table 1.6 MCS test results

Series	MANA	SAND	STX	THETA
Rank	(2)	(3)	(1)	(4)

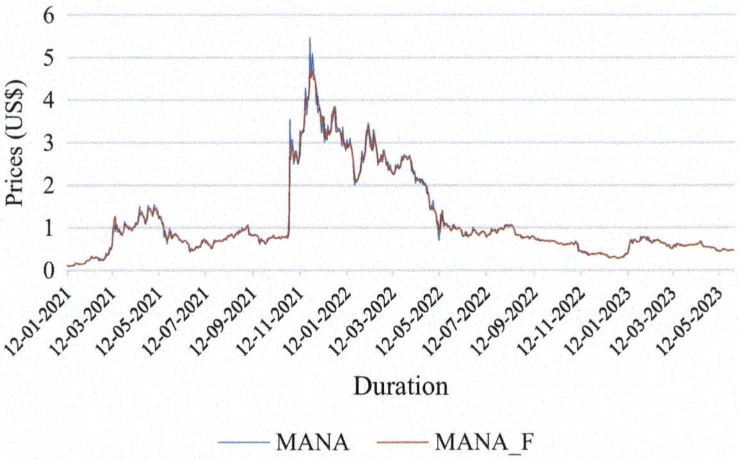

Fig. 1.6 Observed versus forecasted (MANA)

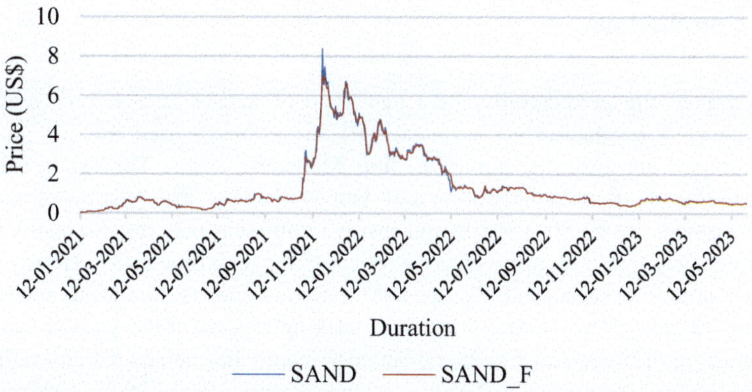

Fig. 1.7 Observed versus forecasted (SAND)

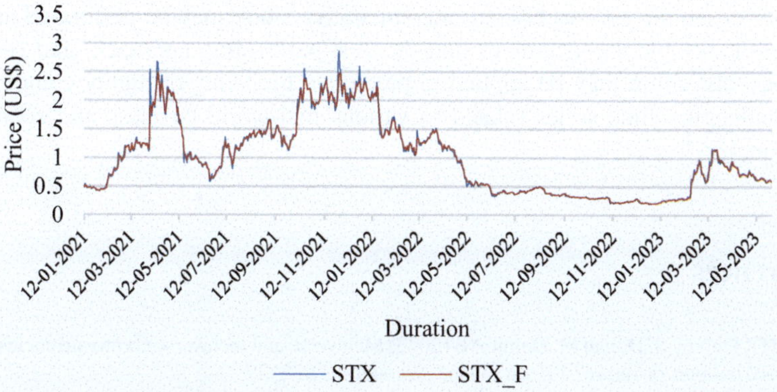

Fig. 1.8 Observed versus forecasted (STX)

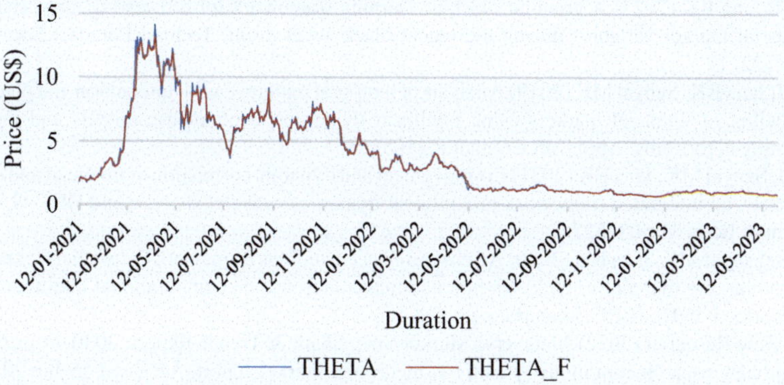

Fig. 1.9 Observed versus forecasted (THETA)

1.5 Conclusions

We examined the predictability of a new type of digital asset known as Metaverse coins. A residual-driven two-stage RF algorithm is used for this purpose. The proposed algorithm is applied to four Metaverse coins. They were selected based on their market capitalization and data availability. The past movements of all these assets are highly volatile and involve nonlinear patterns. Standalone ML algorithms may fail to produce superior forecasts for such time series. However, the proposed approach successfully generated future movements for all four assets with supreme accuracy. This is one of the most striking aspects of the present research. The inclusion of the residual as an explanatory feature has helped to achieve highly accurate forecasts by capturing the unexplained variance. The MCS test confirms that STX is the most predictable and THETA is the least predictable asset. The present work may be extended in the following ways: (i) the RF algorithm may be replaced by other ensemble approaches, (ii) a comparative study may be performed among different ensemble algorithms to find the best-performing approach, and (iii) the proposed algorithm may be applied to predictive modeling of highly complicated and challenging time series related to various real-world business and economic situation.

References

Akkus HT, Gursoy S, Dogan M, Demir AB (2022) Metaverse and metaverse cryptocurrencies (meta coins): bubbles or future? J Econ Financ Account 9(1):22–29

Barrera KG, Shah D (2023) Marketing in the Metaverse: conceptual understanding, framework, and research agenda. J Bus Res 155:113420

Breiman L (2001) Random forests. Mach Learn 45(1):5–32

Ghosh I, Jana RK (2023) A granular machine learning framework for forecasting high-frequency financial market variables during the recent black swan event. Technol Forecast Soc Chang 194:122719

Ghosh I, Jana RK, Sanyal MK (2019) Analysis of temporal pattern, causal interaction and predictive modeling of financial markets using nonlinear dynamics, econometric models and machine learning algorithms. Appl Soft Comput 82:105553

Ghosh I, Sanyal MK, Jana RK (2021) Co-movement and dynamic correlation of financial and energy markets: an integrated framework of nonlinear dynamics, wavelet analysis and DCC-GARCH. Comput Econ 57:503–527

Ghosh I, Sanyal MK, Jana RK (2020) An ensemble of ensembles framework for predictive analytics of commodity market. In: 2020 4th international conference on computational intelligence and networks (CINE). IEEE, Singapore, pp 1–6

Grand View Research (2022) Metaverse Market Size, Share & Trends Report, 2030. https://www.grandviewresearch.com/industry-analysis/metaverse-market-report. Accessed 15 Jun 2023

Hansen PR, Lunde A, Nason JM (2011) The model confidence set. Econometrica 79(2):453–497

Jana RK, Ghosh I (2023) Time-varying relationship between geopolitical uncertainty and agricultural investment. Financ Res Lett 52:103521

Jana RK, Tiwari AK, Hammoudeh S (2019) The inefficiency of Litecoin: a dynamic analysis. J Quant Econ 17:447–457

Jana RK, Ghosh I, Sanyal MK (2020) A granular deep learning approach for predicting energy consumption. Appl Soft Comput 89:106091

Jana RK, Ghosh I, Das D, Dutta A (2021) Determinants of electronic waste generation in Bitcoin network: Evidence from the machine learning approach. Technol Forecast Soc Chang 173:121101

Jana RK, Ghosh I, Goyal V (2022a) Spillover nexus of financial stress during black Swan events. Financ Res Lett 48:102892

Jana RK, Ghosh I, Wallin MW (2022b) Taming energy and electronic waste generation in bitcoin mining: insights from Facebook prophet and deep neural network. Technol Forecast Soc Chang 178:121584

Jana RK, Das D (2020) Did Bitcoin act as an antidote to the Chinese equity market and booster to Altcoins during the Novel Coronavirus outbreak? Available at SSRN 3544794

Jana RK, Ghosh I (2022) A residual driven ensemble machine learning approach for forecasting natural gas prices: analyses for pre-and during-COVID-19 phases. Ann Oper Res 1–22

Kursa M, Jankowski A, Rudnicki W (2010) Boruta—a system for feature selection. Fund Inform 101:271–285

Purdy M (2022) How the Metaverse could change work. Harv Bus Rev. https://hbr.org/2022/04/how-the-metaverse-could-change-work.

Q.ai (2022) What long-term investors need to know about the Metaverse. Forbes. https://www.forbes.com/sites/qai/2022/08/25/what-long-term-investors-know-about-the-metaverse/?sh=45dda8217877.

Schumacher P (2022) The Metaverse as opportunity for architecture and society: design drivers, core competencies. Arch Intell 1(1):1–20

Tiwari AK, Jana RK, Das D, Roubaud D (2018) Informational efficiency of Bitcoin—an extension. Econ Lett 163:106–109

Vidal-Tomás D (2022) The new crypto niche: NFTs, play-to-earn, and metaverse tokens. Financ Res Lett 102742

Chapter 2
DriveHarmony: An AI-Based Safety Assistant Driving System

Khalid Sultan, Aqeel Alsarraf, Ali Alsarraf, Ali Alfeeli, and Ali Alghanemi

Abstract Road safety has been a concern for public and governments across the globe. Statistics show that more than a million people lose their lives annually due to road traffic accidents, and millions more get injured globally. Though different safety technologies have been developed in some modern vehicles to help cope with this issue, most vehicles on roads currently are not equipped with such features. To help mitigate this problem, we propose a solution that is based on machine learning and can fit into any vehicle regardless of its size, age, or model. Our proposed solution includes three subsystems. The first component (AIsys) tries to integrate voice activation towards old machines and acts according to the user's input. The second element (DrowsDet) checks whether the driver is distracted or drowsy while driving as those factors significantly increase the risk of involvement in a crash. The last subsystem (HarmonyCall) allows the system to call the authorities according to the input of the user, or automatically if the system detects an issue with the driver. The device aims to reduce the occurrence and consequences of road traffic accidents by helping drivers stay awake and focused on the road or advise them to stop for rest otherwise. All three subsystems work synchronously together to attain the highest efficiency possible. The proposed solution has been tested and evaluated in real life. The obtained results are promising and prove the effectiveness of the solution.

Keywords Machine learning · Driver fatigue detection · Road safety · Road traffic accidents

K. Sultan (✉) · A. Alsarraf · A. Alsarraf · A. Alfeeli · A. Alghanemi
College of Engineering and Applied Sciences, American University of Kuwait, Salmiya, Kuwait
e-mail: ksultan@auk.edu.kw

A. Alsarraf
e-mail: S00044979@auk.edu.kw

A. Alsarraf
e-mail: S00045654@auk.edu.kw

A. Alfeeli
e-mail: S00038400@auk.edu.kw

A. Alghanemi
e-mail: S00039573@auk.edu.kw

D. K. Sharma et al. (eds.), *Machine Learning for Real World Applications*, Transactions on Computer Systems and Networks, https://doi.org/10.1007/978-981-97-1900-6_2

2.1 Introduction

Distractions happening while driving can lead to major risks and serious accidents which, in turn, may lead to death in many cases. According to a report published by the World Health Organization (WHO) in 2018, the global rate of death caused by traffic collisions is very high and the number of annual road accident-related deaths exceeded 1.3 million which means that the world loses about 3000 deaths per day (or 125 deaths every hour) due to road accidents. In addition, millions of others sustain serious injuries (Global status report on road safety (WHO) 2023). Moreover, statistics show that road traffic accidents have become a prominent killer of people aged between 5 and 29 years (The global health observatory - Road Safety (WHO) 2023). WHO has also indicated that the ratio of serious road traffic accidents for low and middle-income countries is significantly higher than that of high-income countries, marking some African countries at the top of the list in terms of related fatality rate (Kuwait Times homepage 2023).

On a local level, statistics presented by the General Traffic Department in Kuwait show that in 2020 alone there were 352 incidents of death due to road traffic accidents, keeping in mind the lockdown that happened in Kuwait for almost a month and the low population (4.36 million in 2020) (Kuwait Times homepage 2023). Though Fig. 2.1 marks that the number of traffic deaths in Kuwait has decreased compared to previous years, it is considered high for such a small population country.

Driving nowadays is dangerous due to the number of stimuli and distractions a person can be affected by while driving. That is why road safety is a major concern in many countries like Kuwait where the number of accidents per year exceeds the international levels. Literature suggests that a substantial fraction of road accidents is due to human errors including mental fatigue. Statistics also indicate that fatigue-related crashes represent between 10 and 20% of all major road accidents globally (Thomas et al. 2021).

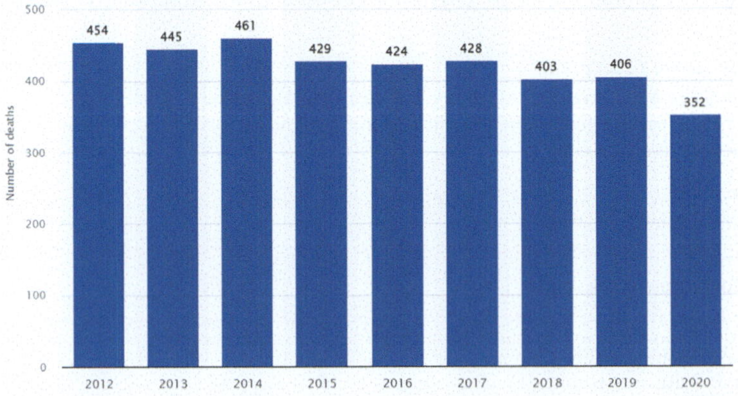

Fig. 2.1 Number of traffic-related deaths in Kuwait between 2012 and 2020 (Saleh 2023)

To cope with the situation, some vehicle manufacturers have incorporated accident-avoidance features including fatigue detection systems in some of their vehicles. These systems offer opportunities to reduce the impact of an accident that may result from severe fatigue or distraction. However, old car models do not have as many safety features -in their systems- as some new ones do. Consequently, many drivers still get into road crashes often because of the lack of safety features in their cars. This holds true especially in situations where people need to drive for long distances or long time such as professional drivers - transport truck drivers and delivery workers for instance. Drivers who drive on highways during the night or for a long time may fall asleep while driving due to drowsiness, fatigue, or other natural phenomena like lack of rest that the human body needs to revive its energy level. Other reasons that could put drivers' safety at risk include, but not limited to, irregular shiftwork, mental and physical stress, fainting while driving, inconsistent sleep schedules, distractions by different factors such as using the phones, interaction with the environment, having a bad mood, etc.

Driver fatigue, distraction, and drowsiness are considered leading factors in road collisions in general, and in heavy vehicle crashes in specific (Ani et al. 2020; Zhao et al. 2016). Studies show that about 80% of traffic accidents are due to human errors including fatigue or drowsiness, and about 20% of road collisions are due to lower level of alertness, and it reaches up to 50% on certain roads (Reshma et al. 2021). Fatigue is a common phenomenon in humans' lives, and defined as a state of tiredness, usually due to long time of driving or lack of sleep, which negatively impacts mental performance and decreases the alertness of drivers (Lal and Craig 2001). In literature, *fatigue* is often used interchangeably with *drowsiness* and *sleepiness*. It deteriorates the driver's concentration and leads to inadequate reaction time. Besides, changes in weather and environmental conditions can also be viewed as important influences that contribute to drowsiness (Reshma et al. 2021).

Unfortunately, many drivers believe that some ad-hoc measures such as opening the window to get some fresh air, turning on the radio, or talking to a passenger could mitigate the fatigue. In fact, these measures have not been proven effective on fatigued driving. Continued use of such ineffective measures implies that drivers tend to avoid appropriate alternatives such as stopping the vehicle and taking a rest (Nordbakke and Sagberg 2007).

Indeed, taking some reliable measures to keep drivers fully attentive on the road is very important to help decrease the number of roadway accidents and consequently save peoples' lives. In this direction, some recent studies have confirmed that increasing the minimum rest limit and reducing the workload for heavy vehicle drivers have successfully improved the fatigue level and lowered the crush rate on roads. Studies also showed that the consumption of caffeine, through tea and coffee for example, not only can help leverage the driver's attentiveness level but also produce faster reaction times and lead to more accurate decisions (Brice and Smith 2002). Figure 2.2 summarizes the different ways of handling fatigue by drivers and innovative solutions to reduce its impact.

However, innovative solutions are still needed to promote safe driving and cope with the issue (Zhao et al. 2016). In some situations, a warning message could be

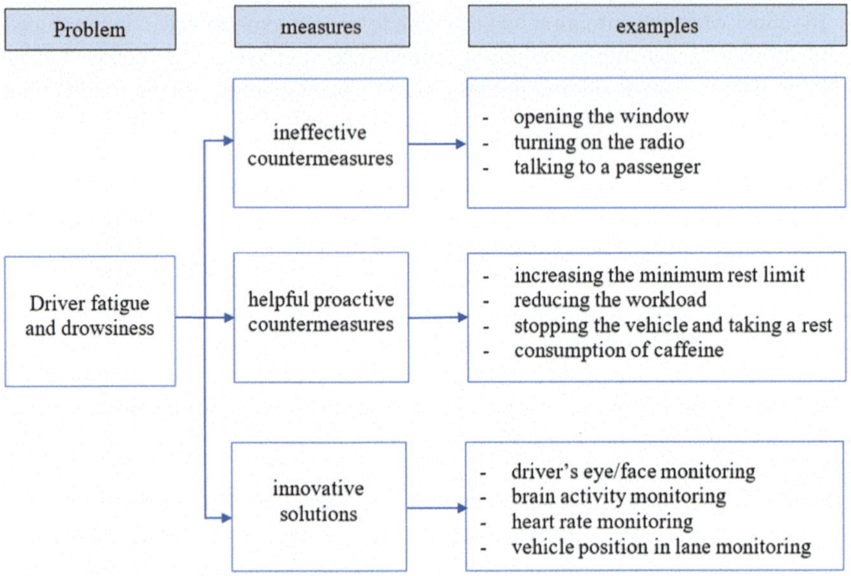

Fig. 2.2 Summary of drive fatigue countermeasures

sufficient to help the drivers get the control back and avoid the occurrence of traffic collisions.

Various techniques and methods have been put forward to eliminate the risk of fatigue and downiness among drivers, and hence prevent road accidents. Among others, the following procedures are the most common adopted ones: (1) vehicle-based measures such as steering pattern monitoring, and vehicle position in lane monitoring; (2) driver's behavioral measures such as yawning detection and driver's eye/face monitoring; (3) physiological measures including brain activity and heart rate; (4) psychophysical measures which include studying the relationship between some physiological signals such as the electrocardiogram, electromyogram, and electroencephalogram (EEG) signals; and (5) biomechanical measures (Ani et al. 2020; Reshma et al. 2021). It is worth mentioning that steering pattern, and vehicle position in lane monitoring works only when the automatic lane-keeping system is not used.

Our main objective is to develop a reliable AI-based monitoring system equipped with a camera to monitor the loss of attention of the drivers during different conditions (i.e., day and night) (Dasgupta et al. 2013). This helps detect drowsiness and fatigue by measuring the alertness level of the driver. So, doing ensures that fatigue-related crashes can be identified and addressed before they occur. The proposed system is called *DriveHarmony*. The main idea of *DriveHarmony* is to continuously monitor the driver's eyes and face to ensure that the driver is always focusing on driving. Whenever the diver closes his/her eyes for 2–3 s or turns his/her face left or right or look down (at the phone for example) for a considerable period, the system will notify the diver by issuing a warning sound so that the driver focuses back on driving, which ultimately helps in avoiding the risk getting into an accident.

The contributions of this article are as follows:

As mentioned earlier, the main aim of this work is to develop an intelligent system to help reduce the effect of driving fatigue and consequently prevent the occurrence of fatal road accidents. The proposed system is called *DriveHarmony* and consists of three main components, namely, *DrowsDet*, *AIsys*, and *HarmonyCall*. The *DrowsDet* is the sub-system responsible for the detection of drowsiness or distraction in a driver. The *AIsys* is the sub-system acting as an interface of the system as it allows the user to communicate with the system via voice. HarmonyCall is the sub-system responsible for sending a distress notification to a nearby hospital in case of an emergency.

2.2 Literature Survey

In this section, we present some of the most recent related proposals for driver fatigue and drowsiness detection. A multi-tasking Convolutional Neural Network model (ConNN) is introduced in Savaş and Becerikli (2020) for driver fatigue and drowsiness detection, leveraging eye and mouth characteristics to model driver behavior. Changes in these features are utilized to monitor fatigue, as both eye and mouth information are simultaneously classified by the proposed Multi-task ConNN model. Fatigue determination involves calculating PERCLOS (Percentage of eye closure) and FOM (frequency of mouth) to assess eye closure duration and yawning frequency, respectively. The study divides driver fatigue into three classes, achieving a 98.81% detection rate on the YawdDD and NthuDDD datasets, highlighting the model's success in comparison. Xu et al. (2023) developed a framework called E-Key, which offers a unified approach for simultaneous personal identification (PI) and driving fatigue detection. The framework utilizes a convolutional neural network and attention structure known as CNN-Attention. To evaluate its effectiveness, they conducted experiments using EEG data obtained from 31 healthy participants who engaged in a 90-min simulated driving task while wearing a dry-sensor system. To compare the performance of their proposed scheme, they benchmarked it against three commonly used models: CNN, CNN-LSTM, and Attention. Remarkably, their framework outperformed all three models in terms of both PI (98.5%) and fatigue detection (97.8%) with statistical significance ($p < 0.01$). Moreover, the spatial–temporal structure of the proposed framework strikes an optimal balance between classification performance and computational efficiency. In Sheykhivand et al. (2022), the researchers presented an automated system that tackles the classification of driver fatigue through a two-stage process. The system combines compressed sensing (CS) theory and deep neural networks (DNNs) while utilizing EEG signals. Initially, the recorded EEG data are compressed using CS theory to reduce computational load. Subsequently, the compressed EEG data are inputted into a deep CNN, specifically designed for not only feature extraction but also feature selection and classification. The devised network architecture comprises seven convolutional layers and three long short-term memory (LSTM) layers. Simulation results demonstrate the effectiveness of the approach at various compression rates (40, 50, 60, 70, 80, and 90) for single-channel recordings. The achieved accuracy rates for these compression rates are as follows: 95%, 94.8%, 94.6%, 94.4%, 94.4%, and 92%, respectively. The work in Shalash (2019) proposes the utilization of transfer learning in a driver fatigue detection system, specifically relying on a single EEG channel to enhance system usability. The process begins by capturing the signal and subjecting it to preprocessing filtering. Subsequently, the signal is transformed into a 2D spectrogram. Finally, the 2D spectrogram undergoes classification using AlexNet through transfer learning, distinguishing between normal and fatigued states. To determine the most accurate channel for classification, this study compares the accuracy of seven EEG channels. The findings reveal that the FP1 and T3 channels demonstrate the highest effectiveness in indicating the driver's fatigue state. These channels achieved an accuracy of 90% and 91%, respectively. The work in Ed-Doughmi et al. (2020) introduces a novel approach for analyzing and predicting driver fatigue and drowsiness. The proposed method involves utilizing a Recurrent Neural Network (RNN) to process sequential frames of the driver's face. To develop and validate the model, authors employed a carefully curated dataset. Furthermore, they incorporated a multilayer model based on 3D Convolutional Networks, enabling us to effectively detect driver drowsiness. Following a rigorous training session, they achieved an impressive accuracy rate, reaching approximately 92%. Fatima et al. (2020) introduced a cost-effective approach to detect driver fatigue by analyzing micro-sleep patterns. They developed two algorithms that enable precise identification of the driver's face and eyes, even when the driver is not directly facing the camera or when eyes are closed. To determine whether the eye is closed or open, they utilized Support Vector Machine (SVM) and AdaBoost algorithms. By analyzing the eye states, authors identified micro-sleep patterns and implemented an alarm system to alert the driver when necessary. The dataset consisted of various subjects of different genders, appearances, and lighting conditions. Their proposed system achieved impressive results, with face detection accuracy reaching 99.9% and eye detection accuracy reaching 98.7%. The average accuracy for SVM and AdaBoost algorithms across all subjects was 96.5% and 95.4%, respectively.

In their study, Karuppusamy and Kang (2020) devised a hybrid multimodal system leveraging neural networks to identify driver fatigue. The system incorporated electroencephalography (EEG) data, gyroscope data, and image processing data. Results demonstrated the efficacy of the hybrid approach, achieving a detection accuracy of 93.91% in accurately discerning the drowsiness level of the driver. The research in Ren et al. (2021) aimed to create a two-level learning hierarchy RBF network (known as RBF-TLLH) that enables the global optimization of crucial network parameters. To accomplish this, EEG data from six healthy individuals in a simulated driving scenario were gathered, capturing both fatigue and alert states. Initially, principal component analysis was employed to extract relevant features from the

EEG signals. Subsequently, the proposed RBF-TLLH model was utilized for classifying the driving status. The findings revealed that the RBF-TLLH approach outperformed other commonly used artificial neural networks in terms of classification performance, with a mean accuracy of 92.71% and an area under the receiver operating curve of 0.9199. Further, an efficient framework was developed in Zhuang et al. (2020) to enhance the accuracy and generalization of eye openness estimation. The framework consists of two components: a segmentation network and a decision network. By employing a light U-Net structure, the segmentation network performs precise pixel-level classification on the eye images, allowing for accurate extraction of pupil and iris features from the video frames. The extracted feature map is then utilized to guide the decision network in estimating eye openness. The effectiveness of the approach was evaluated using the National Tsing Hua University Drowsy Driver Detection (NTHU-DDD) Video Dataset, and the results showed a remarkable 96.72% precision in fatigue detection. Wang et al. (2020) proposed the AMCNN-DGCN model, which combines attention-based multiscale convolutional neural networks (CNNs) with dynamical graph convolutional networks (DGCNs) in order to address two key challenges within a single end-to-end framework. The objective of the model is to effectively handle these issues. The AMCNN-DGCN model begins by employing attention-based multiscale temporal convolutions, enabling automatic learning of frequency filters that extract significant patterns from raw EEG data. Following this, the model utilizes DGCNs to learn spatial filters. In this step, the adjacency matrix is adaptively determined in a data-driven manner, exploiting the inherent relationships between channels. This temporal-spatial structure allows AMCNN-DGCN to capture highly discriminative features. To assess the efficacy of AMCNN-DGCN, the researchers conducted an experiment involving a simulated fatigue driving environment. EEG signals were collected from 29 healthy subjects for analysis and evaluation. Zhou et al. (2020) proposed a method for predicting the transition from a non-fatigued state to fatigue in highly automated driving, utilizing physiological measures. The researchers specifically focused on the PERCLOS (PERcent of time the eyelids CLOSure) measure as an informative indicator of driver fatigue. They identified crucial physiological features associated with fatigue prediction and developed prediction models based on these features. To accomplish this, the researchers employed a technique known as nonlinear autoregressive exogenous network. Remarkably, their model could predict the onset of fatigue with a lead time of at least 13.8 s. The accuracy of the fatigue transition prediction exhibited great promise, yielding an F1 measure of 97.35% and 99.10% for two distinct types of models.

2.3 Methodology

The research team opted for an agile process methodology for the system design and analysis in which progress is analyzed regularly by the entire team, and updates to the system structure are documented, validated, and verified. The system was broken down into three sub-systems and then each sub-system was developed and tested independently. When all sub-systems were completed, they were combined and tested together.

The formulation of the hypothesis was based on questionnaires and relevant literature. The generation of test cases for experimentation was helpful to understand the validity of the hypothesis. The following research questions were formulated to be answered.

- Research Question 1: how to build an AI system that could minimize the effects of fatigue and drowsiness on traffic accidents?
- Research Question 2: how to build a computer system that could act as an emergency protocol in case a driver suddenly experiences a deterioration in health.

2.3.1 Requirement Specification

User Requirements. The driver needs to be able to interact with DriveHarmony by voice when running the AIsys sub-system. Furthermore, a simple UI showing the input from the streaming camera will be available for the user to verify that the DrowsDet sub-system is indeed operating correctly. In case of an emergency or unfortunate event preventing the driver from driving; the system will immediately operate autopilot and HarmonyCall will send a distress voicemail to the nearest hospital.

System Requirements

Non-*functional* Requirements. System dependability and real-time efficiency are beyond important for *DriveHarmony* because its utilities are geared to building a safer driving experience for the user. The system needs to react in real time while the user is driving at high speeds. Furthermore, data processing must be done instantly when the data are still relevant. Such optimization in system efficiency and data processing is to ensure that minimal time has passed from the moment when *DriveHarmony* receives data until it reacts to said data. Thus, the user will not feel that there is a lag in the system that was engineered to ensure his/her safety. Figure 2.3 presents the non-functional requirements of the *DriveHarmony* system.

Functional Requirements Specification. Table 2.1 presents the definitions of the four main functional requirements of DriveHarmony. Tables 2.2, 2.3, and 2.4 list the specifications of REQ1 (*DrowsDet* sub-system requirement), REQ2 (*AIsys* sub-system requirement), and REQ3 (*HarmonyCall* sub-system requirement), respectively.

REQ1

REQ2

REQ3

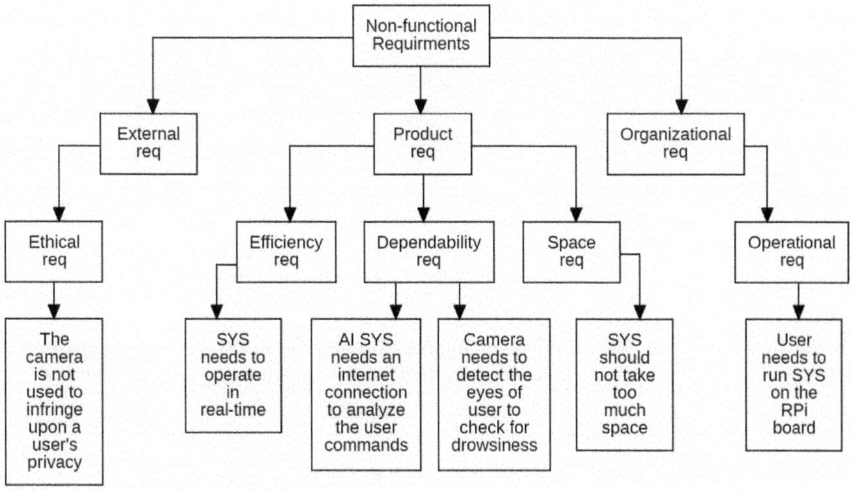

Fig. 2.3 DriveHarmony non-functional requirements

Table 2.1 REQ definitions

REQ id	REQ definition
1	The system shall detect signs of drowsiness or distraction from the user without impeding his/her comfortability
2	The system shall enable the user to activate and deactivate a theoretical state of autopilot using voice command
3	The system shall assist the user in case of an emergency by sending a voicemail to a nearby medical center
4	The system should localize all targets within proximity, even those that are occluded

2.3.2 Design and Analysis

DrowsDet. *DrowsDet* is used to identify if drowsiness/distraction is in effect while the user is driving. The user is not responsible for wearing any special device or mechanism that may impede upon his/her comfortability. We are using computer vision and object detection algorithms to identify whether the user is showing signs of drowsiness or not. For such algorithms, only a camera is needed to be positioned on the front dashboard of the user's car where the camera will input a video stream continuously until the system is stopped. Figure 2.4 shows the sequence model of the *DrowsDet* sub-system.

AIsys. *AIsys* is the sub-system responsible for facilitating user/system communication. It allows a user to control the alert functionality of *DriveHarmony* by voice. *AIsys* is engineered specifically to promote *DriveHarmony* with futuristic functionalities such as voice control which will surely be integrated into many parts of society. The

Table 2.2 DrowsDet REQ specification

Software system	Subsystem/component	Artifact	Ref.
DriveHarmony	*DrowsDet*	FRS	3.3.1.2.2
Function	Drowsiness/distraction detection		
Description	Alerts driver if he/she is showing signs of drowsiness or distraction		
Inputs	Face and eyes		
Source	Camera		
Outputs	Sound alert if drowsiness/distraction detected		
Destination	*DrowsDet* control loop		
Action	Input stream from infrared camera and detect any sign of drowsiness or distraction. Then alert user if drowsiness/distraction is detected		
Requires	Camera detects eyes of driver to detect signs of drowsiness/distraction		
Preconditions	Working, good-resolution IR camera		
Postconditions	None		
Side effects	None		

Table 2.3 AIsys REQ specification

Software system	Subsystem/component	Artifact	Ref.
DriveHarmony	*AIsys*	FRS	3.3.1.2.2
Function	User-System voice interface		
Description	Allows user to control the autopilot state of a car by voice command		
Inputs	User command		
Source	Microphone		
Outputs	System voice reply		
Destination	Speaker		
Action	*AIsys* records at 5-s intervals then analyze the recording to check if a user has any commands targeted to the system		
Requires	Speech-to-text for system to understand recorded user input. Text-to-speech for the system to output a voice reply		
Preconditions	User must say "Harmony" before commanding the system. A connection to the internet		
Postconditions	None		
Side effects	None		

appeal of voice control in automotive applications such as *DriveHarmony* is attractive to both car buyers and manufacturers. *AIsys* can only accept user input during 5-s intervals because it proved to be more stable and accurate than if a user is to input continuously. Therefore, the processes of data acceptance and data processing were separated and implemented in stages, a 5-s interval to accept user input then

Table 2.4 HarmonyCall REQ specification

Software system	Subsystem/component	Artifact	Ref.
DriveHarmony	*HarmonyCall*	FRS	3.3.1.2.2
Function	Calls a medical center/facility if driver shows signs of health deterioration		
Description	Detects probable signs that indicate the driver's inability to continue driving and call a medical center to check the driver in		
Inputs	None		
Source	*DrowsDet*		
Outputs	Voicemail		
Destination	Hospital call center		
Action	A String message is sent to a Twilio server which is then sent to a medical center as a voicemail		
Requires	Concurrency with *DrowsDet*		
Preconditions	Valid Twilio account with Twilio number to access server		
Postconditions	None		
Side effects	None		

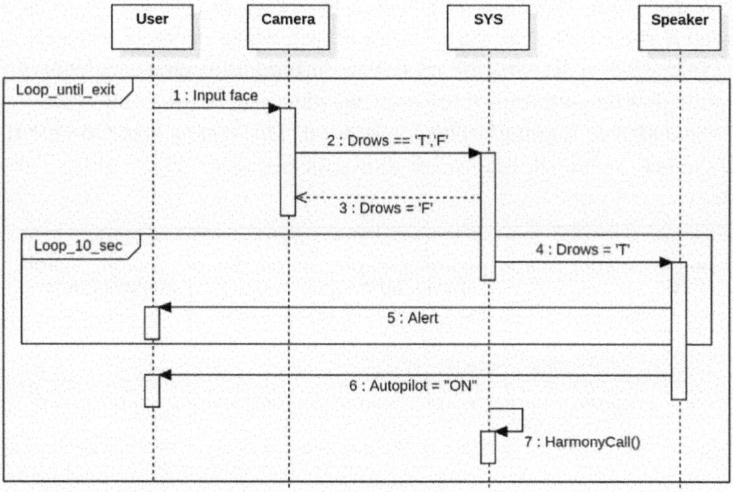

Fig. 2.4 DrowsDet sequence model

processing said input in the following 2-s interval. By partitioning the processes in such a pattern; *AIsys* can focus on one process at a time which resulted in a stable and accurate output during test runs. Figure 2.5 shows the sequence model of the *AIsys* sub-system.

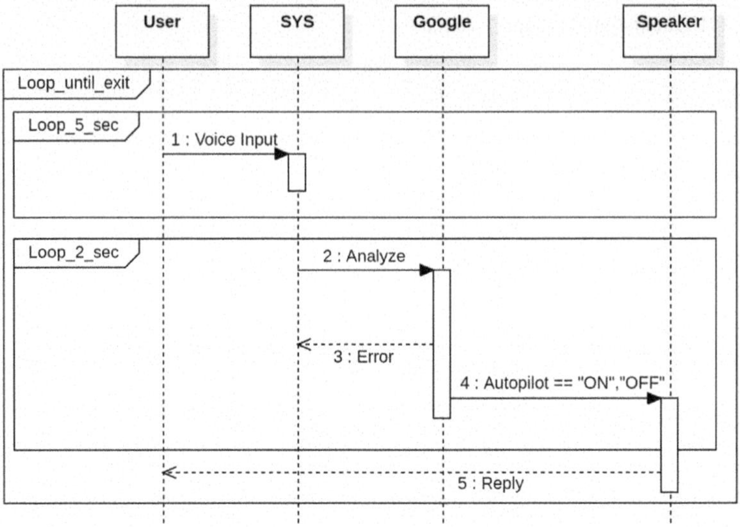

Fig. 2.5 AIsys sequence model

HarmonyCall. HarmonyCall is engineered to act as an emergency protocol that would send a voicemail to a nearby hospital or medical center. *HarmonyCall* is a reliable system that will assist the user even during unforeseen accidents of health deterioration. For the purposes of research, we identified closing one's eyes for more than 10 s as a sign of fainting/health implication. This can be updated later if more accurate data are obtained. Figure 2.6 shows the sequence model of the *DrowsDet* sub-system.

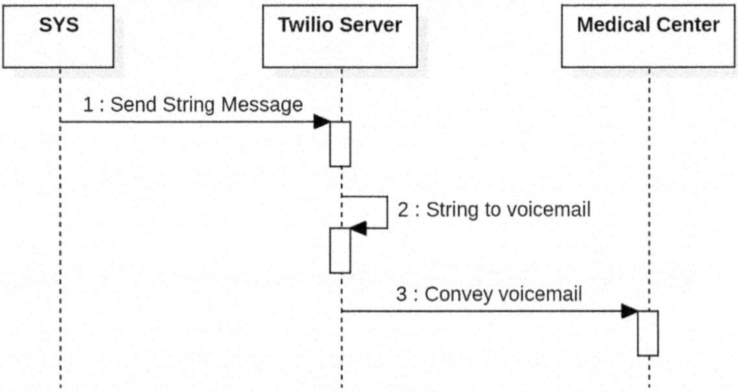

Fig. 2.6 HarmonyCall sequence model

2.4 Implementation

Figure 2.7 shows the block diagram of the proposed system. It also depicts each subsystem and details the data usage of each entity and well as the dependencies among all components of the *DriveHarmony* system. Specifically, the flowchart diagrams of the sub-systems *AIsys*, *HarmonyCall* and *DrowsDet* are shown in Figs. 2.8, 2.9, 2.10, and 2.11, respectively.

2.4.1 Implementation of DrowsDet

DrowsDet uses a camera module connected to a Raspberry Pi (RPi) which keep continuously monitors the driver and hence inputs a video stream of a driver, then matches the stream to a pre-defined face landmark file, as shown in Fig. 2.8, to identify the user's face and the different facial elements such as the eyes, nose, and

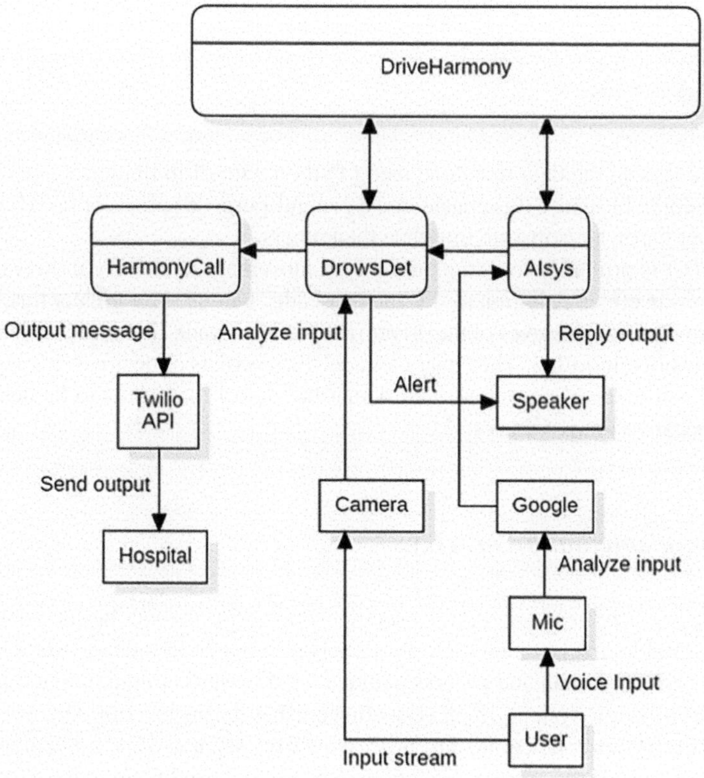

Fig. 2.7 DriveHarmony block diagram

Fig. 2.8 Face landmarks

mouth. Processing the data was done using Python with dLib library. Raspberry Pi has been adopted because it is considered a general-purpose possessor (GPP) with a much easier hardware/software interface than others.

DrowsDet is primarily focusing on the eyes movement to identify if drowsiness/distraction is in effect. Additionally, a "haar cascade" was used for greater functional accuracy in detecting the eyes of the driver. The "haar cascade" is a machine learning object detection algorithm. The "haar cascade" file consists of positive and negative images. Positive images are those that depict the object that needs to be detected, whereas negative images do not.

2.4.2 Implementation of AIsy

The sub-system *AIsys* can only accept user input during 5-s intervals because it proved to be more stable and accurate than if a user is to input continuously. *AIsys* would have to accept input data and process said data if the user is to input continuously. This results in a slow system which very often crashes during test run. Therefore, the processes of data acceptance and data processing were separated and implemented in stages, a 5-s interval to accept user input then processing said input in the following 2-s interval. By partitioning the processes in such a pattern, *AIsys* can focus on one process at a time which resulted in a stable and accurate output during test runs.

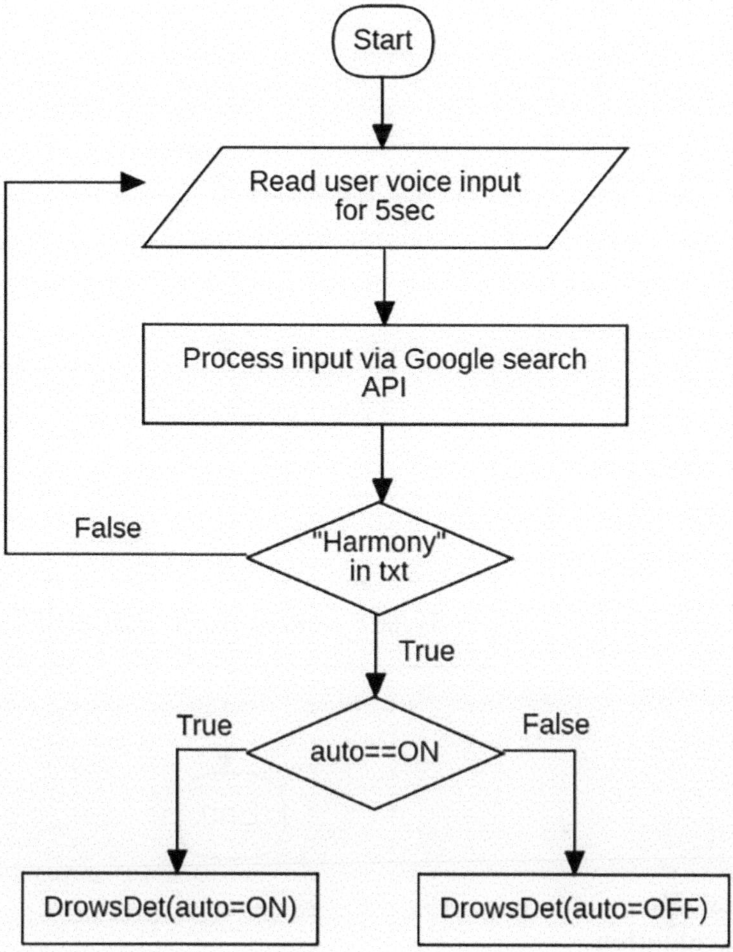

Fig. 2.9 AIsys flowchart

2.4.3 *Implementation of HarmonyCall*

HarmonyCall utilizes the Twilio API to issue a voicemail to a nearby hospital. *HarmonyCall* is initiated and fainting is detected. If so, a string message is sent from *HarmonyCall* to a Twilio server which then proceeds to send the string as a voicemail to a predetermined phone number. It should be noted that for the purposes of this work, the predetermined number was fixed. However, if practically applied; the number should be variable because the aim is to call the nearest hospital at the time of detecting signs of fainting from the user. In this case, the phone number will be determined automatically based on the location of the nearest hospital or medical service provider.

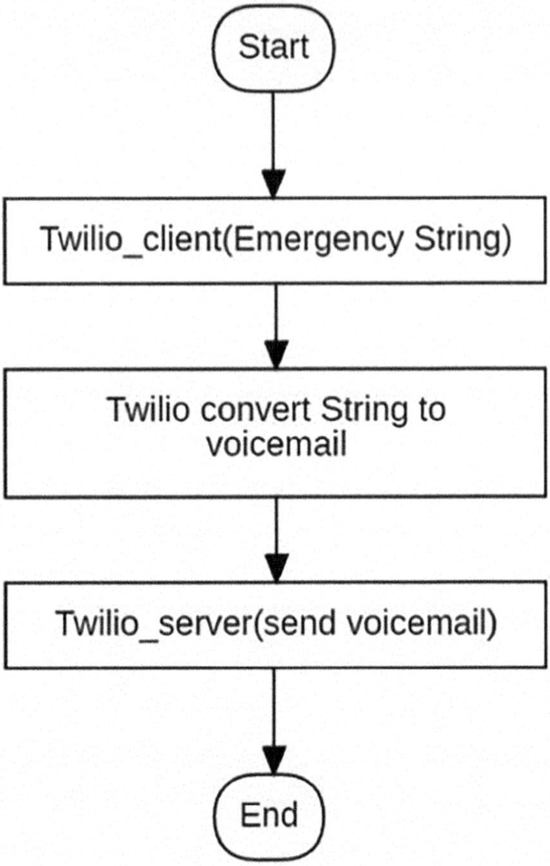

Fig. 2.10 HarmonyCall flowchart

2.5 Evaluation

To evaluate its effectiveness of the *DriveHarmony* system, we have performed a multi-stage testing process in which each component was tested independently and then the entire system was tested to ensure that the three main processes work together effectively. Eventually, the proposed system was integrated into a car and tested on the road.

2.5.1 *Black Box Testing*

For Black Box testing, the main purpose was to analyze the performance of the individual processes (*DrowsDet*, *AIsys*, and *HarmonyCall*) in terms of inputs and outputs as well as to test the *DriveHarmony*, which is the integration of the three

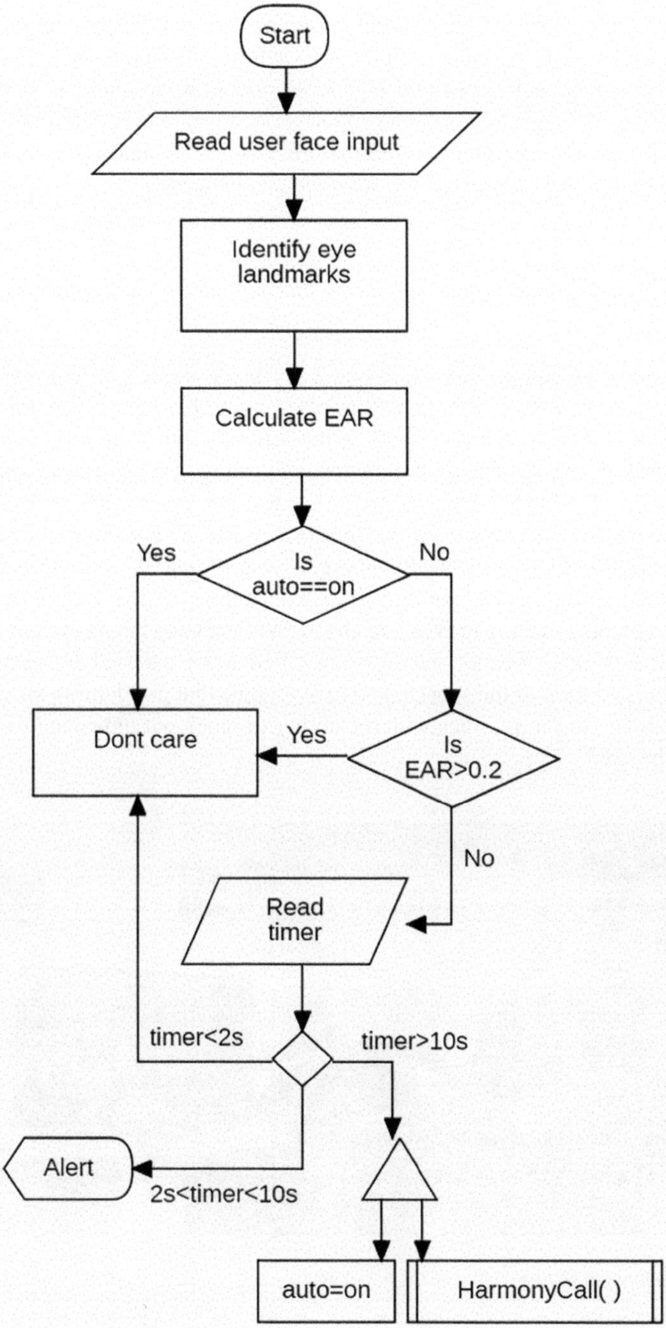

Fig. 2.11 DrowsDet flowchart

previously mentioned processes. Experimentation analysis helps in enhancing the quality and fidelity of the software processes while also simplifying the hardware required. Moreover, during this testing phase, we came to understand the actual behavior versus the expected behavior of DriveHarmony, which is important for the On-Road Testing. Furthermore, system overheads and limitations were analyzed, and the findings were considered for the system enhancement.

2.5.2 Stationary Testing

For the stationary testing, as shown in Fig. 2.12, the developed system was mounted on a phone holder, with a camera held in place by rubber bands. Several locations on the dashboard were utilized, such as the left side and center air conditioning units, before settling on placing the device itself on the dashboard mound right in front of the driver. That decision was made because it was the most optimal location and yielded the most accurate results. Once placed in the most optimal location, the drowsiness detection, as well as the emergency call features, worked as expected, with only minor issues faced regarding the eye detection, which required the driver to sit upright instead of having the seat lean back. While testing the AI system, we came to the realization that when the windows are rolled down, external noises may affect the AI's interpretation of the speech, and to overcome that problem we simply rolled up the windows. In future, mics with the ability to block out unwanted background noise can be used.

Fig. 2.12 Stationary testing

Fig. 2.13 On-road testing

2.5.3 On-Road Testing

The on-road testing, as shown in Fig. 2.13, was conducted first conducted an empty parking lot and then conducted in a quiet and uncrowded area in Kuwait City to avoid any unwanted circumstances and not to put others' safety at risk. During this testing, the emergency call and the fatigue/drowsiness detection worked just as well as they did during the stationary testing. Since the driver is on road and closing his eyes would lead to damage and injuries, the driver instead squinted his eyes just enough to see the road, while allowing the device to detect the eyes as being closed. As expected, on-road testing proved the effectiveness of *DriveHarmony* in detecting the drive fatigue through monitoring the drivers' eyes and the warning message was enabled so that the driver can focus back on driving. When the pattern has repeated for several times, the system advised the driver to stop driving and take some rest. However, when the driver continued -for the sake of testing- to close his eyes again and again and not stop the car, a phone call to the authorities was placed.

2.6 Conclusion

In this chapter, we presented an approach to combat driver's fatigue and dronishness and reduce the risk of having fatigue-related traffic crashes. The developed system, called "*DriveHarmony*", continuously monitors the drivers' eyes and face to detect any fatigue or drowsiness. Once detected, the system issues an audio notification to help the driver focus back on driving. If the pattern of fatigue repeats, the system will request the driver to stop driving and take a rest. Failing to respond to the system instructions results in notifying the authorities about the case and the risk of potential accident. If driver shows signs of health deterioration, medical center/facility will be automatically called. This system helps reduce the crush rate on roads and consequently saves people's lives.

References

Ani MF, Kamat SR, Fukumi MF, Noh NA (2020) A critical review on driver fatigue detection and monitoring system. Int J Road Saf 1(2):53–58

Brice CF, Smith AP (2002) Effects of caffeine on mood and performance: a study of realistic consumption. Psychopharmacology 164:188–192

Dasgupta A, George A, Happy SL, Routray A (2013) A vision-based system for monitoring the loss of attention in automotive drivers. IEEE Trans Intell Transp Syst 14(4):1825–1838

Ed-Doughmi Y, Idrissi N, Hbali Y (2020) Real-time system for driver fatigue detection based on a recurrent neuronal network. J Imaging 6(3):8

Fatima B, Shahid AR, Ziauddin S, Safi AA, Ramzan H (2020) Driver fatigue detection using viola jones and principal component analysis. Appl Artif Intell 34(6):456–483

Global status report on road safety (WHO). https://www.who.int/publications/i/item/9789241565684. Accessed 19 Jun 2023

Karuppusamy NS, Kang BY (2020) Multimodal system to detect driver fatigue using EEG, gyroscope, and image processing. IEEE Access 8:129645–129667

Kuwait Times homepage. https://timeskuwait.com/news/352-traffic-accident-deaths-in-2020/. Accessed 19 Jun 2023

Lal SK, Craig A (2001) A critical review of the psychophysiology of driver fatigue. Biol Psychol 55(3):173–194

Nordbakke S, Sagberg F (2007) Sleepy at the wheel: knowledge, symptoms and behaviour among car drivers. Transp Res Part f: Traffic Psychol Behav 10(1):1–10

Ren Z, Li R, Chen B, Zhang H, Ma Y, Wang C, Zhang Y et al (2021) EEG-based driving fatigue detection using a two-level learning hierarchy radial basis function. Front Neurorobotics 15:618408

Reshma J, Aishwarya B, Farheen Khanam ZM, Sai Vennela G, Lekhana AV (2021) A survey on driver drowsiness detection techniques. Int J Latest Technol Eng Manag Appl Sci (IJLTEMAS) IX(XI), 13–17

Saleh S (2023) Number of deaths due to traffic accidents in Kuwait 2012–2020. Statista homepage. https://www.statista.com/statistics/652679/kuwait-road-traffic-number-of-deaths/. Accessed 19 Jun 2023

Savaş BK, Becerikli Y (2020) Real time driver fatigue detection system based on multi-task ConNN. IEEE Access 8:12491–12498

Shalash WM (2019) Driver fatigue detection with single EEG channel using transfer learning. In: 2019 IEEE international proceedings on imaging systems and techniques (IST). IEEE Press, Abu Dhabi, United Arab Emirates, pp 1–6

Sheykhivand S, Rezaii TY, Meshgini S, Makoui S, Farzamnia A (2022) Developing a deep neural network for driver fatigue detection using EEG signals based on compressed sensing. Sustainability 14(5):2941

The global health observatory - Road Safety (WHO). https://www.who.int/data/gho/data/themes/road-safety. Accessed 19 Jun 2023

Thomas MJW, Gupta CC, Sprajcer M, Demasi D, Sach E, Roach G, Sargent C, Dawson D, Ferguson SA (2021) Fatigue and driving: an international review. Appleton Institute - Central Queensland University, Adelaide, Australia

Wang H, Xu L, Bezerianos A, Chen C, Zhang Z (2020) Linking attention-based multiscale CNN with dynamical GCN for driving fatigue detection. IEEE Trans Instrum Meas 70:1–11

Xu T, Wang H, Lu G, Wan F, Deng M, Qi P, Sun Y et al (2023) E-key: an EEG-based biometric authentication and driving fatigue detection system. IEEE Trans Affect Comput 14(2):864–877
Zhao J, Hao K, Ding Y (2016) Driver fatigue monitoring system using video images and steering grip force. In: 5th international proceedings on measurement, instrumentation and automation. Atlantis Press, Shenzhen, China, pp 610–617 (2016)
Zhou F, Alsaid A, Blommer M, Curry R, Swaminathan R, Kochhar D, Lei B et al (2020) Driver fatigue transition prediction in highly automated driving using physiological features. Expert Syst Appl 147:113204
Zhuang Q, Kehua Z, Wang J, Chen Q (2020) Driver fatigue detection method based on eye states with pupil and iris segmentation. IEEE Access 8:173440–173449

Chapter 3
Effective Integration of Clustering and Classification or Regression Machine Learning Algorithms

Daniel Okunbor

Abstract This chapter includes a good exposition of centroid-based, density-based, distribution-based, and hierarchy-based clustering machine learning algorithms and various supervised machine learning models. K-means remains a popular unsupervised machine learning algorithm, we will discuss how the K-means can build hierarchical clustering. There are several classifiers and regressors in the literature and since the focus of the chapter is on unclassified, unlabeled, or unstructured datasets, the goal will be to start with clustering algorithm to determine the different classes or labels in the data. The resulting labeled data are used to build a classifier or a regressor as appropriate. This approach is important as significant datasets are unclassified or unlabeled. This approach has several applications, especially in natural language processing.

Keywords Centroid-based · Density-based · Distribution-based · Hierarchical-based clustering machine learning · K-means · Unclassified dataset · Unsupervised learning

3.1 Introduction

Machine learning algorithms include supervised, unsupervised, and reinforcement learning. Supervised learning algorithms are referred to as ground truth models since they model datasets that are labeled, where the labels are used as targets in the implementation and to assess the validity of the model. The objective is to learn from the dataset using the targets to develop a predictive model used to predict the label or class or outcome of a new data. Supervised learning can be thought of as a training process that has a teacher (target). Unsupervised learning algorithms model unlabeled or unclassified datasets. Here, the dataset does not include targets, labels,

D. Okunbor (✉)
Fayetteville State University, Fayetteville, NC 28301, USA
e-mail: diokunbor@uncfsu.edu

D. K. Sharma et al. (eds.), *Machine Learning for Real World Applications*, Transactions on Computer Systems and Networks, https://doi.org/10.1007/978-981-97-1900-6_3

classes, or outcomes. The objective is to determine patterns and other useful information from the dataset. It can be an important approach for data segmentation, pattern recognition, sentiment analysis, and many more. These two categories of learning algorithms (supervised and unsupervised) rely on existing data. The reinforcement learning algorithms on the other hand depend on the object of study and the object's environment; the data which represent the behaviors or actions are collected dynamically during training which involves simulating the object (agent) within the environment using a reward system. The simulation involves changing the object's state within the environment.

The supervised learning model is a predictive model; think of it as an interpolation or extrapolation model. It is interpolation if the new data fall within the range of the existing data and extrapolation if the new data are outside of the range. Two types of supervised learning models used in practice are classification and regression (Raschka and Mirjalili 2017; Muller and Guido 2016). Typically, classification techniques model discrete output values while regression techniques model continuous output values. The classification determines a decision boundary dividing classes in the data while the regression finds the fit function that optimally predicts the outcome. Examples of classification algorithms are logistic regression, k-nearest neighbors, support vector machines, kernel support vector machines, Naïve Bayes, decision tree, random forest, and artificial neural networks (Muller and Guido 2016).

As mentioned previously, the unsupervised learning models do not involve any ground truth and learning is based on unlabeled data. The objective is to obtain useful information by grouping the data as in clustering or reducing the dimensionality as in feature extraction. The features (also called attributes) are the causal or independent variables in the dataset.

This chapter focuses on unsupervised learning with the goal of developing clusters from the dataset and using the clustered dataset in a supervised learning to obtain a predictive model. In the process of clustering, dimension reduction techniques, as in other machine learning models, can be used to improve model accuracy. Although we will discuss several clustering models, the k-means clustering, and the density-based clustering will be implemented. The architecture is shown in Fig. 3.1.

This chapter is organized as follows. In Sect. 3.2, different clustering and dimensionality reduction techniques are discussed. While k-means algorithm may be the most popular of all the clustering algorithms, finding the optimal number of clusters remains a computationally hard problem. Although there are heuristic techniques for obtaining a good approximate number of clusters, testing different k values remains a widespread practice. The classification algorithms are discussed next in Sect. 3.3 along with regression techniques. There are a variety of classifiers and regressors in the literature, however, the focus will be on classification. Regression algorithms may be appropriate in certain situations, the integration approach discussed will create classes or outcomes that will be categorical and therefore may not be suitable for regression models. The clusters are assigned labels that will be used in classification. The unsupervised learning served as a prior model for the predictive model.

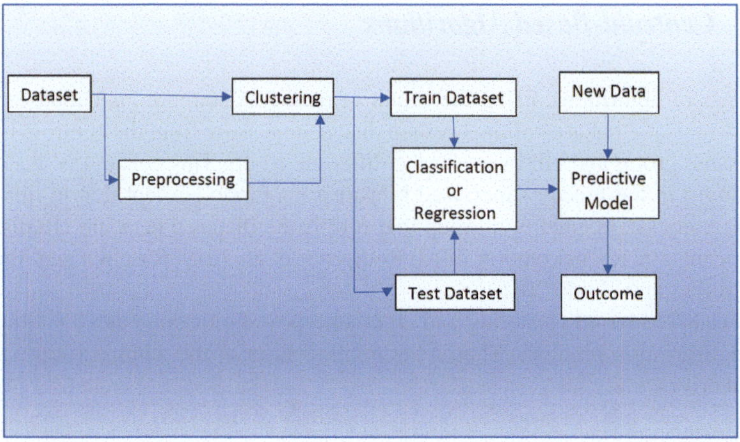

Fig. 3.1 Combined model

Section 3.4 discusses example models based on the integration of clustering and classification models, while Sect. 3.5 focuses on the use of sklearn to build integrative models. Section 3.6 includes concluding remarks and a summary of the chapter.

3.2 Clustering Algorithms

In clustering, the heterogeneous dataset is partitioned into homogeneous groups using pattern extraction and inferences (Brownlee 2020; Shalev-Shwartz and Ben-David 2014b). These groups are called clusters and, in the context of supervised learning, may be considered as classes or labels. Data in each group have similar characteristics, hence the homogeneity. The literature is replete with the definition of clustering (Shalev-Shwartz and Ben-David 2014b; Muller and Guido 2016). In the explanation by Shalev-Shwartz and Ben-David (2014b), data in the same group are like each other, and data in separate groups are dissimilar.

Clustering algorithms are based on similarity functions that represent the degree of similarity of data within the same group or dissimilarity of data from separate groups. The complexity of the algorithm rests on the difficulty in determining the similarity. These similarity functions distinguish the categories of clustering algorithms. The following are types of clustering algorithms found in the literature: centroid-based, density-based, distribution-based, connectivity-based, fuzzy-based, and constraint-based. Note that there may be other clustering algorithms that may fit into any of these types (Aravind 2022; Rodriguez et al. 2019; API Reference. https://scikit-learn.org/stable/modules/classes.html; What is Clustering in Machine Learning: Types and Methods 2022).

3.2.1 Centroid-Based Algorithms

This class of algorithms, using the center of data concentration, is called centroid group data. Data that are within a certain distance measure from the centroid belong to the same group and others are in the different group. This concept is analogous to that used in the greedy supervised k-nearest neighbors, except that in this case, the data point is assigned to a group that fulfills the distance measure (Budisteanu and Mocanu 2021; Chakraborty 2017; Rodriguez et al. 2019; Raschka and Mirjalili 2017; Muller and Guido 2016).

The centroid-based clustering and all other types of clustering can be thought of as an optimization problem. That is the minimization of the within groups sum of squared errors, viz,

$$\min_{w} \varphi(x, \mu, w) \tag{3.1}$$

where μ is the centroid, x is data point and $w = \text{prob}(x \in C)$ is the probability that x is in cluster C. This objective function φ depends on the distance measure. There are a variety of distance measures (Scikit Learn – Quick Guide undated; Brownlee 2020; Shalev-Shwartz and Ben-David 2014b), used in practice, namely,

- Euclidean distance (2-norm): $d^2(x, y) = \sum_{j=1}^{n}(x_j - y_j) = ||x - y||^2$. Here, it is assumed that there are n features. Each group is spherical in shape, that is, each example data is within a spherical distance from the centroid.
- Manhattan distance (1-norm): $d(x, y) = \sum_{j=1}^{n}|x_j - y_j|$. Each example data is within a square that envelopes the centroid. This is a particular case of the Minkowski distance.
- Minkowski distance: $d^p(x, y) = \sum_{j=1}^{n}\left(|x_j - y_j|\right)^p$
- Eisen cosine correlation distance: $d(x, y) = 1 - \frac{|\sum_{j=1}^{n} x_j y_j|}{\sqrt{\sum_{j=1}^{n} x_j^2 \sum_{j=1}^{n} y_j^2}}$. Uses the angle of the two vectors at the origin to calculate the distance between them.
- Pearson correlation measure: $d(x, y) = 1 - \frac{\sum_{j=1}^{n}(x_j - \mu_1)(y_j - \mu_2)}{\sqrt{\sum_{j=1}^{n}(x_j - \mu_1)^2 \sum_{j=1}^{n}(y_j - \mu_2)^2}}$. In this distance measure, we first calculate the mean values of x and y, namely, μ_1 and μ_2, respectively. The distance measure is typically suitable for agglomerative clustering.

K-Means

The popular k-means algorithm is an example of centroid-based algorithms. It is a simple algorithm to implement. However, the number of clusters k must be known a priori. Typically, the initial centroids are randomly selected from the dataset. The distances between the examples and the centroids are computed and the examples are placed in the clusters using the following:

$$x \in C^r, \ \textit{if} \ d(x, \mu_r) = \min_{j}\{d(x, \mu_j), j = 1,2, \ldots, k\} \tag{3.2}$$

The centroids are updated using the means (hence it is called k-means) of the clusters, viz:

$$\mu_j = \frac{\sum_{i=1}^{m_j} x_i \in C^j}{m_j}, m_j = |C^j| \tag{3.3}$$

The assignment of examples and the update to centroids are repeated until the clusters or centroids no longer change or until a specified maximum number of iterations is reached.

The distance measure to use is application-dependent. As indicated, the number of clusters, k must be known ahead of the implementation of the algorithm. Finding an optimal k is a nondeterministic polynomial-time (NP) problem. There are heuristic techniques for estimating a "good" k value. The two most popular techniques are the elbow and the silhouette methods (Finley and Joachims 2010; Raschka and Mirjalili 2017).

There are variants of the k-means such as the bisecting k-means, the mean shift k-means, and the mini-batch k-means, see (API Reference. https://scikit-learn.org/stable/modules/classes.html; What is Clustering in Machine Learning: Types and Methods 2022) for good description of these algorithms.

In the elbow method, start with a small value of k and gradually increase its value to obtain predicted clusters. For each k value, calculate the within the clusters sum of squared errors $\varphi(x, \mu) = \sum_{j=1}^{k}\sum_{i=1}^{m_i} d_i^2(x_i, \mu_j)$. The chosen k value is one in which $\varphi(x, \mu)$ begins to diminish and this point is the elbow.

The silhouette approach uses a special value called silhouette for similarity measure of an instance data to determine how its associated cluster, which is called cohesion, compared to the rest of the separated clusters (this is called separation). It has been shown that the silhouette typically varied values from -1 to 1. A higher silhouette value indicates that the sample has been correctly placed, while a negative silhouette value is an indication of too many or too few clusters. Let $\mu_1(i)$, $\mu_2(i)$ denote the mean distances of the i-th sample in the cohesion and the separation, respectively and let C_K denotes the K-th cluster. The formula for silhouette value is (Raschka and Mirjalili 2017).

$$s(i) = \begin{cases} \frac{\mu_1(i)-\mu_2(i)}{\max\{\mu_1(i),\mu_2(i)\}}, \text{ if } |C_K| > 1 \\ 0, \text{ if } |C_K| = 1 \end{cases} \tag{3.4}$$

where the measure similarity is

$$\mu_1(i) = \frac{1}{|C_K| - 1} \sum_{x_i, x_j \in C_K, i \neq j} d(x_i, x_j) \tag{3.5}$$

and the measure of dissimilarity is:

$$\mu_2(i) = \min_{i \neq j} \frac{1}{|C_K|} \sum_{x_j \in C_K,} d(x_i, x_j) \tag{3.6}$$

The elbow and silhouette techniques are implemented in sklearn (API Reference. https://scikit-learn.org/stable/modules/classes.html; Scikit Learn – Quick Guide undated).

3.2.2 Density-Based Clustering

The density-based clustering is based on continuous regions of high density in the dataset, and it is especially suitable for spatial clustering applications. It relies on two hyperparameters which are the cluster size for the cluster to be considered dense and the distance measure used to ascertain the neighborhood to a given example data is dense.

The density-based can effectively separate outliers and cluster observations in high-density areas and build clusters that have an arbitrary shape or clusters within clusters. There are four density-based clustering algorithms found in the literature, and there are density-based spatial clustering of applications with noise (DBSCAN), **Ordering *Points to Identify Cluster Structure*** (OPTICS), Density Clustering (DENCLUE), and Clustering Quest (CLIQUE). The DBSCAN is the only algorithm described, see (API Reference. https://scikit-learn.org/stable/modules/classes.html; Brownlee 2020; What is Clustering in Machine Learning: Types and Methods 2022; Great Learning Team 2022) for information about the remaining algorithms.

DBSCAN Algorithm

The DBSCAN algorithm is the most widely used of all the density-based clustering algorithms and it is described using two parameters, namely, an epsilon, *eps* or ε and the minimum neighbor samples (or points) denoted by *minPts*. The ε is used to represent the radius of the circular region of density points of a point (these are called the core points) and *minPts* is the allowable minimum number of points within the circular density region. Every dataset sample is assigned one of three labels—core point, border point, and noise point. A core point is at least a specified *minPts* of neighboring points that fall within radius of *eps*. The border point has fewer neighbors than *minPts* within *eps* but lies within the radius *eps* of a core point. Points that are not core or border points are noise points (Raschka and Mirjalili 2017).

The DBSCAN algorithm is in three steps:

1. Determine core, border, and noise points as demonstrated in Fig. 3.2.
2. Form a separate cluster for each core point or connected group of core points.
3. Assign each border point to the cluster of its corresponding core points. All unassigned points are noise points.

The DBSCAN algorithm has the advantage of determining arbitrarily shaped clusters such as shown in Fig. 3.3.

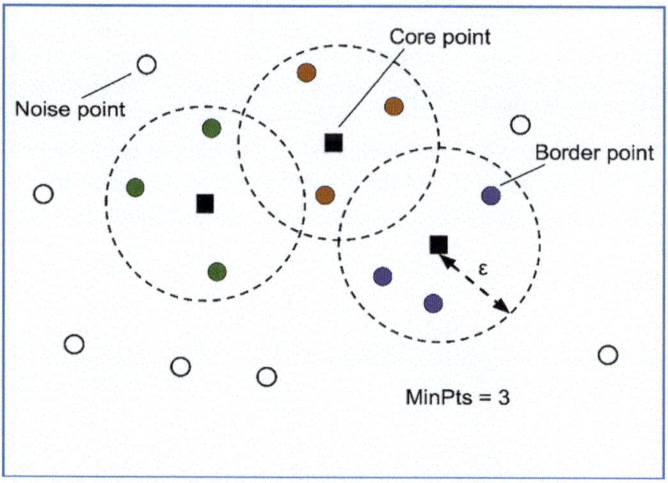

Fig. 3.2 Illustration of DBSCAN parameters (Raschka and Mirjalili 2017)

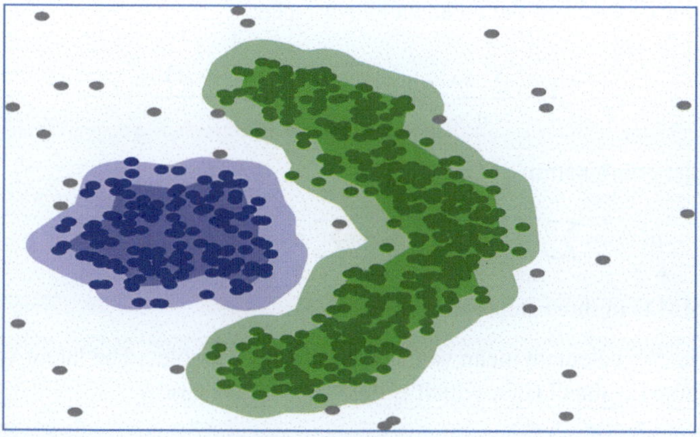

Fig. 3.3 DBSCAN can find non-linearly separable clusters (DBSCAN. https://en.wikipedia.org/wiki/DBSCAN)

3.2.3 Distribution-Based Clustering

This clustering model is also called model-based clustering, and it is closely related to statistics and is based on probability distributions. A cluster **is defined as objects belonging to the same distribution**. These clustering approaches assume data that are composed of distributions, such as Gaussian distributions. The cluster includes data objects that have a higher probability of being in it. Each cluster has a central point, the higher the distance of the data point from the central point, the lesser will be its probability to be included in the cluster (Scikit Learn – Quick Guide undated). One example of distribution-based clustering is the Gaussian mixture model (**GMM**).

Gaussian Mixture Model (GMM)

The GMM model maximizes the likelihood that a sample point belongs a given cluster using both the mean μ and the covariance Σ as described in a probability density function of Gaussian distribution (or normal distribution). The likelihood function is:

$$L(X) = \prod_{i=1}^{n} N(x_i|\mu, \Sigma) \tag{3.7}$$

where

$$N(x|\mu, \Sigma) = \frac{1}{\sqrt{2\pi|\Sigma|}}\exp\{-\frac{1}{2}(x-\mu)\Sigma^{-1}(x-\mu)\} \tag{3.8}$$

The maximum of the log of the likelihood function called the maximum likelihood estimate is obtained from the mean and covariance:

$$\widehat{\mu} = \frac{\sum x_i}{n}, \widehat{\Sigma} = \frac{\sum(x_i - \widehat{\mu})(x_i - \widehat{\mu})^T}{n} \tag{3.9}$$

The linear combination of Gaussians is:

$$p(x) = \sum \pi_k N(x_k|\mu_k, \Sigma_k), \sum \pi_k = 1, k = 1,2, \ldots, K \tag{3.10}$$

The GMM is in three steps:

1. E-phase: We calculate mean values of the latent variables. The latent variable in this context is the cluster, which is unknown in this case.

$$r_{ik} = \frac{\pi_k N(x_k|\mu_k, \Sigma_k)}{\sum \pi_k N(x_k|\mu_k, \Sigma_k)} \tag{3.11}$$

2. M-phase: In this phase, the log-likelihood function is maximized to update appropriate parameters:

$$E = \sum_i \sum_k r_{ik}\{\ln \pi_k + \ln N(x_k|\mu_k, \Sigma_k) \tag{3.12}$$

$$\pi_k = \frac{\sum r_{ik}}{n}, \mu_k = \frac{\sum r_{ik}x_i}{n}, \Sigma_k = \frac{\sum r_{ik}(x_i - \mu_k)(x_i - \mu_k)^T}{\sum r_{ik}} \tag{3.13}$$

3. Phases (1) and (2) are repeated until convergence is reached, that is, when there is longer an increase in the log-likelihood function.

As described above, GMM is an optimization problem like the k-means except that it uses in addition to the mean, the covariance. The k-means algorithm is regarded as hard clustering since every point belongs to exactly one cluster. The GMM is considered a soft clustering because every point is assigned to several clusters with specific probabilities and for this reason, the GMM can be used for non-spherical clusters.

3.2.4 Hierarchy Clustering

A clustering method that is based on hierarchy is called linkage-based clustering and it is based on bottom-up fashion where each data point is considered a cluster initially and in a repetitive fashion, the algorithm merges clusters of the previous clustering until all points are in single large cluster. With this class of algorithms, two parameters are used to define the algorithm, the first parameter being the distance between clusters, while the second parameter is used as a stopping criterion. The merging uses the distance function as explained below:

1. Single linkage clustering: this phase is based on the minimum distance between data points of the two clusters.
2. Average linkage clustering: this phase uses the average distance between a single point in one cluster and points in the rest.
3. Max linkage clustering: uses the maximum distance between points in two different clusters.

The hierarchy clustering as described above is clearly an agglomerative or divisive clustering. The computing process commences from data that is completely fragmented dataset to build larger clusters. When the stopping criteria are employed, the algorithm produces dendrogram-like tree of domain subsets that have a single-member set in its leaves and at its root is a single cluster consisting of all data points. Figure 3.4 illustrates a hierarchy clustering consisting of data points whose values are “A”, “B”, “C”, “D”, “E”, and “F”.

For hierarchical-based clustering, the stopping criteria may include:

1. Fixed number of clusters: In this case, we will stop merging clusters once we have reached a fixed number of clusters.
2. Distance upper bound: In this case, we will stop merging once all distances of points between clusters are within an upper bound.

3.2.5 Grid-Based Clustering

A clustering that is based on grid structure is called grid-based clustering, and it is also called a spectral clustering. According to Data Mining 365 (2020), the first step in a grid-based clustering is to use a uniform grid mesh to represent the problem

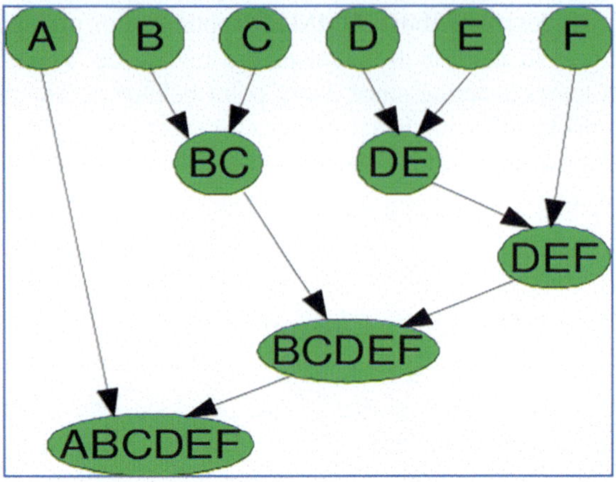

Fig. 3.4 Agglomerative clustering (Patlolla 2018)

domain space. Following that, statistics are collected for objects that are recognized in each individual mesh cell. The next step is clustering, and this is performed on the grid, instead of data objects themselves. It has been shown that grid-based clustering has high computing performance, however, such performance is dependent on grid size (Data Mining 365 2020). There are several grid-based clustering algorithms in practice, including, Statistical Information Grid (STING) developed by Wang, Yang and Muntz in 1997; WaveCluster, a multi-resolution clustering approach using wavelet method and developed by Sheikholesami, Chatterjee and Zhang in 1998; and CLIQUE which combines grid approach and density-based approach. A brief description of the STING algorithm is presented in the next section (Finley and Joachims 2010; Wang et al. 2008; Shalev-Shwartz and Ben-David 2014a; Mahendru 2019; Data Mining 365 2020).

STING Algorithm

As described in Data Mining 365 (2020), the STING algorithm splits the spatial area into rectangular mesh of cells. The resolution approach determines the levels of cells. A cell at a higher level is split into multiple smaller cells at the next lower level. Once this is done, statistics for each cell are calculated. The following statistics, namely, count, mean, standard deviation, min, and max of cells in the lower level are then used to specify cells at the higher level. Additional statistical information such as the probability distribution (normal, uniform) is also computed and used in this grouping of cells.

STING utilizes hierarchy to divide the spatial area into rectangular mesh of cells that resemble a quadtree. According to Data Mining 365 (2020), STING uses spatial database for representing the data points, from which statistical parameters are calculated for each cell.

3.3 Classification and Regression Algorithms

There are numerous machine learning algorithms for supervised learning. As described in Section I, the supervised learning model consists of datasets that are labeled or that have outcome or class or dependent variable. A supervised learning model is either a classification or a regression. Classification is used when the outcome variable is categorical or discrete while regression is used when the outcome variable is continuous. Our focus in the chapter is on classification simply because the clustering algorithm that will be used will result in a categorical outcome variable. While there are many classification algorithms such as the artificial neural networks (perceptron, multi-layer perceptron, adaptive linear neural, backpropagation, deep learning), logistic regression, support vector machines, Naïve Bayes, self-organizing maps, decision trees, k-nearest neighbors, ensemble learning such as random forest; the chapter will only describe the logistic regression and support vector machines. However, artificial neural network (ANN) algorithm has gained wide popularity due to the hype for deep learning.

The nature of supervised learning allows it to be used widely for predictive modeling. Given a dataset, the objective is to train the model to obtain a predictive model that can be used for predicting the outcome of new data. In training the model, the feature variables (x) and the outcome or class variable (y) are related using the linear function:

$$y = \varphi(x) = w^T x \tag{3.14}$$

and the parameter w called the training weight minimizes the cost function (also called the sum of squared errors):

$$c(w) = \frac{1}{2}\sum \left(y^i - \varphi\left(x^i\right)\right)^2 \tag{3.15}$$

The cost function is minimum along the gradient descent, hence the update to the training weight is accomplished using:

$$w = w + \Delta w,\ \Delta w = -\eta \nabla c(w) \tag{3.16}$$

The $\nabla c(w)$ is the gradient of $c(w)$ and η is the learning rate. The update to the learning weight is repeated until the cost function reaches a threshold or specified maximum number of iterations. These new parameters, the learning rate and the maximum number of iterations, are called hyperparameters in machine learning and are tunable at implementation.

3.3.1 Logistic Regression

The logistic regression model is based on conditional probabilities. The linear function is defined as the log of the likelihood function (i.e., the probability that a class or label came from given features). This is written as

$$\ln(p(y = 1|x)) = \ln\left(\frac{p}{p-1}\right) = w^T x = z, \text{ or } p = \frac{1}{1+e^{-z}} = \varphi(z). \tag{3.17}$$

The φ in this case is sigmoid function. The associated cost function for logistic regression is the negative of the log-likelihood function.

$$c(w) = -\sum y^i \ln(\varphi(z^i) + (1 - y^i)\ln(1 - \varphi(z^i)) \tag{3.18}$$

The $\nabla c(w)$ is easily derived from the minimization of the cost function, since $\frac{\partial \varphi}{\partial z} = \varphi(z)(1 - \varphi(z))$. In situations where the model is overfitting (that with high variances) or underfitting (high bias), the process of regularization is employed. Variance measures the variability or consistency of the prediction model when applied numerous times and bias measures how far off the predictions are the correct values. The L1 and L2 regulations are typically utilized in the training to correct for high variance or high bias.

3.3.2 Support Vector Machines (SVM)

Support vector machine algorithm is another widely used machine learning algorithm. It is both for classification and regression. This algorithm uses the concepts of margin and hyperplane to classify the dataset. The algorithm finds an optimal hyperplane that distinctly separates data points, and this is called the decision boundary hyperplane. This hyperplane divides the plane into two planes, the positive plane and negative plane, and hyperplanes in these regions are called positive hyperplane and negative hyperplane, respectively. Hyperplanes of points closest to the decision boundary hyperplane are called support vectors. The margin is the distance between positive support vector and negative support, see Fig. 3.5.

The margin can easily be written as

$$\varphi(w) = \frac{2}{||w||} \tag{3.19}$$

The goal of the support vector machines is to maximize the margin (or minimize $||w||$) subject to some constraints, namely,

$$w_0 + w^T x^i \geq 1 \text{ if } y^i = 1; \; w_0 + w^T x^i \leq -1 \text{ if } y^i = -1 \tag{3.20}$$

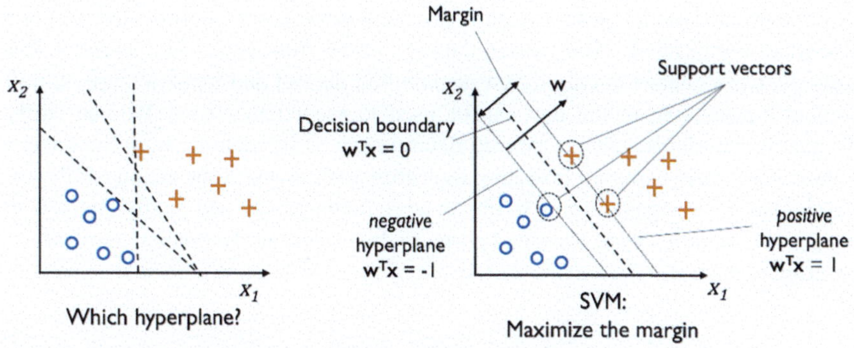

Fig. 3.5 Support vector machines (Raschka and Mirjalili 2017)

or written compactly, it is:

$$y^i\left(w_0 + w^T x^i\right) \geq 1, \forall i \tag{3.21}$$

Using Lagrange formulation, the optimization is equivalent to the minimization of

$$\lambda||w||^2 + \frac{1}{n}\sum\left(1 - y^i\left(w_0 + w^T x^i\right)\right). \tag{3.22}$$

For nonlinear kernels of support vector machines, the features are transformed into a different hyperplane called kernels. Such transformations can be polynomial, Gaussian radial basis or sigmoid.

3.3.3 *Integrated Learning Algorithms*

As discussed in the preceding sections, supervised learning, unsupervised learning, and reinforcement learning are three different approaches for machine learning. Since a considerable number of applications produce data that is unlabeled and sometimes unstructured, supervised learning is not appropriate. While preprocessing techniques such as dimensionality reduction and scaling may result in better sanitized data, they do not create labels for the data if the dataset is unlabeled.

The proposed approach uses an algorithm for unsupervised learning to apportion dataset into separate groups or clusters. These clusters are then assigned as labels, and these labels constitute the outcome or dependent variable making the resultant dataset suitable for supervised learning. The assumption in this approach is that the unsupervised learning serves as an expert system, and the resultant dataset can now be used to train the model to obtain a predictive model. The chosen unsupervised learning model must be application appropriate and therefore application dependent.

The proposed model ensures that the given dataset without labels is used to create new dataset with labels. The labeled dataset is used from supervised learning. For example, use k-means to classify an unclassified dataset and logistic regression to develop a predictive model that can be used to predict new data. We can easily evaluate the model by using a labeled dataset such as Iris dataset, train the dataset without the labels using a clustering algorithm such as the k-means and follow it with logistic regression training to obtain a predictive model and finally check how many of the samples were classified correctly by executing the predictive model on the original dataset.

3.4 Building Learning Model Using Sklearn

Python is becoming the de facto language for machine learning with its rich libraries dedicated to machine learning applications. In particular, the scikit-learn (sklearn) is considered as the most useful software containing robust Python libraries for machine learning. The literature is replete and inundated with the machine learning applications using machine learning algorithms implemented in sklearn. As an open-source Python library container, changes to sklearn modules are made on a regular basis. The first full version of the sklearn was released in 2013, and there have been many versions since then. The sklearn is built upon other libraries, including numpy, scipy, joblib, matplotlib, and pandas. These libraries and many other libraries are available on Jupyter Notebook, which is an integrated development environment (IDE) packaged with Anaconda. Anaconda is also an open-source graphical interface library that is downloaded from anaconda.com. Anaconda comes with many application programming interfaces including Jupyter Notebook.

The sklearn library package implements many learning algorithms including supervised learning, unsupervised learning, preprocessing, performance metrics algorithms. These algorithms are grouped into packages and modules. There are modules for generating synthetic datasets and the list of datasets available in the module can be found in https://en.wikipedia.org/wiki/List_of_datasets_for_machine-learning_research. Table 3.2a and b contain some of the modules and implemented algorithms (API Reference. https://scikit-learn.org/stable/modules/classes.html; Scikit Learn – Quick Guide undated; Raschka and Mirjalili 2017; Muller and Guido 2016).

Several algorithms are implemented as Python classes and, therefore, they must be instantiated to create objects from them. Others are implemented as functions. There are two types of attributes, parameter attributes and outcome attributes. The outcome attribute should not be confused with outcome variable. The parameter attributes represent various input parameters needed for the object and keyword arguments are used to pass values to the parameters. As a convention, all parameter attributes have default values, and the default value is used when one is supplied via keyword argument at the time the object is defined.

Table 3.1 sklearn modules and algorithms

Modules	Algorithms
3.1a	
sklearn.datasets	make_blobs, make_classification, load_iris, load_iris, load_digits, load_wine, fetch_olivetti_faces, fetch_lfw_people, load_breast_cancer, make_moons, make_hastie_10_2, make_friedman1, make_regression, make_circles, load_boston, load_files, etc.
sklearn.linear_model	LogisticRegression, Perpcetron, SGDClassifier, LinearRegression, Ridge, Lasso, SGDRegressor, Perceptron
sklearn.model_selection	KFold, cross_val_score, train_test_split, cross_validate, ShuffleSplit, LeaveOneOut, GroupKFold, GridSearchCV,
sklearn.svm	OneClassSVM, LinearSVR, NuSVR, LinearSVC, NuSVC, SVC, SVR
sklearn.preprocessing	StandardScaler, PolynomialFeatures, MultiLabelBinarizer, LabelBinarizer, MinMaxScaler, normalize, scale, Binarizer, OneHotEncoder, KBinDiscretizer
sklearn.ensemble	AdaBoostClassifier, AdaBoostRegressor, GradientBoostingClassifier, GradientBoostingRegressor, ExtraTreesRegressor, ExtraTreesClassifier, RandomForestRegressor, RandomForestClassifier
sklearn.cluster	KMeans, MiniBatchKMeans, AgglomerativeClustering, DBSCAN
Sklearn.cluster.hierarchy	Dendrogram, ward
3.1b	
sklearn.neighbors	RadiusNeighborsRegressor, KNeighborsRegressor, NearestNeighbors, KNeighborsClassifier, RadiusNeighborsClassifier, IsolationForest, LocalOutlierFactor
sklearn.decomposition	PCA, IncrementalPCA, KernelPCA, NMF, LatentDirichletAllocation
sklearn.tree	DecisionTreeRegressor, DecisionTreeClassifier, export_graphviz
sklearn.neural_network	MLPClassifier,
sklearn.feature_selection	SelectPercentile, SelectFromModel, RFE, f_regression
sklearn.feature_extraction.text	CountVectorizer, ENGLISH_STOP_WORDS, TfidfVectorizer

For functions, keyword arguments are used to pass values to parameters. Table 3.2 shows three examples of sklearn classes and functions; the indicated parameter values are the default values, for explanation of the parameters, see the SciKit-Learn Documentation (Scikit Learn – Quick Guide undated).

Table 3.2 sklearn module and functions

class sklearn.linear_model.**LogisticRegression**(*penalty = 'l2', dual = False, tol = 0.0001, C = 1.0, fit_intercept = True, intercept_scaling = 1, class_weight = None, random_state = None, solver = 'liblinear', max_iter = 100, multi_class = 'ovr', verbose = 0)*
sklearn.datasets.**make_blobs**(*n_samples = 100, n_features = 2, *, centers = None, cluster_std = 1.0, center_box = (-10.0, 10.0), shuffle = True, random_state = None, return_centers*
sklearn.model_selection.**train_test_split**(**arrays, test_size = None, train_size = None, random_state = None, shuffle = True, stratify = None)*

3.5 Datasets

For the purposes of demonstrating the proposed model, the following datasets were generated. Except for the Iris dataset, others have two features only.

Dataset #1: Isotopic Gaussian Blobs

This is synthetic dataset carefully generated to ensure that there are distinct clusters. The cluster standard deviation (cluster_std) is very crucial here—the higher the value, the lesser the distinction between clusters. The parameter **centers = 5** ensures that you have five clustering regions. In this example, the StandardScaler is used as a preprocessor to transform the features so that they are centered around mean within unit standard deviation. See the program listing and the generated plots in Table 3.3 and Fig. 3.6.

Dataset #2: Embedded Two Circles

This is a small circle enclosed in a large circle to create a highly linearly inseparable and an arbitrarily shaped dataset. Datasets of this nature cannot be modeled using linear model such as Perceptron, ADALINE, and LogisticClassifier. Table 3.4 shows the program listing used to generate the circles shown in Fig. 3.7.

Table 3.3 Program listing for dataset generation (blobs)

```
from sklearn.datasets import make_blobs
import matplotlib.pyplot as plt
from sklearn.preprocessing import
StandardScaler
#use keyword arguments to supply the
parameter attributes
X, y = make_blobs(n_samples = 250,cluster_
std = 0.5, n_features = 2,
centers = 5, shuffle = True, random_state = 1)
prep = StandardScaler()
X_pred = prep.fit_transform(X)
```

```
plt.scatter(X_pred[:,0],X_pred[:,1],c =
y,marker = 'o',
edgecolor = 'red')
plt.grid()
plt.xlabel('Feature 1')
plt.ylabel('Feature 2')
plt.title('Sample Dataset with Two Features')
plt.tight_layout()
plt.show()
```

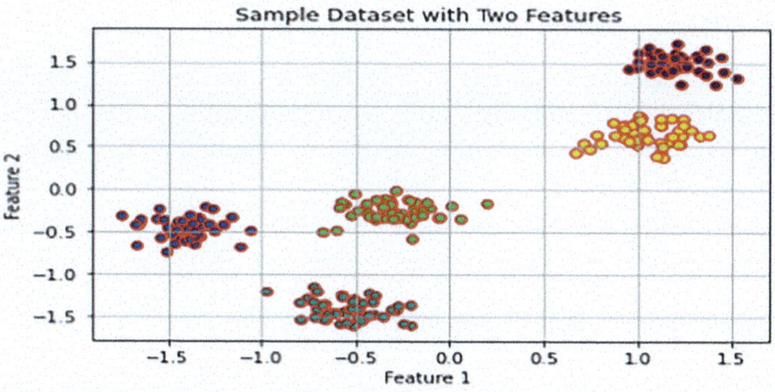

Fig. 3.6 Plot of generated dataset (blobs)

Table 3.4 Program listing for dataset generation (circles)

```
from sklearn.datasets import make_circles
import matplotlib.pyplot as plt
#use keyword arguments to supply the parameter
attributes
#Generate two circles with one inscribed in the
other for a total 1500 points
X,y = make_circles(n_samples = 1500,
random_state = 123, noise = 0.1, factor = 0.2)
plt.scatter(X[:,0], X[:,1], c = y, marker = 'o',
edgecolor = 'red')
plt.grid()
plt.xlabel('Feature 1')
plt.ylabel('Feature 2')
plt.title('Sample Dataset with Two Features')
plt.tight_layout()
plt.show()
```

Fig. 3.7 Plot of generated dataset (circles)

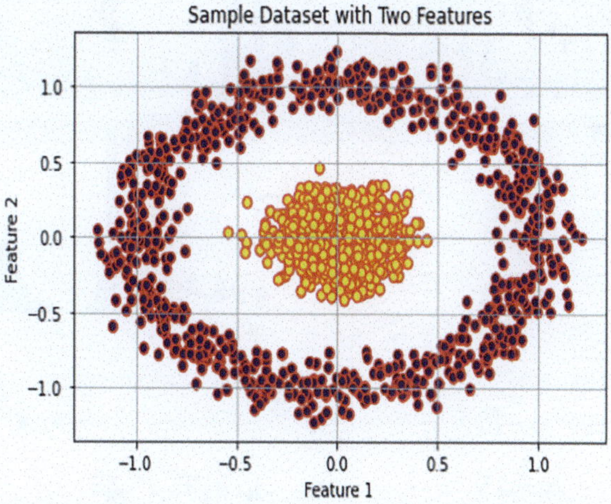

Dataset #3: Iris

This is the Iris dataset that consists of four features and outcome variable (label) and 150 examples. Here, pandas is used to generate DataFrame that is passed to the scatter_matrix function of the plotting module to generate plot of all pairs of the features.

The program listing and plots are shown in Table 3.5 and Fig. 3.8, respectively.

Table 3.5 Program listing for dataset generation (iris)

```
import pandas as pd
from sklearn.datasets import load_iris
iris_dataset = load_iris()
iris_dataframe = pd.DataFrame(iris_dataset.data, columns = iris_dataset.feature_names)
grr = pd.plotting.scatter_matrix(iris_dataframe, c = iris_dataset.target, figsize = (15, 15),
marker = ‘o’,
hist_kwds = {‘bins’: 20}, s = 60, alpha = 0.8)
```

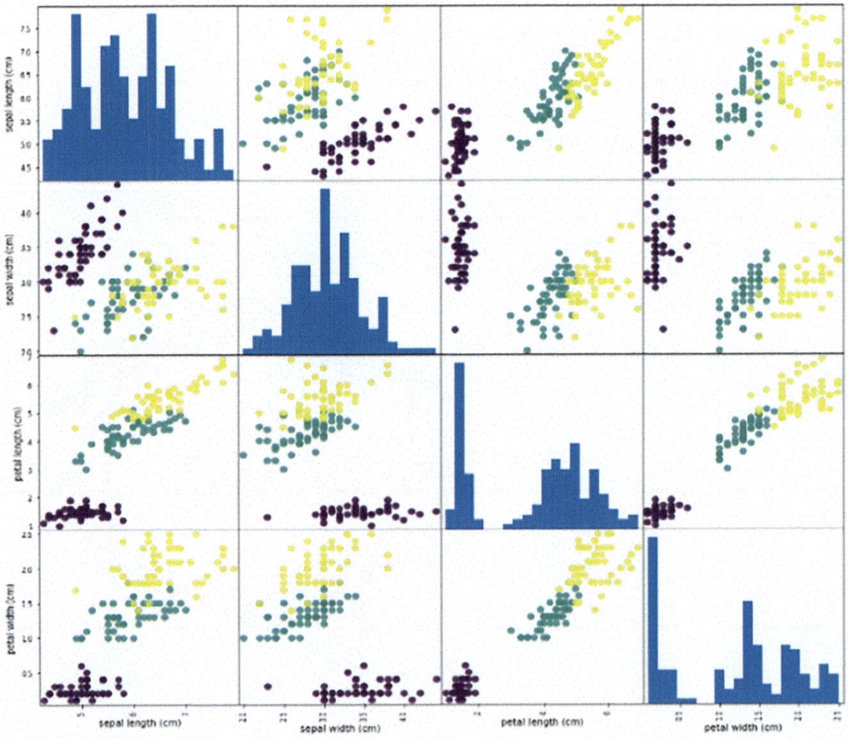

Fig. 3.8 Plots of dataset (iris)

3.5.1 Clustering + Classification using Sklearn

In the first model, K-Means was used to create a labeled dataset. Table 3.6 contains the program listing, and Fig. 3.9 depicts the plots of the K-Means labeled Iris dataset. Since the number of classes was known, there was no need to fine-tune the model, the exact number of clusters (in this case $k = 3$) was specified. A considerable number of samples (89.33%) were correctly labeled.

Table 3.6 Program listing predictive model (K-means)

from sklearn.datasets import load_iris
iris_dataset = load_iris()
from sklearn.cluster import KMeans
iris_kmeans = KMeans(n_clusters = 3)
iris_kmeans.fit_predict(iris_dataset.data)
iris_dataframe = pd.DataFrame(iris_dataset.data, columns = iris_dataset.feature_names)
# create a scatter matrix from the dataframe, color by y_train 	
grr = pd.plotting.scatter_matrix(iris_dataframe, c = iris_kmeans.labels_, figsize = (15, 15), marker = 'o',
hist_kwds = {'bins': 20}, s = 60, alpha = 0.8)
	def compare(y_pred, y):
	count = 0
	for s,t in zip(y_pred, y):
	if s == t:
	count +=1
	return format(count/len(y), '0.2%')
	print(compare(iris_kmeans.labels_, iris_dataset.target))

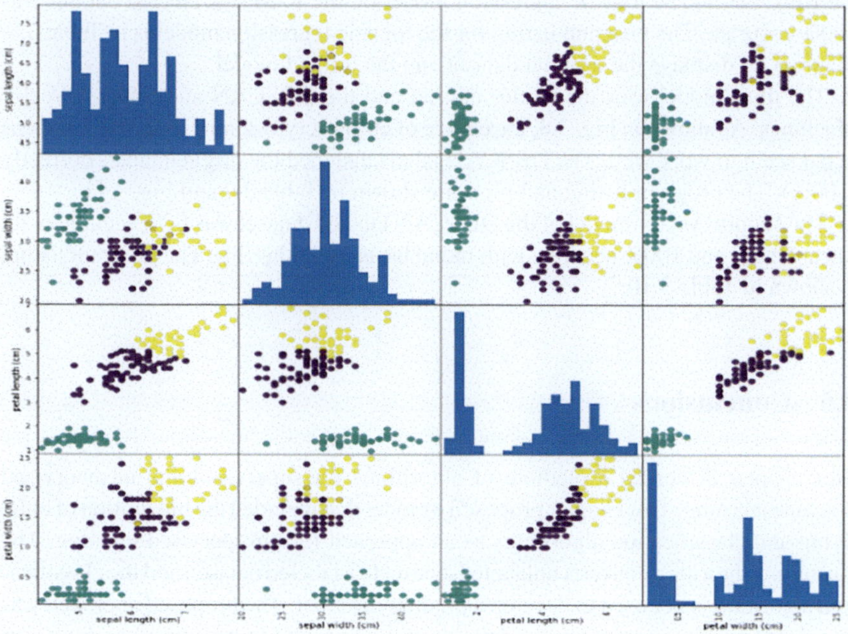

Fig. 3.9 K-means labeled dataset

Table 3.7 Iris dataset: Combined K-means and logistic regression program listing predictive model

from sklearn.datasets import load_iris
iris_dataset = load_iris()
from sklearn.cluster import KMeans
#KMean Learning
iris_kmeans = KMeans(n_clusters = 3)
iris_kmeans.fit_predict(iris_dataset.data)
#Stratify dataset
from sklearn.model_selection import train_
test_split
	X_train, X_test, y_train, y_test = train_test_
	split(iris_dataset.data, iris_kmeans.labels_,
	random_state = 0)
	#LogisticRegression Learning
	from sklearn.linear_model import
	LogisticRegression
	iris_log = LogisticRegression()
	y_pred = iris_log.fit(X_train, y_
	train).predict(X_test)
	print("Accuracy = "+format(iris_log.score(X_
	test, y_test), '0.2%'))

The next step is to use logistic regression to train the K-Means labeled Iris dataset stratifying the dataset into train dataset and test dataset using the train_test_split function of the sklearn.model_selection module. When the train_size (which is between 0.0 and 1.0) is not specified, a default 0.25 is used. Here, the train dataset is 75% of the dataset. Table 3.7 is the program listing for the model. The accuracy of the model is 97.4%.

The second model trains isotopic Gaussian blobs using the K-Means and LogisticRegression. Figure 3.7 represents the original dataset and the K-Means labeled data. The labels represented in colors do not exactly match. Therefore, you cannot exactly compare the K-Means labels with the original labels as done previously in the Iris dataset. The logistic regression model on the K-Means labeled dataset was 100% accurate. The program listing for the logistic regression model is in Table 3.8. Figure 3.10 displays the original dataset and the trained model.

The third model uses the circles dataset and the DBSCAN and Support Vector Machines. As shown in Fig. 3.8, the choice of eps is very essential, eps = 0.3 and eps = 0.4 seem to work well. The DBSCAN labels matched the original labels correctly (100%). The program listing and plots are shown in Table 3.9 and Fig. 3.11.

The support vector model of the DBSCAN labeled dataset was 66% accurate with linear kernel and 100% accurate with radial basis function (rbf). The program listing is shown in Table 3.10.

3.6 Conclusions

This chapter discussed a plethora of algorithms for supervised and unsupervised machine learning. This chapter proposed a model that included the integration of clustering and classification algorithms as an approach for unsupervised learning. The clustering algorithm converts unlabeled dataset to a labeled dataset, and the classification algorithm is not used to develop a predictive model. The proposed approach was tested and evaluated on synthetic datasets and was found to be promising, however, much still is to be done in the analysis for it to be widely acceptable. It is important

Table 3.8 Blobs dataset: Combined K-means and logistic regression program listing for predictive model

From sklearn.datasets import make_blobs from sklearn.preprocessing import StandardScaler X, y = make_blobs(n_samples = 250,cluster_std = 0.5, n_features = 2, centers = 5, shuffle = True, random_state = 1) prep = StandardScaler() X_pred = prep.fit_transform(X) #KMeans Learning from sklearn.cluster import Kmeans blobs_kmeans = Kmeans(n_clusters = 5) blobs_kmeans.fit_predict(X_pred)	plt.scatter(X_pred[:,0],X_pred[:,1],c = blobs_kmeans.labels_,marker = 'o',edgecolor = 'red') plt.grid() plt.show() from sklearn.model_selection import train_test_split X_train, X_test, y_train, y_test = train_test_split(X_pred, blobs_kmeans.labels_, random_state = 0) #LogisticRegression Learning from sklearn.linear_model import LogisticRegression blobs_log = LogisticRegression() y_pred = blobs_log.fit(X_train, y_train).predict(X_test) print("Accuracy = "+format(blobs_log.score(X_test, y_test), '0.2%'))

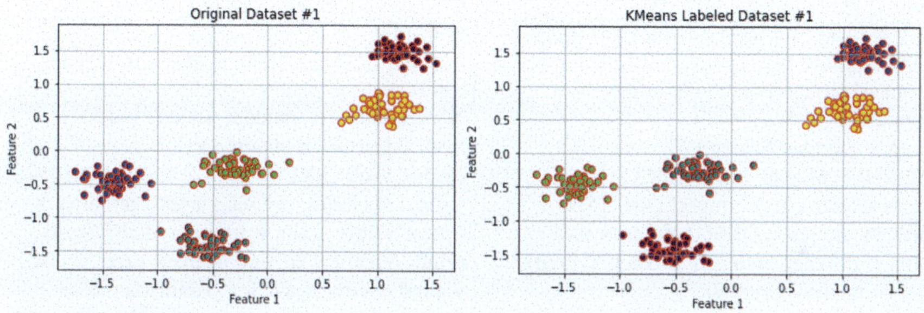

Fig. 3.10 K-Means labeled blobs dataset versus original blobs dataset

Table 3.9 Circles dataset: DBSCAN program listing for predictive model

from sklearn.datasets import make_circles import matplotlib.pyplot as plt X,y = make_circles(n_samples = 1500, random_state = 123, noise = 0.1, factor = 0.2) from sklearn.preprocessing import StandardScaler prep = StandardScaler() X_pred = prep.fit_transform(X) #DBSCAN Learning from sklearn.cluster import DBSCAN dbscan_circles = DBSCAN(eps = 0.3) clusters = dbscan_circles.fit_predict(X_pred)	plt.scatter(X[:,0], X[:,1], c = clusters, marker = 'o', edgecolor = 'red') plt.grid() plt.xlabel('Feature 1') plt.ylabel('Feature 2') plt.title('DBSCAN Labeled Dataset #3 (eps = 0.3)') plt.tight_layout() plt.show()

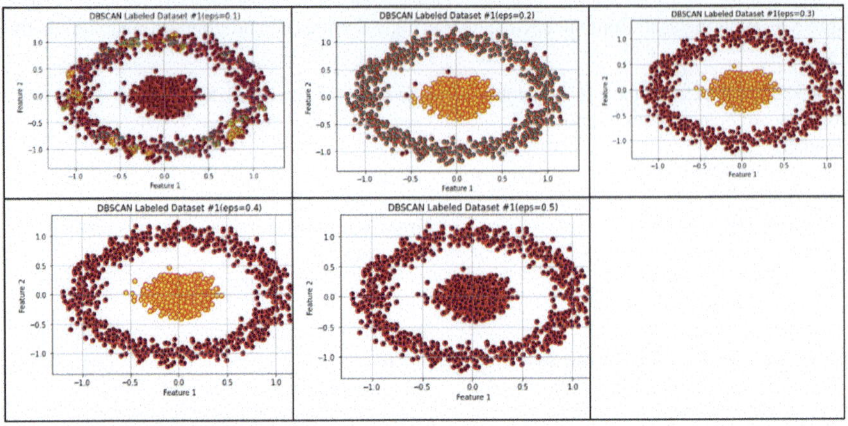

Fig. 3.11 DBSCAN labeled dataset—circles

Table 3.10 Circles dataset: Combined DBSCAN and SVC program listing for predictive model

from sklearn.datasets import make_circles
import matplotlib.pyplot as plt
X,y = make_circles(n_samples = 1500,
random_state = 67, noise = 0.1, factor = 0.2)
from sklearn.preprocessing import
StandardScaler
prep = StandardScaler()
X_pred = prep.fit_transform(X)
from sklearn.cluster import DBSCAN
dbscan_circles = DBSCAN(eps = 0.4)
cltrs = dbscan_circles.fit_predict(X_pred)
def compare(y_pred, y):
count = 0
for s,t in zip(y_pred, y):
if s == t:
count +=1
return format(count/len(y), '0.2%')
	print(compare(cltrs, y))
	from sklearn.model_selection import train_
	test_split
	X_train, X_test, y_train, y_test = train_test_
	split(
	X_pred, clusters, random_state = False)
	from sklearn.svm import SVC
	circles_svc = SVC()
	y_pred = circles_svc.set_params(kernel =
	'linear').fit(X_train, y_train).predict(X_test)
	print("Accuracy = "+compare(y_pred, y_test))
	y_pred = circles_svc.set_params(kernel =
	'rbf').fit(X_train, y_train).predict(X_test)
	print("Accuracy = "+compare(y_pred, y_test))

to note that many applications generate unlabeled datasets and to build a predictive model using such datasets, and integrating unsupervised and supervised learning seems imperative.

References

Aravind CR (2022) Exploring clustering algorithms: explanation and use cases, Neptune labs. https://neptune.ai/blog/clustering-algorithms

API Reference. https://scikit-learn.org/stable/modules/classes.html

Brownlee J (2020) Clustering algorithms with python. In: Python machine learning. https://machinelearningmastery.com/clustering-algorithms-with-python/

Budisteanu EA, Mocanu IG (2021) Combining supervised and unsupervised learning algorithms for human activity recognition. Sensors (basel, Switzerland) 21(18):6309. https://doi.org/10.3390/s21186309

Chakraborty T (2017) EC3: combining clustering and classification for ensemble learning. In: 2017 IEEE international conference on data mining (ICDM), pp 781–786

Data Mining 365 (2020) Grid-based clustering-STING, WaveCluster and CLIQUE. https://www.datamining365.com/2020/04/grid-based-clustering.html

DBSCAN. https://en.wikipedia.org/wiki/DBSCAN

Finley T, Joachims T (2010) Supervised k-means clustering

Great Learning Team (2022) DBSCAN Algorithm how does it work? Contributed by Pavan Kumar Raja. https://www.mygreatlearning.com/blog/dbscan-algorithm/

Mahendru K (2019) How to determine the optimal K for K-Means. https://medium.com/analytics-vidhya/how-to-determine-the-optimal-k-for-k-means-708505d204eb

Muller AC, Guido S (2016) Intro. to machine learning with python, a guide for data scientists. O'Reilly Media, Inc. ISBN 976-1-449-36941

Patlolla CJ (2018) Understanding the concept of hierarchical clustering technique. https://towardsdatascience.com/understanding-the-concept-of-hierarchical-clustering-technique-c6e8243758ec

Raschka S, Mirjalili V (2017) Python machine learning, machine learning and deep learning with python, scikit-learn, and TensorFlow, 2nd. Packt Publishing Ltd. ISBN 978-1-78712-593-3

Scikit Learn – Quick Guide (undated) TutorialPoint simply easy learning. https://www.tutorialspoint.com/scikit_learn/scikit_learn_quick_guide.htm

Rodriguez MZ, Comin CH, Casanova D, Bruno OM, Amancio DR, Costa L, Rodrigues FA (2019) Clustering algorithms: a comparative approach. PLoS ONE 14(1):e0210236. https://doi.org/10.1371/journal.pone.0210236

Shalev-Shwartz S, Ben-David S (2014a) Dimensionality reduction. In: Understanding machine learning: from theory to algorithms. Cambridge University Press, Cambridge, pp. 278–294. https://doi.org/10.1017/CBO9781107298019.024

Shalev-Shwartz S, Ben-David S (2014b) Clustering. In: Understanding machine learning: from theory to algorithms. Cambridge University Press, Cambridge, pp 307–322. https://doi.org/10.1017/CBO9781107298019.024

Wang W, Wang C, Cui X, Wang A (2008) A clustering algorithm combine the FCM algorithm with supervised learning normal mixture model. In: 2008 19th international conference on pattern recognition, pp 1–4

What is Clustering in Machine Learning: Types and Methods (2022) Machine learning, ANALYTIXLABS. https://www.analytixlabs.co.in/blog/types-of-clustering-algorithms/

Chapter 4
Design and Development of Machine Learning-Based Depression Identification Decision Support System

Nilesh Verma and Rishabh K. Hota

Abstract Depression is a prevalent mental disorder characterized by a loss of interest in positive aspects of life, fatigue, and impairment in daily activities. In severe cases, it can significantly affect an individual's quality of life. This research aims to develop a system for identifying individuals with depression based on their activity data. The proposed system, named the Depression Identification Decision Support System (DIDSS), utilizes data collected from wearable sensors to extract statistical features in both Temporal and Frequency dimensions. These features are then used to train an XGBoost classifier. The classifier achieves impressive results, with an accuracy of 99.48% and a sensitivity of 99.88%. Moreover, a user-friendly DIDSS based on the XGBoost model has been implemented, allowing interactive entry of activity data for real-time depression identification.

Keywords Depression Identification Decision Support System (DIDSS) · Machine Learning (ML) · Feature extraction · Feature selection

4.1 Introduction

Depression is a frequent mental condition that is defined by a persistently down mood, a loss of interest or enjoyment, as well as guilt, poor self-worth, and hopelessness (American Psychiatric Association 2013). Depression may be quite crippling and cause a variety of physical and psychological issues. According to the World Health Organization (WHO), health is a state of total physical, mental, and social well-being and not merely a state of disease or infirmity. More than 350 million individuals worldwide experience depression, which can worsen into a major health issue, especially if it lasts for a long time and is moderate to severe in intensity.

N. Verma (✉)
University of Waikato, Hamilton, New Zealand
e-mail: nv78@students.waikato.ac.nz

R. K. Hota
BMS College of Engineering, Bengaluru, India

D. K. Sharma et al. (eds.), *Machine Learning for Real World Applications*, Transactions on Computer Systems and Networks, https://doi.org/10.1007/978-981-97-1900-6_4

Depression can also impair daily life, including professional and social as well as emotional activities. In the worst case it might result in suicide, which accounts for almost one million fatalities each year. The infant and young adult population in Latin America are facing many mental health issues; about 20% of these people have disorders that call for the intervention of health related services; however, this figure is understated because adolescents tend to hide and cover up their own problems from youth and lack the confidence to access therapeutic settings (Rodriguez-Ruiz et al. 2019). Symptoms of depression do not depend upon sex, age are any other factors. Once depression has been diagnosed, it is crucial that the patient receives treatment that reduces the symptoms that this illness brings on. The fact that depression might be resistant to some treatments, though, is one of the difficulties that have emerged. Examined the effects of repetitive transcranial magnetic stimulation (rTMS), particularly on cognitive function, in treatment-resistant depression (Currier et al. 2020). The findings suggest that this non-invasive brain stimulation method can be used to improve cognition in depression that is unresponsive to therapy. The traditional way of monitoring depression relies on patient reports (Faedda et al. 2016) but this form of monitoring is biased. Changes in behavior and perception of the outside world are also noticeable. Ecological Momentary Assessment (EMA) is an alternative to these reports. When possible, EMA captures behavior, emotions, and other types of activities as they occur in real-life circumstances. With the proliferation of wearable technology (such as monitoring bands, smart watches, etc.) and smartphones equipped with motion sensors like gyroscopes and accelerometers, EMA measurements can be performed almost instantly, enabling the population to receive more mental health services without the need for brand-new, specialized devices (Artese et al. 2017). The objective of this research is to develop a Decision Identification Decision Support System (DIDSS) to identify depression states from a patient's motor activity utilizing data from a smartwatch using Machine Learning (ML) techniques. A major contribution of the proposed research is the feature extraction method from the activity data fetched from wearable devices.

4.2 Literature Review

Depression is a major public health problem, and it is estimated that approximately 16.2 million adults in the United States suffer from major depressive disorder in any given year (Fasmer et al. 2016). There are many decision support systems (DSSs) that have been developed to help clinicians identify and treat patients with depression (Fasmer et al. 2016). However, most of these DSSs are based on traditional methods, such as logistic regression, to make predictions. In recent years, ML has become a popular tool for developing DSSs because it can learn complex patterns from data and make more accurate predictions than traditional methods (Zhang et al. 2022).

Stress detection is a crucial element in the development of DIDSS. Several studies have employed ML for stress detection from electrocardiogram (ECG) signals in automobile drivers (Keshan et al. 2015; Leng et al. 2015; Mozos et al. 2017). These studies have shown that ML can be used effectively for stress detection in automobile drivers. There is a growing body of evidence to suggest that circadian rhythm dysregulation is a key factor in the development and severity of bipolar disorder (BD). Studies have shown that BD patients have altered levels of melatonin, a hormone that regulates the sleep–wake cycle and that these alterations are associated with changes in BD symptoms (Alloy et al. 2017; Bayani et al. 2017). Furthermore, recent research has demonstrated that wearable devices that track activity levels can be used to detect changes in circadian rhythms and potentially predict BD episodes (Frogner et al. 2019; Jakobsen et al. 2020).

The development of a DIDSS using ML has the potential to improve the accuracy of predictions for the severity of depression and help clinicians identify and treat it.

There are several methods that have been proposed for the detection of depression episodes, including the use of ML algorithms. ML algorithms are a type of artificial intelligence that can be used to learn from data and make predictions about future events (Ray 2017). Several studies have shown that ML algorithms can be used to detect depression episodes in patients with unipolar and bipolar disorders (Frogner et al. 2019; Garcia-Ceja et al. 2016; Jakobsen et al. 2020; Mozos et al. 2017; Rodriguez-Ruiz et al. 2019) In particular, motor activity has been shown to be a promising information source for the detection of depression episodes (Frogner et al. 2019; Rodriguez-Ruiz et al. 2019). ML algorithms have been used by many authors to detect depression episodes in patients with unipolar and bipolar disorders using motor activity data (Pacheco-González et al. 2019) on the other hand EEG signals have been used to detect depression with promising results. A study by Acharya et al. (2015) used nonlinear features from EEG signals to develop a depression diagnosis index. Another study by Rizwan et al. (2019) used ML techniques to classify depression patients and normal subjects from EEG signals. In another similar work by Galvan-Tejada et al. (2019) authors used Genetic Algorithm to optimize features of the dataset obtained from an Actigraph watch to develop a model for identification of depression. Authors have extracted features using time and frequency domains.

Actigraphy data, which is data collected from wearable activity monitors, has also been used to detect depression. Bhakta and Sau (2016) using ML to predict depression among senior citizens. Author (Artese et al. 2017) used Actigraphy data to examine the relationship between personality and physical activity in older adults.

Textual data, such as social media posts, has also been used to detect depression. Authors (Uddin et al. 2022) have used deep learning algorithms to predict depressive symptoms from a large textual dataset and authors (Garcia-Ceja et al. 2018a; Jakobsen et al. 2020) have used ML to classify depression patients from social media posts.

This research work focuses on the development of a DIDSS using ML algorithms with feature extraction and feature selection methods to predict the severity of depression based on the patient's symptoms. This DIDSS allows an end user to put motor activity data to predict depression either for a single patient or for a group of patients. The user can see the activity through the graph. The entire experiment was carried out on the basis of four different sets of data obtained as per the time periods and using 10 ML algorithms. DIDSS was developed using the best classifier performing better on a particular set of data.

4.3 Material and Methods

Figure 4.1 shows step by step process to carry out the proposed research work with the material and methods used. Data is very important ingredients while using ML for a particular domain. The performance of any ML model is strongly based upon the amount and nature of the data set used to train the ML models. Step 1 explains data selection and data pre-processing for developing ML models. The dipressjon dataset has been identified for the model development from the research work carried out by (Garcia-Ceja et al. 2018a). Data pre-processing is an important part of the ML process, therefore second step depicts the data pre-processing which consists of the sample selection, normalization, shuffling, and removing incomplete cases. As a third step feature extraction was carried out in order to extract a number of statistical features followed by feature selection as the fifth step and finally the data sets prepared after applying feature extraction and feature selection were feeded to 10 ML classifiers including XGBoost. Training of ML models was done using K-Fold cross-validation and finally, models were validated on the basis of various classification metrics. The details of each step are as below:

A. **Data Description**

The Depresjon dataset provides a comparative analysis between patients without depression (referred to as controls) and patients diagnosed with depression (referred to as condition). The dataset includes relevant information for each participant, and it was originally compiled by Garcia-Ceja et al. (2018a).

For data collection, the participants' levels of activity were monitored using Actigraph watches (Cambridge Neurotechnology Ltd., Cambridge, UK, model AW4) with a sampling frequency of 32 Hz. These watches record motion with a sensitivity of 0.05g, storing the number of counts associated with the intensity of movements in the memory. The Actigraph data, along with Montgomery-Sberg Depression Rating Scale (MADRS) scores, were gathered for each individual. The dataset primarily focuses on features across time, including timestamps (Dates and times of measurements) and activity levels, which were used for the research. The "control" and "condition" repositories encompass the depression data utilized to support the findings of this study. The dataset was downloaded from the Simula Datasets (Garcia-Ceja et al. 2018b) source. The original dataset recorded data per minute for a total of 55 patients. It consists of motor activity recordings for 23 unipolar and bipolar depressed patients, as well as 32 healthy controls. The data collection period spanned 693 days, with 402 days from the control group and 291 days from the condition group.

Figure 4.2 depicts the energy level of a control subject and a condition subject over time, using Actiwatch samples. The control subject exhibited higher energy levels

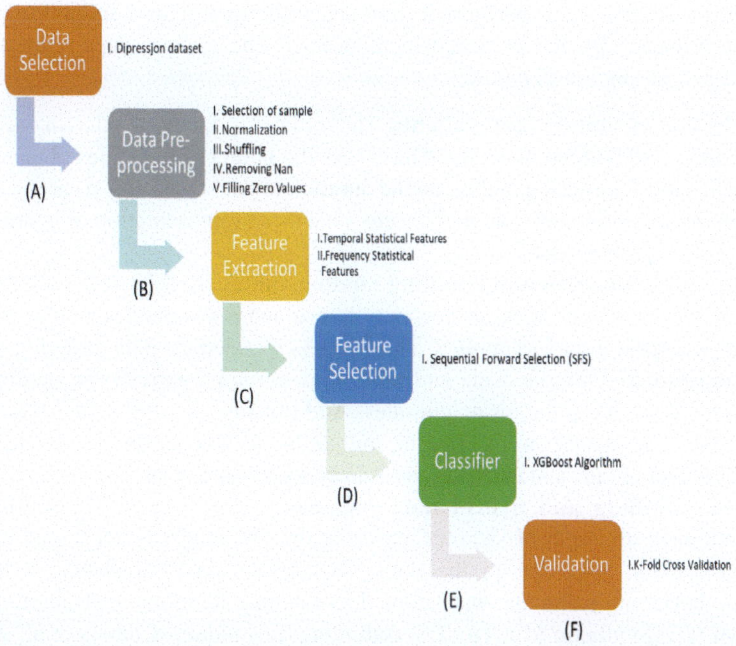

Fig. 4.1 A process flow diagram of proposed research work

compared to the case subject, suggesting a potential correlation between energy levels and depression.

B. Data Pre-processing

Data pre-processing plays a crucial role in the ML process, ensuring that the data is appropriately prepared for analysis. In this study, four primary stages of

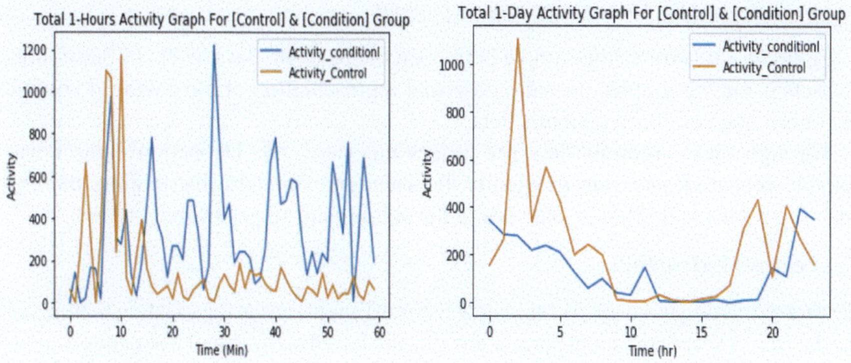

Fig. 4.2 Samples of a control (Left) and a condition (Right) from the Depression database taken from the Actiwatch

data pre-processing were performed to enhance the quality and suitability of the Depresjon dataset for the proposed methodology. The various steps followed for data pre-processing are as below:

i. **Selection of samples and patients**: The initial step involved selecting relevant samples and patients from the dataset. As the number of samples obtained for each subject varied depending on the duration of data recording, a careful selection process was undertaken to ensure a balanced representation of control and condition patients data.
ii. **Data normalization and handling missing values**: To achieve a standardized distribution of data with an average of zero and a standard deviation of one, data normalization was applied. This process rescaled the data, bringing all the features onto a similar scale and preventing certain features from dominating the analysis due to larger magnitudes. Additionally, missing values (NaN and zeros) in the dataset were filled in, mitigating the impact of incomplete cases and reducing bias and uncertainties related to missing data.
iii. **Data shuffling and hybrid data balancing**: The dataset was shuffled to randomize the order of samples, ensuring that the model is not biased toward any specific sequence in the data. To address class imbalance, a hybrid data balancing technique was adopted, combining Random Oversampling and SMOTE (Synthetic Minority Oversampling Technique) (Chawla et al. 2002). This approach created synthetic samples for the minority class (depressive patients) and balanced the volume of data for the suggested methodology.
iv. **Creation of three different datasets**: To cater to various experimental scenarios, three distinct datasets were prepared based on the time duration of recordings:

   - Dataset-1 (1-Hour Data): Comprised of 17,015 samples for depressive patients and 9,212 samples for non-depressive patients.
   - Dataset-2 (Top 1-Hour Data): Included 1,920 samples for depressive patients and 1,379 samples for non-depressive patients.
   - Dataset-3 (1-Day Data): Contained 739 samples for depressive patients and 405 samples for non-depressive patients.

Figure 4.3 illustrates the sample counts on the basis of class levels of these three datasets using bar graphs, providing a visual representation of the balanced volume of control and condition patient's data.

Through these rigorous data pre-processing steps, the Depresjon dataset was suitably prepared, ensuring its quality, balance, and standard distribution, thereby enabling robust and reliable analyses in the subsequent stages of the research.

C. Feature Extraction

Feature extraction is a way to obtain new feature dimensions using existing features of the dataset. A set of statistical parameters that are based on time and frequency dimensions were obtained after performing the feature extraction procedure (Galván-Tejada et al. 2019). These features were taken from the database, which was gathered from

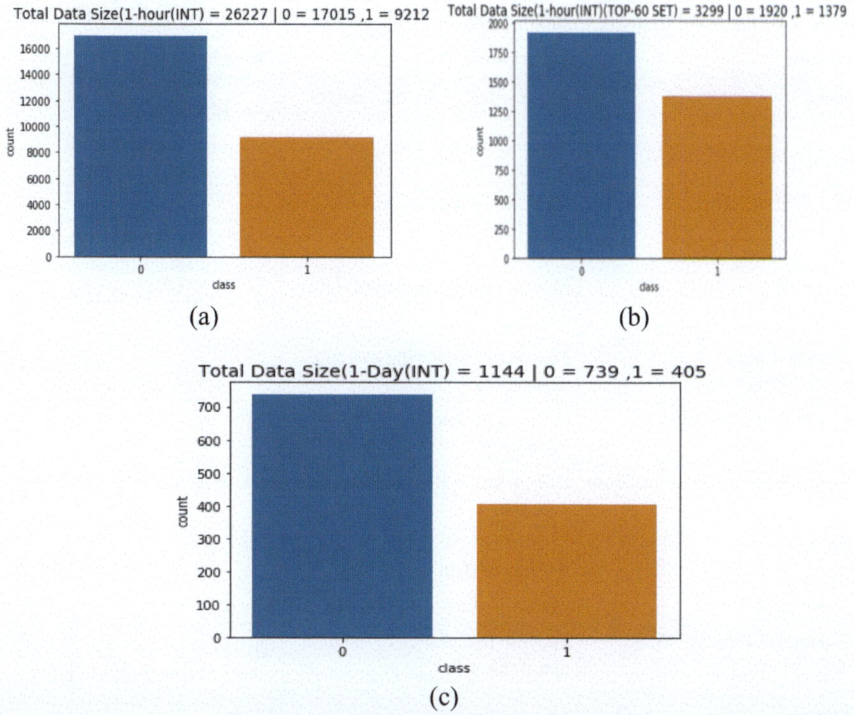

Fig. 4.3 Samples recorded for **a** 1-Hour **b** Top 1-Hour and **c** 1-Day of patient activity data

the individuals' activities through Actigraphy. Table 4.1 shows features of frequency and time domains with feature names and their corresponding mathematical formula.

The fast Fourier transform, abbreviated as FFT, is applied to the data in order to convert them from the time domain into the frequency domain. This transformation can be computed using the equation:

$$y[k] = \sum_{n=0}^{N-1} e^{-2\pi j \frac{kn}{N}} x[n]$$

where x(n) represents each motor activity gathered every minute on an hour, N represents the total observations on an hour lapse, k indicates the current frequency, and x(k) represents the spectrum components of the samples.

In the formulas shown in Table 4.1, various symbols represent specific mathematical operations and parameters. For clarity, the following explanations are provided:

- n represents the sample size.
- N represents the population size (N − 1 is used for sample variance calculation).
- μ denotes the mean value of the data.

Table 4.1 Time and frequency domain features

Feature	Description	Formula
Mean	Mean value of the data points	$\mu = 1/n \sum_{i=1}^{n} x_i$
Median	Median value of the data	$x_{(n+1)/2}$
Standard deviation	Measure of data dispersion from the mean	$sd = \sqrt{\sum_{i=1}^{n} (xi - \mu)^2/(N-1)}$
Variance	Measure of the spread of data	$sd^2 = 1/n \sum_{i=1}^{n} (xi - \mu)^2$
Trimmed mean	Mean after removing extreme data points	$\overline{\chi} = \frac{\sum_{i=p+1}^{n-p} \chi_{(i)}}{n-2p}$
Coefficient of variation	Relative measure of data variability	$CV = sd/\mu$
Inverse coefficient of variation	Reciprocal of the coefficient of variation	$ICV = \mu/sd$
Kurtosis	Measure of the peakness or flatness of data	$K = \mu/\sigma$
Skewness*	Measure of data asymmetry	$S = (\mu - \upsilon)/\sigma$
Quantile * 1, 5, 25, 75, 95, 99%	Specific percentiles in the data distribution	$Q[i](p) = (1 - \gamma x[j] + \gamma x[j+1]$
Minimum	The smallest value in the dataset	Min(n)
Maximum	The largest value in the dataset	Max(n)
Absolute Energy	Measure of total energy in the data	$U = -\frac{GMm}{r}$
Interquartile range	Data dispersion between the third and first quartiles	Q3 − Q1

- σ represents the standard deviation of the data.
- χ_((i)) denotes the data point at the ith position after ordering the dataset.
- υ represents the median value in the dataset.
- p is a proportion used to calculate the trimmed mean and quantiles.
- γ is a function of j and g, where j = floor(np + m) and g = np + m − j.
- m represents a constant determined by the sample quantile type.

Features of all the prepared datasets (Dataset-1, Dataset-2, and Dataset-3) are extracted using the above frequency and time dimensions.

D. **Feature Selection**

Feature selection is a process to choose the best feature subset for which the ML model will perform well. In order to get the best feature subset using 10 cross fold validation, a Sequential Feature Selector (SFS) strategy is used by combining the three sets of features (1-Hour, Top 1-Hour, and 1-Day). SFS evaluates each feature separately before choosing the ones that produce the highest accuracy. Table 4.2 shows features under the time domain and frequency domain along with the feature ID.

E. **Classification Techniques**

During the stage of classification, ten distinct ML algorithms were used to differentiate between depressive and non-depressive patients based on the features that were chosen for each dataset. More specifically, the stage makes use of the features that were determined to have the highest rank in the prior stage.

F. **Validation**

In order to assess the performance of the developed ML models, various classification measures were utilized as shown in Table 4.3.

These classification measures serve as important evaluation metrics for the models developed using the Depresjon dataset. The accuracy, sensitivity, specificity, ROC curve, precision, and F-score provide insights into the model's performance in distinguishing between control and condition subjects.

Table 4.2 Frequency and time domain features along with feature ID

Feature ID	Time domain feature	Feature ID	Frequency domain feature
1	Mean	20	Mean
2	STD	21	STD
3	Variance	22	Variance
4	Kurtosis	23	Kurtosis
5	Skewness	24	Skewness
6	Quantile * 1	25	Quantile * 1
7	Quantile * 5	26	Quantile * 5
8	Quantile * 25	27	Quantile * 25
9	Quantile * 75	28	Quantile * 75
10	Quantile * 95	29	Quantile * 95
11	Quantile * 99	30	Quantile * 99
12	Trim mean	32	Trim mean
13	Coefficient of variance	32	Coefficient of variance
14	Inverse coefficient of variation	33	Inverse coefficient of variation
15	Minimum	34	Minimum
16	Maximum	35	Maximum
17	Median	36	Median
18	Absolute energy	37	Absolute energy
19	Interquartile range	38	Interquartile range

Table 4.3 Classification measure used

Accuracy (ACC)	True–false sample identification rate
Sensitivity (SEN)	This measure is also termed sensitivity, probability of detection, and true positive rate. It's the proportion of positive samples correctly detected as positive
Specificity (SPE)	This measure shows the proportion of real negatives (e.g., the fraction of frames without an endoscopic finding are correctly identified as a negative result)
ROC	It measures how well a model can differentiate classes
Precision (PRE)	This measure displays the proportion of returning samples accurately recognized as positive (the fraction of retrieved samples that are relevant)
F-score	The harmonic mean of a test's precision and recall

4.4 Experiment and Result Discussion

The experiment of the current research work is divided into four parts as four different experiments. The first three experiments are based on three different datasets prepared and discussed in data pre-processing sub-section while the last experiment is based on feature selection. The purpose of each experiment is to determine the best ML model among the various models developed. Experiments were carried out with a number of well-known ML algorithms such as Nearest Neighbors, Linear kernel Support Vector Machine (SVM), Gradient Boosting, Decision Tree, Logistic regression, Naive Bayes, Random Forest, Neural Network (MLP), AdaBoost, and XGBoost. All these ML algorithms are frequently utilized in various ML processes. Models were measured on the basis of accuracy, sensitivity, specificity, ROC curve, precision, and F-score. Python is used as the programming language for the entire experimental setup and dynamic partition of data using tenfold cross-validation was used to train the models. The details of each experiment are explained in more detail below:

A. **Experiment No. 1**

In the first experiment, 10 ML models were developed using Dataset-1 which consists of original data with samples in one-minute duration. The results are shown in Table 4.4 which shows that the XGBoost model with accuracy = 99.4%, sensitivity = 99.8%, and specificity = 99% outperformed as compared to other models. The XGB achieves the highest levels of performance in terms of all matrices. The results are also shown in a graphical manner for a clear appearance in Fig. 4.4.

B. **Experiment No. 2**

In order to check the performance of ML models on the basis of data, another experiment with Dataset-2 was set up which consisted first 60 samples of each patient, and hence number of samples was reduced. The experimental results with 10 ML models are shown in Table 4.5 which again reflects the robustness of XGBoost and Gradient Boosting models with accuracy = 98.8%, sensitivity = 100%, and specificity = 97.4% outperformed as compared to other models but the performance of

Table 4.4 Performance of various ML models using Dataset-1

Model	ACC	SEN	SPE	ROC	PRE	F-Score
Random Forest	0.987	0.994	0.980	0.987	0.980	0.987
XGBoost	**0.994**	**0.998**	**0.990**	**0.994**	**0.990**	**0.994**
K-Neighbors	0.968	0.980	0.956	0.968	0.958	0.969
SVM	0.748	0.835	0.658	0.746	0.716	0.771
Decision Tree	0.969	0.982	0.955	0.969	0.957	0.970
MLP	0.870	0.877	0.863	0.870	0.869	0.873
Multinomial NB	0.636	0.563	0.711	0.637	0.668	0.611
Logistic Regression	0.732	0.791	0.671	0.731	0.713	0.750
AdaBoost	0.927	0.940	0.913	0.926	0.917	0.929
Gradient Boosting	0.955	0.963	0.947	0.955	0.949	0.956

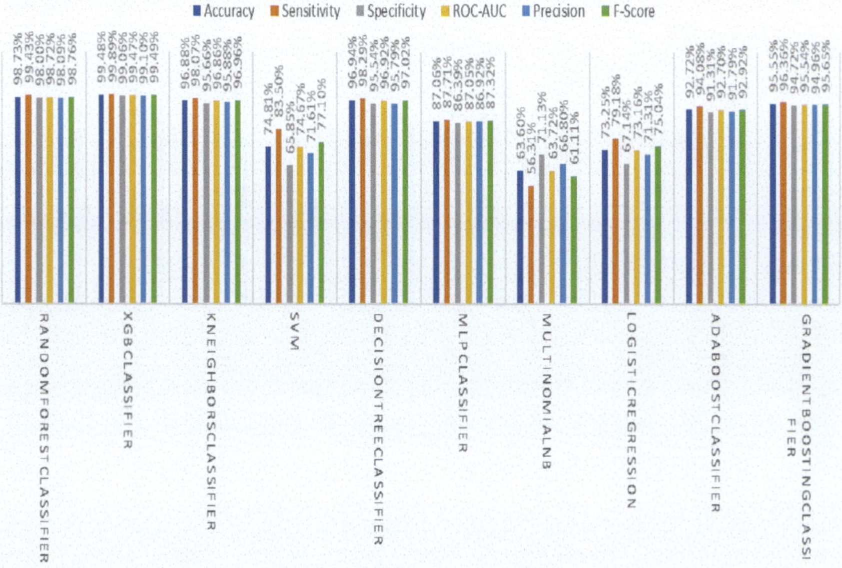

Fig. 4.4 Comparative performance of 10 ML models using Dataset-1

XGBoost in this experiment is no better than performance done in experiment No. 1. Results of Classifiers are also shown in form of comparative bar graph in Fig. 4.5.

C. **Experiment No. 3**

Samples of an entire day of each patient as Dataset-3 are used in this experiment, again with the same 10 ML algorithms. The experimental results are shown in Table 4.6

Table 4.5 Performance of various ML models using Dataset-2

Model	ACC	SEN	SPE	ROC	PRE	F-Score
Random Forest	0.961	0.990	0.923	0.956	0.944	0.966
XGBoost	**0.988**	**1.000**	**0.974**	**0.987**	**0.980**	**0.990**
K-Neighbors	0.895	0.961	0.807	0.884	0.868	0.912
SVM	0.867	0.961	0.743	0.852	0.831	0.891
Decision Tree	0.966	0.980	0.948	0.964	0.961	0.971
MLP	0.889	0.970	0.782	0.876	0.854	0.909
Multinomial NB	0.701	0.951	0.371	0.661	0.666	0.784
Logistic Regression	0.784	0.932	0.589	0.760	0.750	0.831
AdaBoost	0.972	0.990	0.948	0.969	0.962	0.976
Gradient Boosting	**0.988**	**0.998**	**0.974**	**0.987**	**0.980**	**0.990**

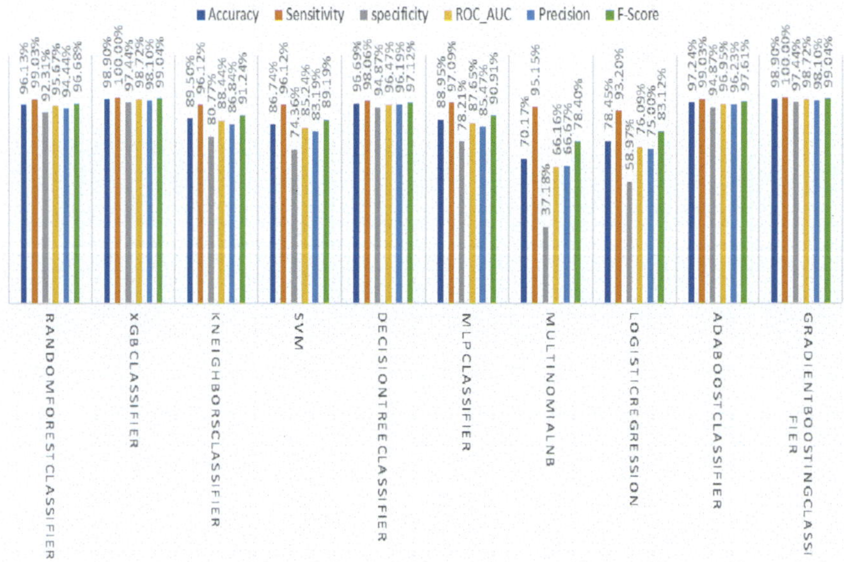

Fig. 4.5 Comparative performance of 10 ML models using Dataset-2

which reflects that XGBoost with accuracy = 95.3%, sensitivity = 97.1%, and specificity = 93.2% is performing better than other ML models. Comparative results of classifiers are also shown in the form of a comparative bar graph in Fig. 4.6.

A comparative performance of all three experiments is shown in Table 4.7 which clearly reflects that the XGBoost model developed with Dataset-1 is performing better than others and hence this model is selected for further process as feature selection in experiment No. 4.

Table 4.6 Performance of various ML models using Dataset-1

Model	ACC	SEN	SPE	ROC	PRE	F-Score
Random Forest	0.930	0.929	0.932	0.930	0.942	0.936
XGBoost	**0.953**	**0.971**	**0.932**	**0.952**	**0.945**	**0.958**
K-Neighbors	0.938	0.943	0.932	0.937	0.943	0.943
SVM	0.723	0.774	0.661	0.717	0.733	0.753
Decision Tree	0.907	0.943	0.864	0.904	0.893	0.917
MLP	0.769	0.760	0.779	0.770	0.805	0.782
Multinomial NB	0.653	0.802	0.474	0.638	0.647	0.716
Logistic Regression	0.738	0.802	0.661	0.731	0.740	0.770
AdaBoost	0.876	0.873	0.881	0.877	0.898	0.885
Gradient Boosting	**0.953**	0.957	0.949	0.953	0.957	0.957

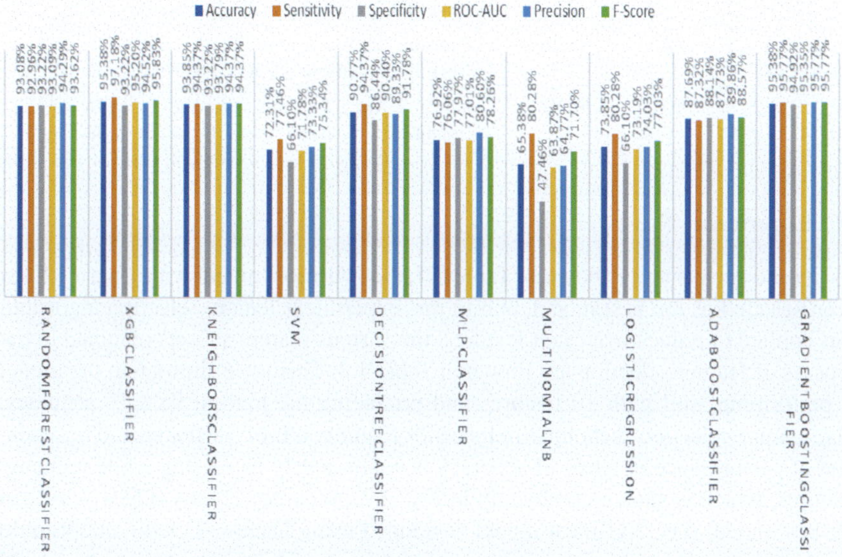

Fig. 4.6 Comparative performance of 10 ML models using Dataset-3

Table 4.7 Performance of XGBoost model with all three datasets

Dataset	ACC	SEN	SPE	ROC	PRE	F-Score
Dataset-1	**0.994**	**0.998**	**0.990**	**0.994**	**0.990**	**0.994**
Dataset-2	0.988	**0.998**	0.974	0.987	0.980	0.990
Dataset-3	0.953	0.971	0.932	0.952	0.945	0.958

Table 4.8 Results of various feature subsets after applying the feature selection technique

Feature subset with ID	No. of feature	Accuracy (%)	Feature name
(5)	01	96.93	('quantile1')
(0, 5)	02	97.93	('mean', 'quantile1')
(0, 5, 10)	03	98.39	('mean', 'quantile1', 'fmean')
(0, 5, 6, 10)	04	98.53	('mean', 'quantile1', 'quantile3', 'fmean')
(0, 4, 5, 6, 10)	05	98.68	('mean', 'skewness', 'quantile1', 'quantile3', 'fmean')
(0, 4, 5, 6, 10, 14)	06	98.69	('mean', 'skewness', 'quantile1', 'quantile3', 'fmean', 'fcofvar')
(0, 3, 4, 5, 6, 10, 14)	07	98.71	('mean', 'kurtosis', 'skewness', 'quantile1', 'quantile3', 'fmean', 'fcofvar')
(0, 3, 4, 5, 6, 10, 11, 14)	08	98.71	('mean', 'kurtosis', 'skewness', 'quantile1', 'quantile3', 'fmean', 'fstd', 'fcofvar')
(0, 3, 4, 5, 6, 10, 11, 12, 14)	09	98.76	('mean', 'kurtosis', 'skewness', 'quantile1', 'quantile3', 'fmean', 'fstd', 'fskewness', 'fcofvar')
(0, 2, 3, 4, 5, 6, 10, 11, 12, 14)	10	**98.82**	('mean', 'var', 'kurtosis', 'skewness', 'quantile1', 'quantile3', 'fmean', 'fstd', 'fskewness', 'fcofvar')

D. **Experiment No. 4**

As per the performance shown in Table 4.7, the XGBoost model with Dataset-1 has been selected as the best model, and in this experiment feature selection algorithm was applied to reduce irrelevant features, the various feature subsets obtained in the process of feature selection are shown in Table 4.8. This table shows that the model is performing well with 10 features and producing the highest 98.82% accuracy. The feature selection technique reduces 28 features out of 38 features. The variation of performance in terms of accuracy is continuously increasing while reducing features, which is shown graphically in Fig. 4.7. Accuracy is only 0.58% less than the best model (i.e., XGBoost model developed using Dataset-1) with an optimum number of features with a feature subset of 10 features. This feature selection method employed in this research work will minimize the number of features to be utilized for depression identification.

4.5 Depression Identification Decision Support System

In order to provide an interactive facility for doctors and patients to identify depression, DIDSS was developed on the basis of the best ML model obtained after applying feature selection and feature extraction techniques as discussed in previous sections.

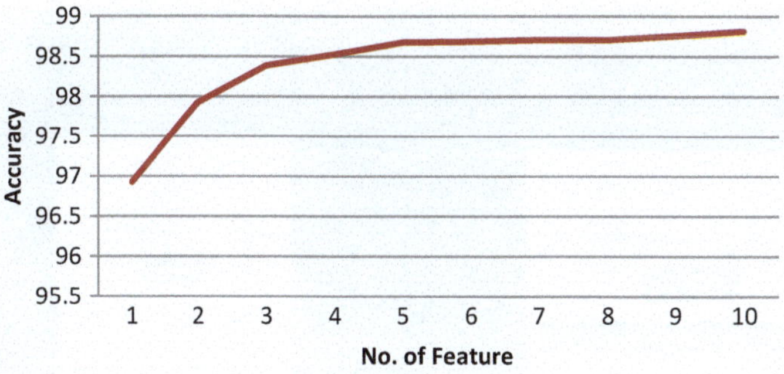

Fig. 4.7 Number of features with respect to accuracy

A graphical view of various screens of the developed system is shown in Fig. 4.8. The first screen of the system provides the facility to login into the system. One can also register as a new user by providing personal information. The system also shows a developer information page with photographs of the developer. An interactive Graphical User Interface (GUI) provides users to enter patient data either of individual patients or a group of patients (Fig. 4.9). Results obtained as depressive or non-depressive will be shown on the screen and can be imported in an Excel file.

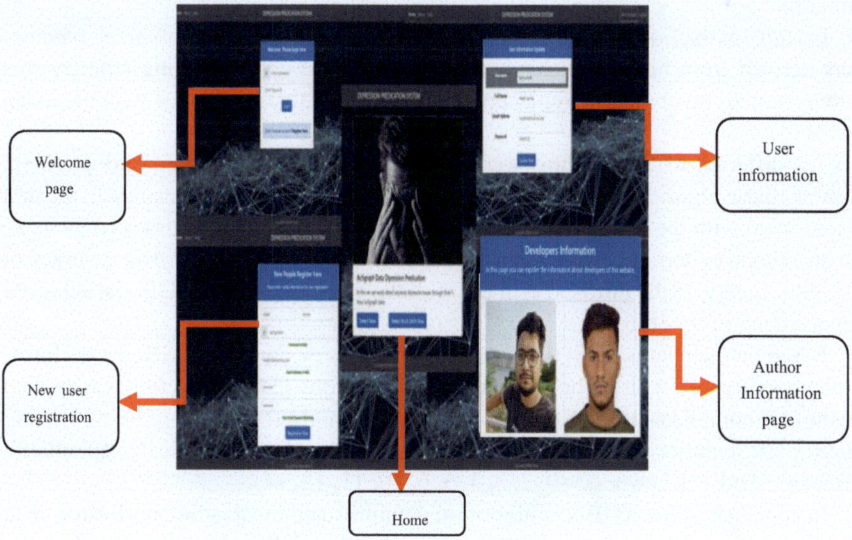

Fig. 4.8 Different screenshots of DIDSS

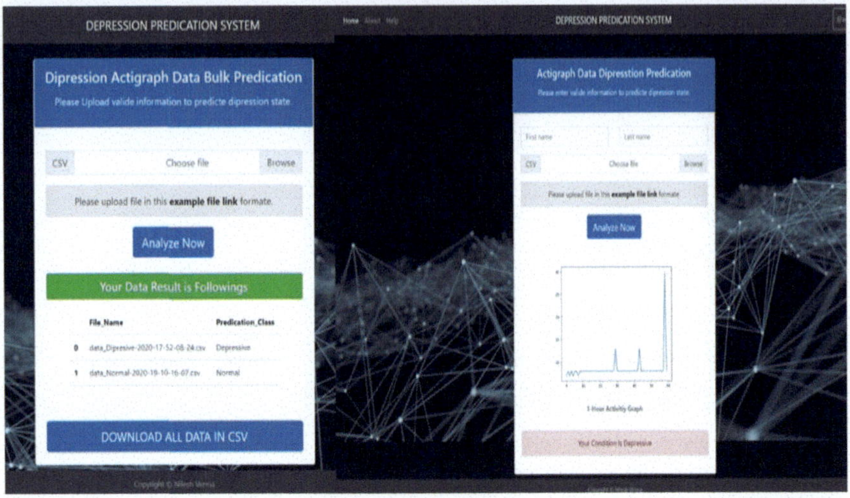

Fig. 4.9 Screenshots related to the identification of Depression using individual (Right) and group (Left) patient data

4.6 Conclusion

In this project, the Depresjon Dataset, consisting of Actigraph data from 22 patients diagnosed with depression and 32 healthy control individuals, was utilized for analysis. Three different datasets were created from the original data, each with a different time duration: one-hour, one-day, and the top 60 patients based on their dataset duration.

Feature extraction was then conducted to obtain a set of statistical parameters derived from both frequency and time. A hybrid data balancing strategy was implemented, involving the replacement of missing values and zeros with the mean value.

Ten different ML algorithms were applied to the datasets to predict depression. Among these algorithms, XGBoost consistently produced the best results. In the first experiment with the one-minute dataset, XGBoost achieved an accuracy of 99.48%. In the second experiment with the one-hour dataset, XGBoost attained an accuracy of 98.89%. Lastly, in the third experiment with the one-day dataset, XGBoost achieved an accuracy of 95.38%.

Based on the results of all three experiments, it is evident that XGBoost performs exceptionally well with the one-hour dataset. Therefore, feature selection was applied to the one-hour dataset, and XGBoost produced the highest accuracy of 98.82% using the top 10 features: 'mean', 'var', 'kurtosis', 'skewness', 'quantile1', 'quantile3', 'fmean', 'fstd', 'fskewness' (0, 2, 3, 4, 5, 6, 10, 11, 12, 14).

In conclusion, the XGBoost algorithm demonstrated outstanding performance in predicting depression based on Actigraph data, especially when utilizing the one-hour dataset and the selected features. These findings showcase the potential of ML in aiding depression diagnosis and treatment and could contribute to advancements in the field of mental health research and clinical practice.

References

Acharya UR, Sudarshan VK, Adeli H, Santhosh J, Koh JEW, Puthankatti SD, Adeli A (2015) A novel depression diagnosis index using nonlinear features in EEG signals. Eur Neurol 74(1–2):79–83

Alloy LB, Ng TH, Titone MK, Boland EM (2017) Circadian rhythm dysregulation in bipolar spectrum disorders. Curr Psychiatry Rep 19(4):21

American Psychiatric Association (2013) Diagnostic and statistical manual of mental disorders. American Psychiatric Association

Artese A, Ehley D, Sutin AR, Terracciano A (2017) Personality and actigraphy-measured physical activity in older adults. Psychol Aging 32(2):131–138

Bayani A, Hadaeghi F, Jafari S, Murray G (2017) Critical slowing down as an early warning of transitions in episodes of bipolar disorder: a simulation study based on a computational model of circadian activity rhythms. Chronobiol Int 34(2):235–245

Bhakta I, Sau A (2016) Prediction of depression among senior citizens using machine learning classifiers. Int J Comput Appl 144(7):11–16. https://doi.org/10.5120/ijca2016910429

Chawla NV, Bowyer KW, Hall LO, Kegelmeyer WP (2002) SMOTE: synthetic minority oversampling technique. J Artif Int Res 16(1):321–357

Currier D, Lindner R, Spittal MJ, Cvetkovski S, Pirkis J, English DR (2020) Physical activity and depression in men: increased activity duration and intensity associated with lower likelihood of current depression. J Affect Disord 260:426–431

Faedda GL, Ohashi K, Hernandez M, McGreenery CE, Grant MC, Baroni A, Polcari A, Teicher MH (2016) Actigraph measures discriminate pediatric bipolar disorder from attention-deficit/hyperactivity disorder and typically developing controls. J Child Psychol Psychiatry 57(6):706–716

Fasmer OB, Hauge E, Berle JØ, Dilsaver S, Oedegaard KJ (2016) Distribution of active and resting periods in the motor activity of patients with depression and schizophrenia. Psychiatry Investig 13(1):112

Frogner JI, Noori FM, Halvorsen P, Hicks SA, Garcia-Ceja E, Torresen J, Riegler MA (2019) One-dimensional convolutional neural networks on motor activity measurements in detection of depression. In: HealthMedia 2019—proceedings of the 4th international workshop on multimedia for personal health and health care, co-located with MM 2019, pp 9–15

Galván-Tejada CE, Zanella-Calzada LA, Gamboa-Rosales H, Galván-Tejada JI, Chávez-Lamas NM, Gracia-Cortés MDC, Magallanes-Quintanar R, Celaya-Padilla JM (2019) Depression episodes detection in unipolar and bipolar patients: a methodology with feature extraction and feature selection with genetic algorithms using activity motion signal as information source. Mob Inf Syst

Galvan-Tejada CE, Zanella-Calzada LA, Gamboa-Rosales H, Galvan-Tejada JI, Chavez-Lamas NM, Gracia-Cort´es MDC, Magallanes-Quintanar R, Celaya-Padilla JM (2019) Depression episodes detection in unipolar and bipolar patients: a methodology with feature extraction and feature selection with genetic algorithms using activity motion signal as information source. Mob Inf Syst 2019:1–12

Garcia-Ceja E, Osmani V, Mayora O (2016) Automatic stress detection in working environments from smartphones' accelerometer data: a first step. IEEE J Biomed Health Inform 20(4):1053–1060

Garcia-Ceja E, Riegler M, Jakobsen P, Torresen J, Nordgreen T, Oedegaard KJ, Fasmer OB (2018a) Motor activity based classification of depression in unipolar and bipolar patients. In: Proceedings—IEEE symposium on computer-based medical systems, pp 316–321

Garcia-Ceja E, Riegler M, Jakobsen P, Tørresen J, Nordgreen T, Oedegaard KJ, Fasmer OB (2018b) Depresjon: a motor activity database of depression episodes in unipolar and bipolar patients. In: Proceedings of the 9th ACM multimedia systems conference, MMSys, pp 472–477

Jakobsen P, Garcia-Ceja E, Riegler M, Stabell LA, Nordgreen T, Torresen J, Fasmer OB, Oedegaard KJ (2020) Applying machine learning in motor activity time series of depressed bipolar and unipolar patients

Joshi ML, Kanoongo N (2022) Depression detection using emotional artificial intelligence and machine learning: A closer review. Mater Today: Proc 58:217–226

Keshan N, Parimi Pv, Bichindaritz I (2015) Machine learning for stress detection from ECG signals in automobile drivers. In: 2015 IEEE international conference on big data (Big Data), pp 2661–2669

Leng LB, Giin LB, Chung WY (2015) Wearable driver drowsiness detection system based on biomedical and motion sensors. 2015 IEEE Sens 1–4

Mozos OM, Sandulescu V, Andrews S, Ellis D, Bellotto N, Dobrescu R, Ferrandez JM (2017) Stress detection using wearable physiological and sociometric sensors. Int J Neural Syst 27(02):1650041

Pacheco-González SL, Zanella-Calzada LA, Galván-Tejada CE, Chávez-Lamas NM, Rivera-Gómez JF, Galván-Tejada JI (2019) Evaluation of five classifiers for depression episodes detection. Res Comput Sci 148(10):129–138

Ray S (2017) Commonly used machine learning algorithms (with Python and R Codes). Analytics Vidhya

Rizwan MF, Farhad R, Mashuk F, Islam F, Imam MH (2019) Design of a biosignal based stress detection system using machine learning techniques. In: 2019 international conference on robotics, electrical and signal processing techniques (ICREST), pp 364–368

Rodriguez-Ruiz JG, Galvan-Tejada CE, Vazquez-Reyes S, Galvan-Tejada JI, Gamboa-Rosales H (2019) Nighttime depression episodes classification using a formal method: knowledge discovery in databases. In: Proceedings—2019 7th international conference in software engineering research and innovation, CONISOFT 2019, pp 207–212

Simula Datasets—Depresjon (n.d.) Retrieved November 2, 2022, from https://datasets.simula.no// depresjon/

Tyrväinen L, Ojala A, Korpela K, Lanki T, Tsunetsugu Y, Kagawa T (2014) The in fl uence of urban green environments on stress relief measures : a field experiment

Uddin MZ, Dysthe KK, Følstad A, Brandtzaeg PB (2022) Deep learning for prediction of depressive symptoms in a large textual dataset. Neural Comput Appl 34(1):721–744

Zanella-Calzada LA, Galván-Tejada CE, Chávez-Lamas NM, Gracia-Cortés MDC, Magallanes-Quintanar R, Celaya-Padilla JM, Galván-Tejada JI, Gamboa-Rosales H (2019) Feature extraction in motor activity signal: towards a depression episodes detection in unipolar and bipolar patients. Diagnostics 9(1):1–13

Zhang W, Wang T, Wang A (2022) Impact of physical activity intensity on longitudinal trajectories of cognitive function and depressive symptoms in middle-aged and older Chinese adults: eight-year prospective study. J Affect Disord 315:64–69

Chapter 5
Machine Learning Algorithmic Model for Pairs Trading

R. Sivasamy, Dinesh K. Sharma, Sediakgotla, and B. Mokgweetsi

Abstract This chapter uses a regression modeling approach and machine learning algorithms to investigate two correlated stocks, Pepsi, and Coca-Cola, during the same trading session. We divided the observed data into training (75%) and testing (25%). We follow the simple linear evolution of the response variable Y (=Pepsi prices) with the predictor X (=Coke prices) in the training data using both the ordinary least squares (OLS) method and the neural network. The goal is to obtain appropriate estimates of the fitted model. After determining the stationary property of the residuals determined by the "Augmented Dickey-Fuller" (ADF) test of this fit, we get forecast values $\widehat{Y}$ of Y. We then develop a trading strategy to examine the joint performance of Y and X processes for future trading using two different co-integrated stationary processes (i) the spread = (Y $-\widehat{Y}$) and (ii) the ratio = ($\widehat{Y}$/ X). An error correction model (ECM) with a lag of 1 is also included to compare its performance with that of the spread and ratio models. Real data sets are used for demonstration, and optimal performance is determined for each case.

Keywords Pairs trading · Stationary series · Spread · Ratio · Positions · Total profits

5.1 Introduction

Fundamental analysts study everything from the general economy to the dynamics of individual industries. Technical analysts use stock charts to identify patterns and trends that predict future stock performance. The approach proposed focuses primarily on technical analysis using machine learning (ML) algorithms. ML models are built using mathematical and statistical methods. This type of modeling process

R. Sivasamy (✉) · Sediakgotla · B. Mokgweetsi
Department of Statistics, University of Botswana, Gaborone, Botswana
e-mail: Sivasamyr@ub.ac.bw

D. K. Sharma
University of Maryland Eastern Shore, Princess Anne, USA

D. K. Sharma et al. (eds.), *Machine Learning for Real World Applications*, Transactions on Computer Systems and Networks, https://doi.org/10.1007/978-981-97-1900-6_5

learns from data over time and improves the accuracy of model-based predictions. Such models are widely used in artificial intelligence, biological sciences, agricultural sciences, stock market analysis, etc. Unsupervised, semi-supervised, reinforcement, and supervised learning are all types of ML models. Each branch can manage high data quantities, precise prediction and categorization, and validation processes.

Many factors usually influence stock prices. Based on the historical price movements of the stock, numerous techniques, such as fundamental analysis and technical analysis, are used in the stock price analysis. A stock's supply and demand are crucial factors in its price changes, which causes it to fluctuate quickly. Investors and stock traders aim to increase their profits by purchasing and reselling stocks. To predict the stock price, various statistical and ML approaches are applied. Statistical tools only analyze historical data of stock prices to forecast future stock prices. General trend charts and graphs, as well as time series techniques like regression analysis and auto-regressive integrated moving averages (ARIMA), are examples of statistical tools.

The book chapter offers a pairs trading tool for analyzing and forecasting stock price movements using various open-source R software packages. We focus on three specific methods: (i) fitting a linear model to real data sets of Pepsi and Coca-Cola stocks, (ii) a neural network approach to fitting non-linear models, and (iii) an error correction model (ECM). Applying these technical questions (i) to (iii) to all other stock price pairs can also help traders make good profits. First, only co-integrated securities that make up a correlated pair are chosen. Due to its ease of use, the Engle-Granger test (1987) is used to test this condition.

For prediction tasks of time series, researchers often employ Box-Jenkins methods (1994) by fitting an ARIMA (p, d, q) model to an observed historical data. Hagan (1996), and Hill (1996) discussed various neural network designs and models for time series forecasting, while Wong (1991), and Wedding and Cios (1996) discussed a combined network with Box-Jenkins Model for forecasting time series data sets. Faraway and Chatfield (1998) discussed some more methods for building, classifying, testing, and analyzing stochastic models in time series analysis. Dacorogna et al. (2001) examined the best mathematical models for analyzing high-frequency data to help us understand the microstructure of markets. Elliott et al. (2005) and Gatev et al. (2006) provided an analytical framework for how many hedge funds use pairs as an investment strategy and how two similar stocks could be traded at a certain distance to make appropriate investment decisions. The distance method proposed by Gatev et al. (2006) chooses pairs where the sum of the historic distances between the price series of the two assets is as small as possible.

High data dimensionality can create the likelihood of finding irrelevant features in all problems. According to Berkhin (2006), such a problem produces severe consequences for dimensions greater than 15. Taking this into consideration, a feasible solution is that the number of dimensions may be upper bounded at 15.

The outcomes of ML techniques are often extremely encouraging. Recurrent Neural Networks (RNN) are used by Huck (2010) and Huck and Afawubo (2015) to produce a one-week forecast from which the projected returns are derived. How to get pairs trading gains that are resistant to trading costs was covered by Do and

Faff (2010, 2012). Artificial neural networks (ANN) are used by Dunis et al. (2009, 2015) to forecast the change in three well-known spreads.

The investor should first choose the securities of interest and then look for potential combinations among the chosen securities, as stated in Caldeira and Moura (2013), in order to choose the proper pair of equities. The investor must next select a pair that is judged suitable for trading among the candidates found. The distance, correlation, and co-integration techniques are the most used ones. With the use of S&P 500 stocks, Krauss (2017) examined the performance of deep neural networks, gradient-boosted trees, and random forests in the context of statistical arbitrage. When compared to more conventional approaches, Kim and Kim's (2019) exploration of the possibilities of employing deep reinforcement learning in a pairs trading scenario produced excellent results.

Krauss (2017) examined the expanding literature on frameworks for pairs trading across five groups. Pereira et al. (2020) investigated a neural ARIMA model for predicting time series. For a broadly diversified portfolio of Chinese equities, Xiang and He (2022) explained how the factors chosen explained the cross section of stock returns and how these variables contributed to the dynamics of the equity risk premium. This explanation was based on the pairs trading method. Han et al. (2023) created a long-short portfolio using an unsupervised learning pairs trading method that was applied to the US stock market from January 1980 to December 2020. Additionally, Sabino da Silva et al. (2023) provided an alternative pairs trading method based on calculating a mispricing index in a novel manner using an optimal linear combination of copulas, and they assessed the statistical and economic performances by looking at the daily returns of the S&P 500 stocks between 1990 and 2015.

5.1.1 Pairs Trading

The main strategy for trading pairs of stocks is to find out if the joint movements of two stocks are related. When the joint movement turns bad, the strategy is to buy worse and sell better, if the joint movement will recover, and profits can be made. When the joint movement resumes, the positions are reversed, resulting in an arbitrage profit from the price difference between the two stocks. Pairs trading has been widely used in the financial industry for over twenty years as a popular method of quantitative statistical arbitrage (Dacorogna et al. 2001) and (Gatev et al. 2006).

Principle of a pair swap

Pairs trading is a trading technique that is neutral to all market conditions. This method involves analyzing the performance of two related securities. When the correlation between stocks weakens, stocks that trade at a higher price are sold or stocks that trade at a lower price are bought. The process of buying and selling stocks occurs with the hope that the spread of risks of two correlated securities will converge. In this article, the author focuses on the roles of correlation and co-integration indicators between two strains.

Consider a pair of stocks X and Y, such as Pepsi and Coca-Cola. When do we expect the ratio of two prices or the difference between prices to remain the same over time? However, there may be differences between the two pairs from time to time due to market factors. For example, the price of one stock may rise at different times, while the price of another may fall relatively. If this difference normalizes over time, we can trade in pairs: sell the best performer (the bullish stock) or buy the worst performer (the bearish stock).

We expect the gap between the two stocks to eventually flatten out with the better performing stocks going down or the underperforming stocks going up or both—our trades can make money in either case. If two stocks rise or fall together without changing the difference between them, then we make a loss, not a profit. Therefore, pairs trading is a market-neutral trading strategy that allows traders to profit from almost any market condition, such as uptrends, downtrends, or sideways movements.

Suppose a trader trades in pairs based on a high statistical correlation between the two securities. Most trades in the pair assume a correlation of 0.80 plus, which can be difficult to determine. Profits are made when the value of a low-yielding security recovers and the price of a higher-yielding security falls. The net result is the total return of the two investments. Second, while historical trends can be accurate, past prices are not always indicative of future trends. Simply asking for a correlation of 0.80 can also reduce the likelihood of the expected result.

5.1.2 Example of Pairs Trade

To illustrate the potential returns of a pairs trading strategy, here we look at Stock A, such as Pepsi, and Stock B, such as Coca-Cola, which have a high correlation of 0.95. Both stocks deviate from their historical short-term trend correlation with a correlation of 0.50. Arbitrage is taking the dollar equivalent of a long position in Stock A, which is underperforming, and a short position in Stock B, which is performing better. The stock has approached and returned to a 0.95 correlation over time. Investors can now profit from long and closing short positions.

5.2 Engle-Granger Procedure for Co-integration Testing

Robert Engle and Clive Granger developed the central focus of co-integration testing in their seminal article Engle and Grange (1987). The two-step procedure is easy to follow and gives a general idea of co-integration.

- First make sure that the individual series are indeed integrated (non-stationary) of the first order I(1). We then regress sequentially using standard OLS and test whether the series of residuals is integrated of order I(0) (leading to stationarity).

- Second, if the residuals are stationary, we can extract the coefficients of the OLS model and use them to form a stationary linear combination of the two-time series.

The prices of financial shares X and Y can behave unpredictably and neither form a stationary process, but w = spread (X, Y) can reach a stationary state over time. First, we use the conventional ADF test to verify each set to make sure they are not perfectly robust (a root of unity exists). Then we either form a spread or move to ratio-based pairs trading.

5.2.1 Hedging and Cointegration

Ideally, this means that if we take a similar trade between two highly correlated pairs, we will make money if both go in the predicted direction. On the other hand, if the price moves in the opposite direction, we will suffer more losses. We should take proactive measures to reduce the loss, using only correlation, but using another measure of asset similarity, such as "co-integration".

Cointegration is a useful econometric tool for identifying common equilibrium assets. Cointegration trading is a trading strategy that aims to profit when parallel related assets deviate from equilibrium. The price of co-integrated assets is held due to the stability of their spreads.

Therefore, it is necessary to maximize the returns of investors and eliminate risks. One such approach to using a pairs trading strategy is hedging. In a highly correlated or perfectly correlated currency pair, we can make opposite trades because of the price difference between the two. Suppose the prices of the two series Y (=Pepsi) and X (=Coca-Cola) follow the pattern as stated in (5.1):

$$Y = \alpha X + e;\ e \sim N(0, 1),\ r = (Y/X) \tag{5.1}$$

where 'α' is the true constant ratio and 'e' is the white noise. For such a pair to perform best between two-time series Y and X, the expected value of the ratio r = Y/X must agree with the mean, i.e., they must be integrated. Cointegration means that the relationship between the two prices varies around the mean, or the distribution of the linear combination (5.1) between Y and X must form a stationary series.

5.2.2 Testing of Cointegration

Although correlation and co-integration are theoretically similar, they are not the same. A simple example of co-integration is a series from a normal distribution. Because two composite time series (like X and Y above) drift toward and away from each other, there are times when the dispersion is large, and the difference is small.

When the ratio (Y/X) moves around the average α, we make money on the ratio of average returns. To do this, we observe when X and Y are separated, i.e., α is too high or too low:

- If the ratio α is lower than usual and we expect it to rise, then we can bet to buy Y and sell X.
- If α is large, we expect it to decrease. So, you can bet on it by selling Y and buying X.

We only make or lose money when stocks X and Y move relative to each other. Threshold levels can thus be easily recommended to monitor the spread between Y and X or the ratio (Y/X).

5.2.3 *What is a Stationary Series?*

A stationary series is one whose changes do not depend on the passage of time. Two price series can be combined into a portfolio in such a way that their linear combination is stationary. The individual price series are said to be co-integrated with each other if the stationary linear combination of the two price series.

For example, if the above ratio (Y/X) forms a stationary series, it enables our statistical models to predict the value of "α" accurately and efficiently. Equivalently, let the spread of the linear combination (5.1) of these two stocks is given by

$$\begin{aligned} &\text{Spread} = \widehat{Y} - \alpha X, \\ &\text{where } Y = \log(\text{Prices of Pepsi}), \text{ and } X = \log(\text{Prices of Coca-Cola}) \end{aligned} \tag{5.2}$$

If the spread series of the above Eq. (5.2) is stationary, then we can say that stocks Y and X are co-integrated with each other. In general, two securities are said to be co-integrated if there is a linear relationship between them, which has a constant mean and variance i.e., there must exist a stationary relationship over time that ensures a constant mean and variance.

The spread portfolio defined in (5.2) oscillates around a statistical equilibrium. Based on this relationship, we can formulate a trading rule, and thus we go long one unit of Y_t and short α units of X_t. On the other hand, when the residual term is getting large, we long α units of X_t and short one unit of Y_t. Hence, we can implement a mean reversion strategy by going long on one stock and going short on the other stock. We can use the 'Augmented Dickey-Fuller' test to check.

5.2.4 *Dickey-Fuller and Augmented Dickey-Fuller Tests*

Strong assumptions are made regarding our data throughout statistical tests. They can only be employed to demonstrate the extent to which the null hypothesis can be rejected. For this issue, significant outcomes must be attained.

The following model equation's null hypothesis that $\theta = 1$ is tested using a Dickey-Fuller test, a unit root test. The first lag's coefficient on Y is α (alpha). Null Hypothesis (H0): (Unit Root Test) $\theta = 1$.

To incorporate a higher order regression into the model and build upon Dickey-Fuller, the Augmented Dickey-Fuller test was created. The null hypothesis, however, is the same as it was for the Dickey-Fuller test. The calculated p-value must be below significance (for example, 0.05) in order to reject the null hypothesis because the null hypothesis implies the existence of a unit root, i.e., $\theta = 1$.

Trading pairs often hedge to some extent against adverse market movements, i.e., making the strategic market neutral. In some cases, however, the spread does not return to the average. To avoid such an unexpected situation, we need to place a stop loss order.

5.2.5 *Algorithm for Pairs Trading*

According to Elliot (2005), the spread between two financial assets can be modeled as a mean reversion process. Such a scatter series can also be fitted using the observed values of two-time series X and Y without a log. If spread = Y-α X changes in place, its average return over the period would be approximately zero. We will explain one such case in the following discussions using real and secondary data of Y and X. For this pair, we suggest the best strategy based on the forecast indicators to make a good profit in the following trading days.

5.2.5.1 Trading with Co-integrated Prices of Pepsi Versus Coca-Cola

(i) We collected prices from the Yahoo website using the following R programming commands.

```
YY<- getSymbols("PEP",start="2017-01-01", auto.assign=F)
XX<- getSymbols("KO",start="2017-01-01",auto.assign=F)
```

We then stored the prices of Pepsi in the YY-matrix with column names "PEP.Open PEP.High PEP.Low PEP.Close PEP.Volume PEP.Adjusted" and Coca-Cola in the XX-matrix with column names "KO.Open KO.High KO.Low KO.Close KO.Volume KO.Adjusted":

Then we extracted the column PEP.Close of YY in Y and the column of XX in X. Thus, each of the two vectors Y and X has included 1437 prices (Table 5.1).

Table 5.1 Data on closing prices of Y = Pepsi and X = Coca-cola for a sample of size = 1437 (=N)

	Y	X		Y	X
[1,]	104.65	41.80	[1432,]	173.22	62.32
[2,]	104.85	41.65	[1433,]	173.90	62.50
[3,]	104.71	41.75	[1434,]	167.41	60.47
[4,]	104.56	41.74	[1435,]	168.68	60.79
[5,]	103.46	41.32	[1436,]	165.88	59.53
[6,]	101.97	41.04	[1437,]	166.97	59.54

Those plots in Fig. 5.1 suggest that each data set on *Pepsi and Coca-Cola* is slightly trending. In addition, these two-time series are positively correlated, and no clue about co-integration. We build a trading rule of "sell, buy, or don't trade" with three conditions:

We now present a few descriptive statistics denoting the price sequence of the Pepsi by Y and that of the Coca-Cola by X (Table 5.2):

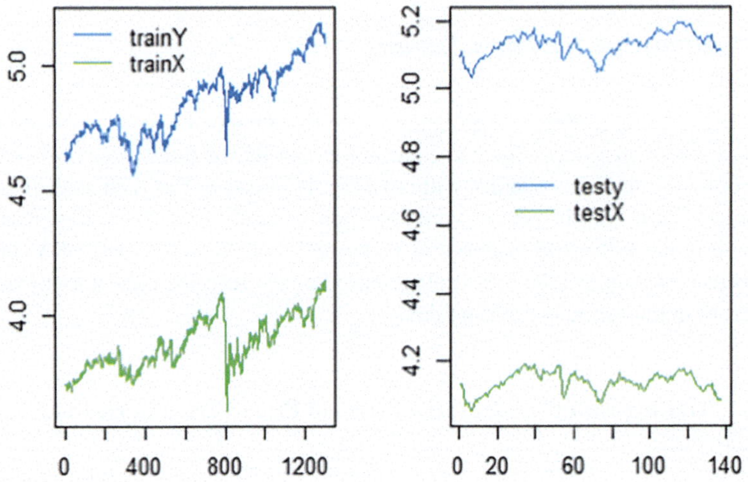

Fig. 5.1 Left plot displays the price movements of Pepsi and Coca-Cola for a training sample of size 1300 days. Right plot displays the price movements of Pepsi and Coca-Cola for a testing sample of size 137 days

Table 5.2 Prices series of Pepsi and Coca-Cola

Price	Min	Mean	Max	S.D.	Correlation
Pepsi (Y)	96.16	132.9	180.4	21.0	0.925249
Coca-Cola (X)	37.56	50.6	66.21	6.23	

5.2.5.2 Trading charts

Let r(t) = stock 1/stock 2 form a stationary and mean reverting around the mean 'm of r(t)' with standard deviation 's = S.D(r(t))'.

(i) Sell \$1 value of stock 1 and buy \$1 value of stock 2 when r(t) just passes 'm + k s' (i.e., r > m + k s) is called 'open position'. Wait for the string 'r' to drop and just cross 'm', at that point close the position and buy stock 1 for \$1 and sell stock 2 for \$1.

(ii) Sell stock 2 for \$1 and buy stock 1 for \$1 because 'r' just crosses r > (m − k s) is called 'open position'. Wait for the string 'r' to go up and just cross 'm', at that point, close the position and buy stock 2 for \$1 and sell stock 1 for \$1, and

(iii) no trade at any other time point.

Those 'm + k s' and 'm − k s' values are called threshold levels where 'k', is a positive value to be chosen by the trader.

Divide the sample of size = 1437 collected on each of the closing prices of Y = Pepsi and X = Coca-Cola into two categories: training data sets, V = trainY = $\{Y_1, Y_2, \ldots, Y_{1300}\}$ and W = trainX = $\{X_1, X_2, \ldots, X_{1300}\}$, each of size $N_1 = 1300$, and testing data sets V_0 = testY = $\{Y_{1301}, Y_{1302}, \ldots, Y_{1437}\}$, W0 = TestX = = $\{X_{1301}, X_{1302}, \ldots, X_{1437}\}$, each of size $N_2 = 137$, Let $N_1 + N_2 = N = 1437$.

5.2.5.3 Algorithm 1: Maximizing the Profit of Pairs Trading

Step 1: Select two stock prices for pairs trading, say X and Y, each of size N. Split pooled series into two disjoint sets for training and testing purposes, say T_1 = (trainY, trainX) and T_2 = (testY, testX). Select the initial number of N_1 items of the chronological series (Y, X) in the pair T_1 and the remaining N-N_1 (=N_2, say) items in the pair T_2.

Step 2: Fit the spread equation for the training set T_1, i.e., spread = log(trainY) − α log(trainX) under the assumption that the hedge ratio is a constant 'α' by regressing the stock process T_1 and thus estimate 'α'.

Step 3: Apply the Augmented Dickey-Fuller Test on the spread values inducting the estimated value of 'α'. If the p-value of the test is less than 0.05 or 0.01, we conclude, at 95% or 99% confidence that the spread is stationary for this pair and go to step 4 or otherwise ignore the further analysis with this pair.

Step 4: When the spread series is stationary, we can try to locate the extreme positions outside the threshold levels to trigger trading orders for the underlying pairs. If the spread series is not stationary, we must find a stationary error correction series using a differencing technique.

Step 5: Use the estimated spread model of Step 4 (which yields the stationary spread), and forecast the spread for the testing series T_2.

Step 6: For indicating the threshold levels, a statistical control graph can be constructed for monitoring of profitable trading positions for the testing pair T_2.

Table 5.3 Results of ADF test applied to Pepsi and Coca-Cola series. V = trainY and W = trainX

PepsiAdf <- ur.df(V)		ColaAdf <- ur.df(W)	
PepsiAdf@cval	PepsiAdf@teststat	ColaAdf@cval	ColaAdf@teststat
1pct 5pct 10pct	Statistic	1pct 5pct 10pct	Statistic
−3.96 −3.41 −3.12	−3.047302	−3.96 −3.41 −3.12	−3.442958

The objective is that the best model for the training set and forecast the values for the testing set. Then we apply all the 6 steps of the procedure described in Algorithm 1.

Let us apply the ADF test to confirm that the series are non-stationary (Table 5.3).

As the observed statistic value in each testing is larger than the 1%, critical value −3.96, we can't reject the null hypothesis of having a unit root (non-stationary) at 1% level of significance. Hence both Pepsi and Coca-Cola stock series are non-stationary processes.

Remark: We can reject the null hypothesis of the ADF test if we find that our series is integrated of order 0, i.e., I(0). If the Null Hypothesis is not rejected, we take the first difference dY_t and test the new series for a unit root. If the new series dY_t is stationary, we can say that it is of order 1, i.e., I(1) or dY_t is of order I(0).

Model selection and estimation: The 'ts.union' command can be used to build a bivariate object for the two-time series. Let V = trainY and W = trainX. We then estimate long-run regression with a constant using the 'stats' package's 'lm' command. However, we are primarily concerned with the residuals from this equation, i.e., spread, which would imply that the variables are co-integrated if the residuals are stationary. Execution of the command in the R programming Y.eq <- lm(V ~ W, data = ts.union (V,W)) gives us

$$\text{Y.eq} = 0.18236 + 1.19919\,\text{W} \tag{5.3}$$

Thus, the spread = V-Y.eq. To confirm that these residuals are stationary we subject them to a unit root test. The output of this exercise produces the statistic value −2.941825 which is less than the critical value of −1.96 at the 5% level. These would suggest that the residuals are indeed stationary, and they may be co-integrated also or they have a long-run relationship or long-run equilibrium relationship between them.

5.2.6 *Pairs Trading Between TestY and TestX Stocks*

Using the regression equation Y.eq of (5.3), we predict a set of 137 values of the testX quantities and store them in a vector set V1.

$$V_1(j) = 0.18236 + 1.19919\,\text{testX}(j)\ \text{for}\ j = 1, 2, \ldots 137$$

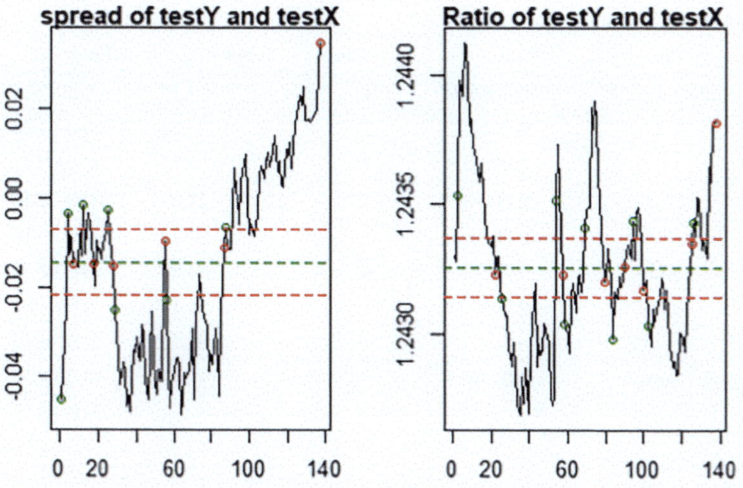

Fig. 5.2 Left describes the graph of the spread estimated between testY and testX. Right plot is that of the graph of the ratio = testY/testX, of size 137 days when k = 0.35. In each case, the favorable positions for selling and buying the stocks are indicated by circles

Therefore, the set of differences, i.e., spreadTest = (testY-V_1), is a stationary series.

We have also constructed a ratio series r = r(t). Let W_1 = testY-0.18236, W_2 = 1.19919 (testX).

and r = (W_1 − W_2)/W_2. and used these r values to trade between Pepsi and Coca-Cola series, based on R-code in four stages:

Step 1: To describe the distribution, we write the R-code 'Plotratio' and use it to plot with threshold levels 'm + k s', 'm', and 'm − k s' for a given value of k, see Fig. 5.2.

Step 2: We develop another R-code called "FindingPositions" to buy and sell stocks by tracking the opening and closing positions.

Step 3: Another R-code i.e., 'ShowPositions' is then written to circle the positions on the graph drawn by the scatter or 'Plotratio' graph, i.e., Fig. 5.2.

Step 4: Finally, an R-code named "PositiveProfit" is developed to calculate the profit or loss at each position shown in the chart and its total.

With these codes written in R programming, results can be obtained from advantageous positions for testY and testX trading pairs. Diagrams showing such details are shown in Fig. 5.2.

We now highlight positions where either selling or buying can occur, and the corresponding profit generated for each position of the spread and the ratio category in Table 5.4.

A simple look at Table 5.4 best illustrates the spread series approach versus the ratio series approach, as the latter produces negative returns for many positions. In addition, the total profit based on the spread method (antilog (0.021611) =

Table 5.4 Profit on spread profit on ratio

S.No	Pos	Profit	S.No	Pos	Profit
1	1,4	0.008619	1	3,22	0.001623
2	4,7	0.002132	2	26,54	0.001620
3	12,17	0.002808	3	54,57	−0.004181
4	25,28	0.002463	4	58,69	0.002533
5	29,55	0.003536	5	69,79	−0.000817
6	56,87	0.002053	6	84,90	0.006590
7	88,137	−0.008531	7	94,100	0.001967
8			8	102,125	−0.005217
	Sum =	0.021611	9	126,137	−0.002615
				Sum =	0.004118

1.05102) is also higher compared to the total profit produced by the ratio analysis (antilog (0.004118) = 1.009527). In any case, the last position is not considered when calculating the total profit, because the last trading day is not a closing position.

5.2.7 *Error Correction Model (ECM/ecm)*

According to Granger's theorem, if two series are non-stationary (i.e., I (1)), then the linear combination of the two series can be stationary, in which case the two variables can be said to be co-integrated. From an economic perspective, two variables are co-integrated when there is a long-term or balanced relationship between them.

Let Y = trainY and X = trainX, each series has a length of 1300 prices. The next part of the procedure involves creating an error correction model (ECM or ecm). Now we choose the lag is one for the dependent variable (Y = trainY, of size 1300).

By taking the first difference diff(Y) of the variable Y, we lose the first observation. In addition, the first lag of the first difference diff(Y)[−1] would shorten the time series by two observations. Let the necessary variable transformations as follows:

$$
\begin{aligned}
\text{Y.d} &= \text{diff(Y)}[-1] \\
\text{X.d} &= \text{diff(X)}\,[-1] \\
\text{error. ecm1} &= \text{Y.eq\$residuals}[-1:-2] \\
\text{Y.d1} &= \text{diff(Y)}\big[-(\text{length(Y)} - 1)\big] \\
\text{X.d1} &= \text{diff(X)}\big[-(\text{length(silver)} - 1)\big]
\end{aligned}
$$

We can ensure that all variables such as Y.d, error.ecm1, Y.d1, and X.d1 have the same size. In this case, we use one dependent variable Y.d to fit the model (5.4) to the observed data set:

$$Y.d = \beta_0 + \beta_1\,\text{error.ecm1} + \beta_2 Y.d1 + \beta_3 X.d1 \tag{5.4}$$

"To estimate the 'ecm.Y' with a linear regression model, we specify a few commands in R programming:

$$\begin{aligned} \text{Data} &= \text{cblind}(Y.d, \text{error.ecm1}, Y.d1, X.d1) \\ \text{ecm.Y} &= \text{lm}(Y.d \sim \text{error.ecm1} + Y.d1 + X.d1,\ \text{data} = \text{Data}) \end{aligned} \tag{5.5}$$

The output of this command of (5.5) estimates the ecm.Y, i.e., (5.6) for the training set:

$$\text{ecm.Y} = 0.011369\,\text{error.emc1} - 0.287029 Y.d1 + 0.044201\,X.d1 \tag{5.6}$$

We now use test.X of size 137 information withinside the envisioned ecm.Y of (5.6) and the expected ecm.Y to gain the envisioned residuals.

All residuals ecm.Y$residuals = (Y.d (Test.Y) – ecm.Y) and ecm.Y shape a desk bound collection around mean 0. If the collection evaluated by R-code is (Y.d – ecm.Y) therefore will become stationary. We can now advocate the nice purchase or promote positions or non-buying and selling days withinside the pairs trading study.

Figure 5.3 is a graph showing the number of positions in circles based on the ecm-model approach that can be traded over a 135-day period. What is special is that Fig. 5.3 has more trading days to sell or buy stocks compared to Fig. 5.2.

Thus, traders who want to trade frequently may prefer 'ecm' method over the spread method discussed earlier and summarized in Fig. 5.2. In this case of ecm-model, the total profit is $1.01772 (=antilog (0.00763)). However, the total profit based on the spread method, i.e., antilog (0.021611) = 1.05102) is higher compared to the total profit from the 'ECM/ecm' analysis of $1.01772 (=antilog (0.00763)).

Future co-integrating sets are expected to change the prices of Pepsi (i.e., Y) and Coca-Cola (i.e., X) according to the changes in Fig. 5.3. Therefore, the trader can

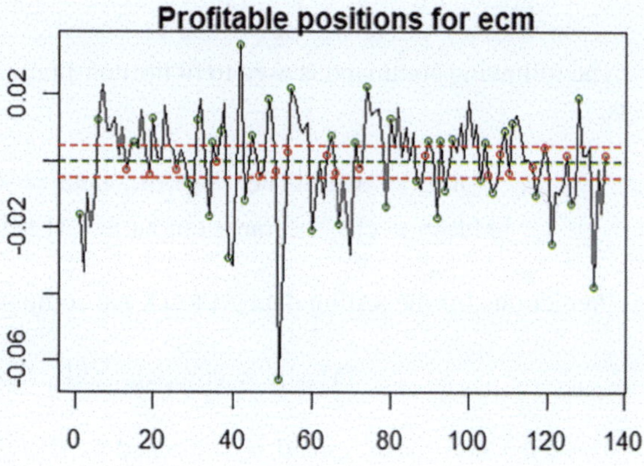

Fig. 5.3 Describes the graph of the spread estimated between testY and testX under ecm when k = 0.35. Since the lag value is 1, ecm model's size is 135 days. The favorable positions for selling and buying the stocks are indicated by circles

choose a subset of the positions marked with these circled positions to get more profit as well.

5.3 Neural Network Approach to Pairs Trading Based on Co-integrated Spread

Neural networks consist of a single input layer of neurons that receive input responses and pass through several hidden layers to evaluate the connections between neurons, and finally reach an output layer that shows us the desired output. The number of hidden layers can be chosen randomly, and such layers use forward and backward methods to estimate the weight of inter-layer neuronal connections and thus the learning process.

Thus, in the learning cycle of model matching, after moving forward, a loss function is obtained, which should be minimized, and the corresponding weights of the neurons in the hidden layer of the gradient descent output are gradually updated by an activation function to obtain a better result. If the activation functions are linear, there is no going back and figuring out what weights to give the input neurons to get a better prediction.

This section considers the same paired data sets, Y = closing prices of Pepsi and X = closing prices of Coca-Cola, each set has size 1437, and the analysis section examines the performance of the co-integrated distributed trading method, i.e., "neural network (NN)" approach. The goal is to check if the neural network method can provide a high, stable, and robust performance compared to the linear models discussed in the previous sections.

Using the same data sets to train the network with 1300 values and test the network, NN model was fit and thus predicted 137 values as before. Readers may refer to an application of NN in time series forecasting (Sivasamy and Omolo 2021). Let,

$$\text{PeL} = \log(\text{trainY}),\ \text{CoL} = \log(\text{trainX}),\ \text{PeL2} = \log(\text{testY}),\ \text{CoL2} = \log(\text{testX})$$

The following command is used to fit the non-linear model using the training data sets:

$$\begin{gathered}\text{nnL} = \text{neural (Pel} \sim \text{CoL, data} = \text{cbind(Pel, CoL)},\\ \text{hiddesn} = \text{c}(5, 3), \text{linear.output} = \text{FALSE})\end{gathered}$$

Predictions for the testing data set testX are computed using the command:

$$\text{PrtestX} = \text{compute (nnL, CoL2)}$$

Thus, the non-linear spread is computed as (PeL2-PrtestX). According to the neural method (Pel ~ CoL, data = cbind(Pel,CoL), hiddesn = c(5,3), linear.output

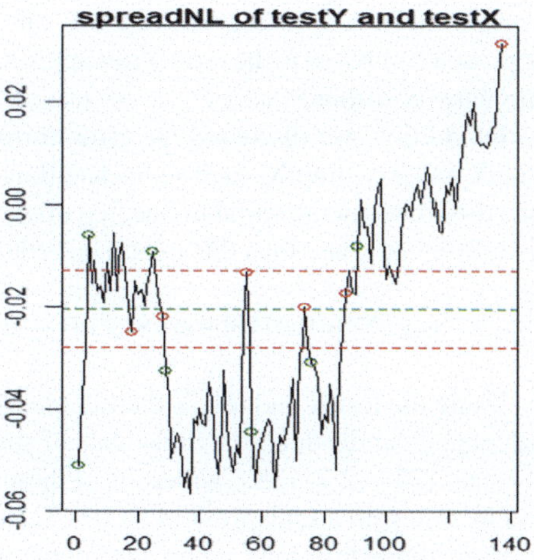

Fig. 5.4 Describes the control chart of the estimated non-linear relation between testY and testX using the neural network method when k = 0.35. Favorable positions for selling and buying shares are marked with circles

= FALSE), there is an excellent number of opening and closing positions for this spread, the series is shown in Fig. 5.4. The corresponding profit per dollar invested is 1.042395.

It should be noted that due to 'Y.eq andlt;- lm(V ~ W, data = ts.union (V,W))', the estimated spread's profit (obtained by both linear model approaches) is 1.05, which is smaller than the general benefit 1.049395 yielded by the neural method.

5.4 Conclusion

Mathematical and statistical aspects of both correlated and integrated stock price pairs are explored. Secondary data on closing price changes for Y = Pepsi and X = Coca-Cola were collected over 1,437 days. A training sample size of 1300 days was used to fit the linear model (i.e., V = trainY and W = trainX);

$$\widehat{V} = 0.18236 + 1.19919\,W$$

Using this estimated regression relationship of Y on X, we predicted a set of vectors V_1 from the 137 values corresponding to the "testX" data set. Then it turned out that the amount of differences or spread = (testY-V1) forms a stationary series, the total profit using the spread method was $1.05102.

During these 137 days, profitable positions were also found for the ratio series r = V_1/testX and are represented as circles in Fig. 5.2, with a total profit of $1.009527 from the report analysis.

A calculation method (Algorithm 1) was developed in 6 steps to create a chart of profitable positions. The expected profit for each rounded position shown in the chart was calculated and presented in Table 5.4 for the spread method and the proportional or ratio method. Looking at Table 5.4, one can see the merits of the spread series approach compared to the ratio series approach, as the latter produces negative returns for many positions.

In addition, we discussed the error correction model (ECM or ecm) using Y = trainY and X = trainX, each series length is 1300 prices. This method considers the first difference of the variable (the first observation disappears) and the first lag of the first difference, and thus this treatment reduced the time series by two observations.

$$\text{ecm.Y} = 0.011369\,\text{error.ecm1} - 0.287029\text{Y.d1} + 0.044201\,\text{X.d1}$$

Using the test.X and text.Y dataset of size 137, we predicted the ecm,Y data and calculated the residuals, i.e., residuals of spread or ecm.Y$ = (Y.d (Test.Y)-ecm.Y) for size. $137 - 2 = 135$. The results of these promotional positions were then shown in Fig. 5.3, with a total profit of $1.01772. It is recommended that traders who want to trade frequently may prefer the "ecm" method with a lag of 1 over the spread method.

Using the NN method, a non-linear model between TextY and TextX was fitted and then the prediction corresponding to testX series was computed. For the spread series, both opening and closing positions were identified in Fig. 5.4. Profit for each pair of opening and closing positions was found and the total profit was also computed.

Comparing the $1.05102 total profit of the lm() method with the $1.049395 profit of the NN model, we can conclude that the NN performs the same as the 'lm() method. Here, NN can perform better than some other methods only if the optimal number of neurons is used in the hidden layers.

The near future of the actual series is expected to match our predicted values with 95% confidence. The main conclusion is that the proposed pairs trading algorithms can achieve very promising results for all types of pairs in real data sets.

References

Berkhin P (2006) A survey of clustering data mining techniques. In: Kogan J, Nicholas C, Teboulle M (eds). Springer, Berlin, pp 25–71

Box G, Jenkins G, Reinsel G (1994) Time series analysis. Prentice Hall, Forecasting and Control

Caldeira J, Moura G (2013) Selection of a portfolio of pairs based on cointegration: a statistical arbitrage strategy. Rev Bras Finanças. Rio de Janeiro 11(1):49–80

Do B, Faff R (2010) Does simple pairs trading still work? Finan Anal J 66(4):83–95

Do B, Faff R (2012) Are pairs trading profits robust to trading costs? J Finan Res 35(2):261–287

Dunis CL, Laws J, Evans B (2009) Modelling and trading the soybean-oil crush spread with recurrent and higher order networks: a comparative analysis. In: Artificial higher order neural networks for economics and business. IGI Globa, pp 348–366

Dunis CL, Laws J, Middleton PW, Karathanasopoulos A (2015) Trading and hedging the corn/ethanol crush spread using time-varying leverage and nonlinear models. Eur J Finan 21(4):352–375

Dacorogna MM, Gencay R, Muller U, Olsen RB, Olsen OV (2001) An introduction to high frequency finance. Academic Press, New York

Engle RF, Grange CWJ (1987) Co-integration and error correction: representation, estimation, and testing. Econometrica 55:251–276

Elliott R, Van Der Hoek J, Malcolm W (2005) Pairs trading. Quant Finan 5:271–276

Faraway J, Chatfield C (1998) Time series forecasting with neural networks: a comparative study using the airline data. Appl Stat 47(2):231–250

Gatev E, Goetzmann WN, Rouwenhorst KG (2006) Pairs trading: performance of a relative value arbitrage rule. SSRN eLibrary

Han C, He Z, Toh A (2023) Pairs trading via unsupervised learning. Eur J Oper Res 307:929–947

Hagan M, Demuth H, Beale M (1996) Neural network design. PWS Publishing Company

Hill T, O'Connor M, Remus W (1996) Neural network models for time series forecasts. Manage Sci 42(7):1082–1092

Huck N (2009) Pairs selection and outranking: an application to the S&P 100 index. Eur J Oper Res 196(2):819–825

Huck N (2010) Pairs trading and outranking: the multi-step-ahead forecasting case. Eur J Oper Res 207(3):1702–1716

Huck N, Afawubo K (2015) Pairs trading and selection methods: is cointegration superior? Appl Econ 47(6):599–613

Krauss C (2017) Statistical arbitrage pairs trading strategies: review and outlook. J Econ Surv 31(2):513–545

Kim T, Kim H (2019) Optimizing the Pairs-trading strategy using deep reinforcement learning with trading and stop-loss boundaries. Complexity 3:1–20

Pereira E, Scarpin CN, Teixeira Júnior LA (2020) Time series forecasting by using a neural arima model based on wavelet decomposition. Indep J Manage Prod 7(1). http://www.ijmp.jor.br

Sabino da Silva F, Ziegelmann F, Caldeira J (2023) A pairs trading strategy based on mixed copulas. Q Rev Econ Finan 87:16–34

Sivasamy R, Omolo B (2021) Intelligent computing and mathematical modelling. Easy Chair June 20–21

Wong F (1991) Time series forecasting using backpropagation neural networks. Neurocomputing 2:147–159

Wedding D, Cios K (1996) Time series forecasting by combining RBF network, certainty factors, and the Box-Jenkins model. Neurocomputing 10:149–168

Xiang Y, He J (2022) Pairs trading and asset pricing. Pac Basin Finan J (72)1–20

Chapter 6
Evaluation of Machine Learning Models for Optimized Crop Recommendation

Manish Kumar Shukla, Ritu Raj Sondhiya, and Vikash Ku. Singh

Abstract In this research paper, we evaluate the performance of several machine learning algorithms, including Random Forest, Decision Tree, Logistic Regression, and KNN, for soil data classification to enhance crop recommendations. The findings reveal that the Random Forest algorithm excels in accuracy and specificity compared to alternative methods. Notably, optimized crop recommendations derived from the Random Forest algorithm exhibit a significant enhancement in both crop yield and quality when contrasted with traditional approaches. This study endeavors to identify the most efficacious machine learning algorithms for predicting soil types based on their properties and recommending optimal crops for cultivation. Utilizing a dataset encompassing soil properties such as type, pH level, organic matter content, and nutrient composition from diverse farms and agricultural fields, we conducted thorough preprocessing and cleaning procedures to eliminate outliers and ensure data quality. The study underscores that the Random Forest algorithm surpasses other algorithms in soil data classification, achieving notable accuracy, precision, and recall scores. This performance demonstrates its capability to accurately predict crops based on their properties. Furthermore, hypothesis analysis revealed strong evidence of a significant relationship between temperature and rainfall with crop labels, suggesting the necessity of including these factors in predictive models or further analyses. However, selecting the ideal model without considering the real-world environment risks becoming a matter of ill-supported opinion, as noted in prior research (Cortes and Vapnik, 1995). This study provides farmers and agricultural practitioners valuable insights on utilizing machine learning algorithms in soil data classification and crop recommendation to enhance agricultural production optimization.

Keywords Predictive machine learning · Soil data classification · Crop recommendation

M. K. Shukla (✉) · R. R. Sondhiya · V. Ku. Singh
Department of Computer Science, Indira Gandhi National Tribal University Amarkantak (M.P), Amarkantak 484887, India
e-mail: mks12ds@gmail.com

D. K. Sharma et al. (eds.), *Machine Learning for Real World Applications*, Transactions on Computer Systems and Networks, https://doi.org/10.1007/978-981-97-1900-6_6

6.1 Introduction

As the global population continues to expand, there is a rising need to meet the growing demand for food production. To meet this demand, farmers must maximize their crop yields. One approach to achieving this is by implementing optimized crop recommendations. India, being an agricultural country where farming is a key component of the economy, relies heavily on agriculture for livelihoods and overall development. However, various factors such as climate change, land degradation, and water scarcity have impacted crop production and yield, necessitating the adoption of precision agriculture techniques. Precision agriculture involves leveraging advanced technologies, including machine learning algorithms, to collect and analyze soil, climate, and crop growth data. This data analysis serves as a catalyst for delivering optimized crop recommendations to farmers. The pivotal role of machine learning algorithms in classifying soil data based on parameters such as pH, organic matter content, and nutrient levels cannot be overstated. These algorithms play a vital role in providing tailored crop recommendations and fertilizer guidance, leveraging the outcomes of the classification process.

India boasts an extensive agricultural sector, with over 1.6 million square kilometers of land dedicated to cultivation, ranking it as the second-largest globally. The country's economy heavily relies on this sector, given the substantial involvement of the population in agriculture. The potential for India to emerge as an agricultural superpower holds promise for rural development and poverty reduction. Notably, small farms dominate the Indian agricultural landscape, with over 75% of land holdings being less than 5 acres. The majority of crops are dependent on rainfall, as only approximately 45% of the land is irrigated. An estimated 55% of India's population depends on farming for their livelihood.

Regrettably, there is no universal system in place to aid farmers in agriculture. However, India possesses a rich repository of agricultural data from the past, which can be harnessed for recommendations using machine learning, data mining techniques, and algorithms to suggest suitable crops, irrigation methods, and predictions. Agricultural practices have undergone significant transformations in recent decades, primarily due to technological advancements. This includes the integration of sensors, devices, machines, and information technology. Contemporary agriculture relies on sophisticated technologies such as robots, temperature and moisture sensors, GPS technology, aerial imagery, and various complex Internet of Things (IoT) devices. The implementation of these advanced devices in agriculture enables farmers and businesses to increase profitability, efficiency, safety, and environmental sustainability. The emergence of digital agriculture and its associated technologies has created numerous opportunities for data collection. Remote sensors, cameras, and other connected devices can collect data on a 24-h basis across an entire farm or land. Traditional farming practices often result in suboptimal crop yield and inefficient irrigation, leading to lower profits for farmers. This paper proposes a Crop Recommendation system that leverages machine learning algorithms to address these challenges.

This research aims to evaluate the performance and effectiveness of various machine learning algorithms in optimizing fertilizer recommendations for agricultural practices. Specifically, the focus will be on assessing the predictive capabilities of these algorithms in soil data classification to enhance crop recommendations in the context of India. Soil data will be collected from different geographical locations by leveraging the expertise of farmers and agricultural experts.

Machine learning models will be employed to predict and recommend the most suitable crops and corresponding fertilizer requirements based on the analyzed soil characteristics. The accuracy of recommended crops and fertilizers will undergo evaluation through a comparative analysis of classification results against ground truth data. Collaborative efforts with farmers and agricultural experts ensure the research aligns with the practical needs of the agricultural community. Their insights and feedback become integral for validating the effectiveness of machine learning algorithms and refining the models' recommendations. This study systematically assesses various predictive machine learning algorithms for soil data classification, gauging their efficacy in delivering optimized crop recommendations.

Initiating the research process, soil data is gathered from diverse regions, considering parameters such as pH, organic matter, nutrient content, and texture. Subsequent to data collection, preprocessing procedures are implemented to eliminate extraneous information and address outliers, ensuring the accuracy and reliability of the data central to the study. The subsequent phase involves the identification of suitable machine learning methods for soil data classification. Performance metrics, as detailed by Pasquel et al. (2022), are employed to evaluate the algorithms' effectiveness. The best-performing algorithm is then employed to develop a crop recommendation system. This system, leveraging soil data as input, provides recommendations for optimal crop selection, including guidance on the best fertilizers, ideal planting times, and anticipated yield.

This research will contribute to developing a robust and reliable system for soil data classification and crop recommendation, benefiting farmers, agricultural experts, and the economy as a whole.

6.1.1 Background

India, renowned as an agricultural country, heavily relies on farming as the primary source of livelihood for a significant portion of its population. The Ministry of Agriculture and Farmers' Welfare reports that agriculture contributes around 17–18% to the country's GDP, and approximately 60% of the population is directly or indirectly involved in agricultural activities (Analytical Reports n.d.). Despite its prominence, the Indian agriculture sector encounters numerous challenges, including fragmented landholdings, low productivity, insufficient irrigation facilities, the impact of climate change, and soil degradation.

The National Bureau of Soil Survey and Land Use Planning reveals that approximately 30% of India's total land area suffers from soil degradation, leading to diminished crop yields and reduced incomes for farmers. In response to these challenges, the Indian government has introduced various initiatives like the Soil Health Card scheme and the Pradhan Mantri Fasal Bima Yojana. These programs aim to foster sustainable agriculture practices and enhance farmers' livelihoods. Land degradation remains a pressing environmental issue in India, affecting significant portions of the country's total geographical area (TGA). In 2018–19, about 97.85 million hectares, equivalent to 29.7% of India's TGA of 328.72 million hectares, underwent land degradation. This trend has been on the rise over the years, with 94.53 million hectares (28.76% of TGA) affected in 2003–05 and 96.40 million hectares (29.32% of TGA) in 2011–13 (Land Degradation and Desertification in India n.d.). The desertification situation is equally concerning, with approximately 83.69 million hectares undergoing desertification in 2018–19. This figure surpasses the land affected in previous years, as 81.48 million hectares experienced desertification in 2003–2005 and 82.64 million hectares in 2011–13. Desertification poses a significant threat to the sustainability of agriculture and ecosystems in arid and semi-arid regions of the country (Land Degradation and Desertification in India n.d.).

Using predictive machine learning algorithms in soil data classification can empower farmers to optimize their crop recommendations and enhance their yields (Arooj et al. 2018). By leveraging these algorithms, farmers can gain insights into the characteristics and health of their soil, enabling them to make informed decisions regarding suitable crops, fertilizer application, and resource allocation. Machine learning algorithms can provide accurate predictions of crop yields and suggest suitable crop varieties for specific soil types. According to the Ministry of Agriculture and Farmers' Welfare, the average landholding size in India is about 1.08 hectares, making it difficult for small and marginal farmers to adopt modern technologies and improve their incomes.

The Indian government has set a target to double farmers' incomes by 2022. The country has experienced impressive growth in agri and allied exports, soaring from 32.81 billion USD in 2015–16 to 50.24 billion USD in 2021–22—an impressive 53.1% increase. This remarkable performance is a result of agricultural reforms, technology adoption, and favorable climatic conditions. Strategic trade policies and international collaborations have also contributed to success, elevating the nation's status as a key player in global agri-trade. With a commitment to sustainability and continued investments, the country is poised for further growth, bolstering its position in the global agricultural market and ensuring economic prosperity for the agricultural sector (Doubling of Farmers Income 2022). However, achieving this goal requires significant improvements in agricultural productivity and profitability. Using predictive machine learning algorithms in soil data classification can help farmers achieve this objective by providing them with accurate crop recommendations and reducing production costs.

6.1.2 Dataset Description

In the research methodology, the first step was to collect soil data from farmers in two regions of Madhya Pradesh, Annupur, Shadhol, and Bhopal districts. Soil testing experts collected a total of 7439 soil samples from these regions. After the data collection, the next step was to perform data preprocessing, exploratory analysis, and crop prediction based on the soil features. The data features included Soil_ID, pH, Ec dS/m, Organic Carbon (%), Nitrogen, Phosphorus, Potash Sulfur, Iron, Manganese, Zinc, Copper, Temperature (°C), Rainfall, Region, crops_label, and Season. Based on the soil data features, the analysis predicted the suitable crop for a particular soil type. For example, wheat is ideal for soil with a pH ranging 6.0–7.5 and rainfall ranging 500–750 mm. Soybean is suitable for soil with pH ranging 6.0–7.5 and rainfall ranging 750–1000 mm. Rice is ideal for soil with pH ranging 5.0–6.5 and rainfall ranging 1250–1500 mm. Watermelon is suitable for soil with pH ranging 6.0–7.5 and rainfall ranging 600–800 mm. Cucumber is ideal for soil with pH ranging 5.5–7.0 and rainfall ranging 500–750 mm. The research methodology included the collection of soil data, analysis of the data using predictive machine learning algorithms, and crop prediction based on the soil data features. The features of the dataset are described follows:

- Soil_ID: This is an identifier that is unique to each soil sample in the dataset.
- pH: The pH of the soil indicates its acidity or alkalinity.
- Ec dS/m: This is a measurement of the soil's ability to conduct electricity, commonly expressed in decisiemens per meter (dS/m), representing the soil's salinity.
- Organic Carbon: The percentage of carbon in the soil originating from plant and animal matter is called organic carbon.
- Nitrogen (Kg/ha): It is a necessary nutrient for plant growth and is often measured in kilograms per hectare.
- Phosphorus (Kg/ha): Another critical nutrient for plant growth, phosphorus, is measured in kilograms per hectare.
- Potash (Kg/ha): Potassium is another crucial nutrient for plant growth, measured in kilograms per hectare.
- Sulfur (mg/kg): Sulfur is a plant's secondary nutrient, measured in milligrams per kilogram.
- Iron (mg/kg): Iron is a plant micronutrient measured in milligrams per kilogram.
- Manganese (mg/kg): Manganese is another plant micronutrient measured in milligrams per kilogram.
- Zinc (mg/kg): Zinc is also a micronutrient for plants, measured in milligrams per kilogram.
- Copper (mg/kg): Copper is another plant micronutrient measured in milligrams per kilogram.
- Temperature (°C): Temperature is the measure of heat or coldness of the soil in degrees Celsius.
- Rainfall (mm): Rainfall is the measure of precipitation in millimeters.

- Region: Region denotes the location of the soil sample.
- Crops_label: Crops_label denotes the soil sample's recommended crop(s).
- Season: Season denotes the recommended season(s) for the recommended crop(s).

These features represent the various characteristics of the soil samples in the dataset, and the notations used are commonly accepted abbreviations and symbols used to denote these characteristics.

6.2 Literature Survey

Agriculture holds substantial significance in India's economy and job market. One of the prevailing challenges faced by Indian farmers is the absence of informed crop selection based on soil requirements, leading to a notable negative impact on productivity. Despite being one of the leading global producers of agricultural goods, India struggles with low farm productivity, necessitating efforts to enhance output while minimizing labor (Reddy et al. 2019). India, which boasts the second-largest global population, heavily relies on agriculture for employment (Peraka et al. 2020). Unfortunately, farmers often cultivate the same crops repeatedly without experimenting with new varieties. Additionally, indiscriminate application of fertilizers, lacking knowledge about nutrient and quantity requirements, not only hampers crop output but also contributes to soil acidity and top-layer damage (Peraka et al. 2020).

In addressing these challenges, Reddy et al. (2019) employ an ensemble model featuring majority voting techniques, random tree, CHAID, K-Nearest Neighbor, and Naive Bayes. This collaborative approach aims to recommend suitable crops based on soil parameters with high accuracy and efficiency Reddy et al. (2019). Peraka et al. (2020) propose an intelligent irrigation system utilizing IoT technology, incorporating soil moisture sensors to inform watering decisions. The system not only provides real-time updates to farmers via a mobile app but also contributes to water conservation by reducing irrigation water wastage, which can reach up to 30% Peraka et al. (2020).

Optimizing crop yield is crucial in meeting the increasing global demand for food, and one practical approach is utilizing machine learning techniques to recommend the most suitable crop fertilizer based on soil parameters. Fertilizer recommendation is vital in crop management as it directly impacts yield and quality. However, traditional methods of fertilizer recommendation can be time-consuming and costly, often relying on soil tests and crop nutrient requirements. Machine learning techniques have demonstrated immense potential in optimizing fertilizer recommendations by leveraging soil parameters to predict the optimal fertilizer needs for crops. In their study, Li et al. (2019a, 2019b) proposed a machine learning-based fertilizer recommendation system designed explicitly for tea plantations. Utilizing the Random Forest algorithm and soil data, they achieved an accuracy rate of 92.5% in forecasting the ideal fertilizer demand for tea plants.

Similarly, Zhang et al. (2020a, 2020b) researched apple orchards and presented a machine learning-based fertilizer recommendation system. They applied the Gradient Boosting method and soil data to accurately predict the optimal fertilizer requirement for apple trees, achieving an accuracy rate of 91.5%. In the context of corn fields, Zhang et al. (2021a, 2021b) developed a machine learning-based fertilizer recommendation system employing the Support Vector Machine (SVM) algorithm. By utilizing soil data, they successfully predicted the best fertilizer demand for maize with an impressive accuracy rate of 94.2%. Li et al. (2021a, 2021b) proposed a machine learning-based fertilizer recommendation system using the ANN technique for rice fields. Through soil data analysis, they achieved a remarkable accuracy rate of 95.6% in forecasting the ideal fertilizer requirement for rice. In their study on cotton fields, Liu et al. (2021a, 2021b) introduced a machine learning-based fertilizer recommendation system. By utilizing the Decision Tree algorithm and considering soil factors, they accurately determined the optimal fertilizer requirement for cotton, achieving an accuracy rate of 93.4%.

Chen et al. (2021a, 2021b) developed a machine learning-based fertilizer recommendation system targeting vegetable fields. Using the KNN algorithm and soil data, they successfully forecasted the ideal fertilizer requirement for veggies with an accuracy rate of 91.7%. The studies mentioned highlight the efficacy of machine learning algorithms in providing accurate fertilizer recommendations for various agricultural contexts. By leveraging soil data and applying advanced algorithms, these systems optimize fertilizer usage and enhance crop yield in tea plantations, apple orchards, corn, rice, cotton, and vegetable fields. Machine learning algorithms have demonstrated significant promise in optimizing crop fertilizer recommendations.

Gopi and Karthikeyan (2022) introduced the Multimodal Machine Learning-Based Crop Recommendation and Yield Prediction (MMML-CRYP) model, focusing on the Indian agricultural context. The study emphasizes the challenges in crop recommendation due to various variables and presents a two-stage model. The initial stage employs the equilibrium optimizer with kernel extreme learning machine (EO-KELM) for crop recommendation, while the random forest (RF) technique is utilized for crop yield prediction. The MMML-CRYP model demonstrated superior performance compared to other methods, achieving a maximum accuracy of 97.91%. The R2 score analysis revealed its effectiveness in outperforming other methodologies, showcasing a maximal R2 score of 98.54%. The study suggests future exploration of deep learning (DL) models for crop classification.

Oikonomidis et al. (2022) conducted a systematic literature review (SLR) on the application of deep learning (DL) for crop yield prediction. The study identified 456 relevant studies and narrowed them down to 44 primary studies for in-depth analysis. CNN emerged as the most common and effective algorithm, particularly in dealing with images of maize. Challenges such as the lack of large training datasets and the risk of overfitting were highlighted. The study calls for future research to address data limitations and enhance the interpretability of DL models.

Nehra et al. (n.d.) focused on fertilizer and crop recommendations using the Internet of Underground Things (IoUT) and machine learning. The study aims to provide a cost-effective and automated system for soil analysis. The research utilizes unsupervised machine learning and a CNN to process soil images captured by sensors. The proposed model achieved an accuracy of 91%, showcasing its potential for soil analysis and crop recommendation.

Kuradusenge et al. (2023) conducted a comprehensive study examining the impact of climate change on crop production, with a specific focus on Irish potatoes and maize in Musanze, Rwanda. Utilizing data mining techniques, including random forest, polynomial regression, and support vector regression, the researchers predicted crop harvests based on historical weather and yield data. The study highlighted Random Forest as the preferred model, identifying optimal weather conditions for each crop. The research methodology involved collecting extensive datasets related to weather conditions and crop yields over a significant time span. This data played a crucial role in training and validating machine learning models, allowing the researchers to establish patterns and correlations between environmental factors and crop productivity. The study's findings hold significance for agricultural practices in the region, as the predictive models, particularly the Random Forest algorithm, demonstrated effectiveness in anticipating crop harvests. Notably, the study identified optimal weather conditions for each crop, offering valuable insights for farmers and agricultural decision-makers to make informed choices regarding crop cultivation strategies and resource allocation.

Senapaty et al. (2023) present an innovative study in precision agriculture, focusing on the development of an IoT-enabled soil nutrient analysis and crop recommendation (IoTSNA-CR) model. Published in Computers, the research integrates Internet of Things (IoT) technology, cloud computing, and machine learning tools to enhance agricultural practices. The primary objective is to provide continuous and real-time support for farmers through advanced technologies. IoT sensors were incorporated into the agricultural landscape to monitor and analyze soil nutrient levels in real time—a critical aspect of precision agriculture enabling informed decision-making based on the current state of fields.

The proposed IoTSNA-CR model goes beyond simple data collection, integrating machine learning techniques to provide tailored crop recommendations based on analyzed soil nutrient information. The study employs a multi-class support vector machine (MSVM) with a directed acyclic graph (DAG) structure, further optimized using the fruit fly optimization method (FFO). The results are noteworthy, with the IoTSNA-CR model achieving a high accuracy rate of 0.973. This underscores the model's effectiveness in offering precise and reliable recommendations for crop selection based on soil nutrient analysis. The incorporation of IoT technology ensures continuous monitoring of soil conditions, allowing for timely interventions and adjustments in agricultural practices. Cloud computing enhances accessibility and scalability, enabling farmers to access analyzed data and recommendations remotely, aligning with the trend of digital agriculture. The authors contribute significantly to precision agriculture by introducing an IoTSNA-CR model that integrates IoT, cloud computing, and machine learning. The high accuracy achieved by the model has the potential to revolutionize farming practices, providing real-time insights into soil nutrient conditions and personalized crop recommendations to address current agricultural challenges and promote sustainable and efficient farming practices in the future.

Iniyan et al. (2023) comprehensively explored machine learning methods for predicting crop yields. The study emphasized the complexity of farming systems and the importance of machine learning in decision-making. The study concluded that a feature engineering-based LSTM model exhibited the highest efficiency with an accuracy of 86.3%. The study delves into the intricate nature of farming systems, underscoring the complexity inherent in predicting crop yields and highlighting the pivotal role that machine learning plays in the decision-making processes of agriculture. The researchers employed various multiple regression models, showcasing the diversity of approaches to address the multifaceted challenges of crop yield prediction. The wide array of algorithms reflects the researchers' commitment to a thorough investigation, considering different aspects of the data and exploring various methodologies to enhance predictive accuracy. The study's findings revealed that among the multitude of models tested, a feature engineering-based Long Short-Term Memory (LSTM) model emerged as the most efficient. LSTM, a type of recurrent neural network (RNN), is well-suited for sequence data and can capture long-term dependencies in time-series data, making it particularly relevant for crop yield prediction's dynamic and temporal nature. The LSTM model exhibited an impressive accuracy of 86.3%, signifying its ability to discern intricate patterns within the dataset and make accurate predictions. Feature engineering, a process of selecting, modifying, or creating features from the available data, played a crucial role in enhancing the performance of the LSTM model, showcasing the significance of thoughtful feature selection in the context of crop yield prediction.

Mamatha and Kavitha's (2023) research, featured in the Measurement: Journal of the International Measurement Confederation, contributes to the literature by addressing the critical issue of global food scarcity. Focused on hydroponics farming techniques within a smart greenhouse environment, the study proposes an innovative solution to enhance crop growth management. The researchers advocate for an automated greenhouse system designed to optimize conditions for crop growth. This approach aligns with the burgeoning field of precision agriculture, where technological advancements are leveraged to maximize resource efficiency. The utilization of the KNN algorithm is a notable aspect of the study, demonstrating its application in predicting absolute crop growth rates. By incorporating machine learning into hydroponics farming, Mamatha and Kavitha aim to offer precise and tailored insights, particularly for leafy vegetables. The research showcases a convergence of agricultural science and technology, emphasizing the potential for automated systems to revolutionize crop management and contribute to addressing global food security challenges.

These studies highlight the diverse applications of machine learning in agriculture, from fertilizer recommendation to crop yield prediction and intelligent farming systems. The findings emphasize the potential of machine learning to revolutionize agriculture, offering precise insights for informed decision-making and addressing global food security challenges. Further research is warranted to explore the full potential of machine learning techniques in optimizing fertilizer recommendations for diverse crops and geographies.

6.3 Assumptions and Notations

6.3.1 Data Assumptions

The data assumptions in this research assert the accuracy and representativeness of the data for India's soil and crop conditions. Additionally, the assumptions involve correctly implemented and optimized machine learning algorithms, consideration of pertinent soil and climate parameters, and the suitability of recommended crop varieties without accounting for socio-economic or market factors.

6.3.2 Notations

In the notation system employed, 'X' symbolizes the input dataset, with each row representing specific soil and climate conditions and each column denoting different parameters. 'Y' is utilized to signify the target variable, indicating the recommended crop variety. The research assumes that the evaluation of recommended crop varieties focuses on yield potential and does not consider assessments related to quality or other attributes.

6.3.3 Data Reliability and Crop Suitability

The data used in the research is assumed to be accurate and reliable. The recommended crops for each soil type are considered suitable and effective for the given conditions. The weather conditions during the crop growth period are assumed to be favorable and will not deviate significantly from historical patterns.

6.3.4 Hypothesis Analysis

To perform a hypothesis analysis, we must first state our null and alternative hypotheses to check whether a significant relationship exists between temperature and rainfall with crop labels. To test this hypothesis, we used the statistical ANOVA test.

Null hypothesis: No significant relationship exists between temperature and rainfall with crop labels.

Alternative hypothesis: A significant relationship exists between temperature and rainfall with crop labels.

6.4 Process Flow Diagram

A process flow diagram illustrates the sequential steps of our proposed method to build machine learning models:

I. **Collect raw data from various sources**

II. **Preprocess the data**

Clean, transform, and normalize the collected raw data.

III. **Perform feature extraction**

Discover essential features crucial for training the machine learning model.

IV. **Examine data for insights and patterns**

Explore the dataset for valuable insights and patterns that can enhance the model.

V. **Divide the dataset**

Split the dataset into training and testing datasets.

VI. **Train the machine learning model**

Utilize various techniques to train the machine learning model on the prepared training dataset.

VII. **Evaluate model performance**

Employ ROC score and accuracy evaluation metrics to assess the model's performance on the testing set.

VIII. **Validate the model**

Put the model through its paces with new data, comparing the results to the actual outcomes.

IX. **Continuous improvement**

Continuously improve the model by collecting new data, retraining the model, and repeating steps 7–9.

We can see the flow diagram to predict crop labels in Fig. 6.1.

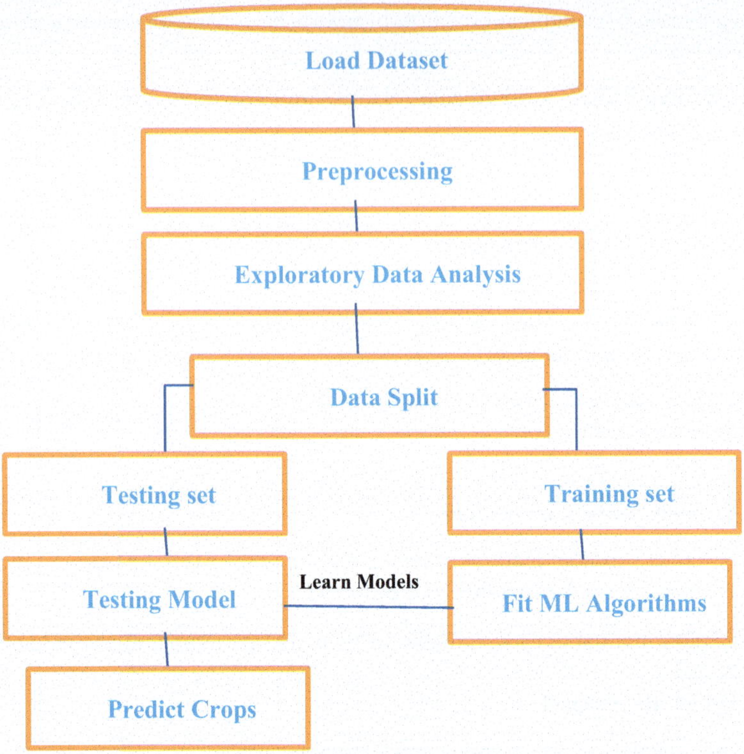

Fig. 6.1 Flow diagram to predict Crops label

6.5 Mathematical Model

Mathematical models are pivotal in crop recommendation systems, aiding agricultural decision-making and optimizing crop yields. The commonly utilized models are Random Forest, Logistic Regression, KNN, and Gradient Boosting (Rashid et al. 2021). Lastly, Gradient Boosting combines weak learners into a strong predictor, offering high accuracy in crop prediction. These diverse mathematical models provide valuable insights into the agricultural domain, helping farmers make informed choices and maximize their harvests.

6.5.1 Random Forest

This algorithm is a widely recognized technique for classification and regression tasks. During the training phase, it generates an ensemble of decision trees. It determines the final output class by aggregating the individual predictions of these trees, either through the mode of categories for classification or the mean prediction for regression (Breiman 2001). During prediction, the algorithm calculates the average prediction from all the trees in the forest, providing the mean forecast for classification or regression tasks. It offers several advantages, including high accuracy, robustness against noise and missing data, and the capability to handle high-dimensional datasets with frequent features. However, massive datasets may incur computational expenses. Additionally, if the relationship between the features and the target variable exhibits a high degree of non-linearity, the random forest algorithm may yield suboptimal results (Liaw and Wiener 2002).

The mathematical representation of the random forest model is as follows

Let X be a (n x m) input matrix, where n is the number of samples and m is the number of features. Let y represent the vector of target values. The random forest technique generates a set of T decision trees designated by f1, f2,…, fT. A random subset of k features is chosen at each tree node from the given set of m features, where k << m. The optimal split is determined based on a criterion for impurity reduction, such as Gini impurity or information gain.

The prediction of the random forest model for a new input vector x is given by

$$f(x) = (1/T) * \text{sum}(i = 1 \text{ to } T)\, fi(x) \tag{6.1}$$

where fi(x) is the prediction of the ith decision tree for the input vector x; this unpredictability ensures that each tree is unique, lowering the risk of overfitting.

6.5.2 *Logistic Regression*

It is a supervised learning algorithm designed for binary classification tasks where the dependent variable is categorical. This logistic function, commonly known as the sigmoid function, converts input values into a probability range of 0 to 1, signifying the likelihood of the event occurring (Hosmer et al. 2013). The mathematical representation of the logistic function is as follows:

$$P(Y = 1|X) = 1/(1 + e \wedge (-z)) \tag{6.2}$$

Here

- P(Y = 1|X) represents the conditional probability of Y’s binary outcome is 1, given the predictor variables X.
- e denotes the mathematical constant e, which is approximately equal to 2.71828.
- z represents the linear combination of the predictor variables and their respective coefficients. It is calculated using the formula:

- $z = b0 + b1X1 + b2X2 + \ldots + bpxp$. Here, b0 is the intercept term and b1, b2, …, bp are the coefficients of the predictor variables X1, X2, …, Xp, respectively (Hosmer et al. 2013).

The logistic regression model can make accurate probability-based predictions for binary classification tasks by iteratively optimizing the coefficients. To estimate these coefficients, iterative optimization procedures such as gradient descent are commonly used.

6.5.3 KNN Model

This model is a non-parametric classification method that does not rely on assumptions about the data distribution. Instead, it focuses on identifying similar instances or locations within a dataset. The KNN algorithm achieves this by identifying the nearest neighbors to a given point and assigning a class label to that point based on those neighbors' classes (Cover and Hart 1967). By considering the classes of the nearest neighbors, the KNN model determines the most likely class for a new data point. This approach allows the KNN model to make predictions based on the local structure of the data, making it suitable for tasks where the decision boundary between classes is non-linear and complex to define. The KNN model is especially helpful when it is difficult to determine the non-linear class boundary.

The KNN algorithm is renowned for being straightforward to comprehend and use. It does, however, have some restrictions. For instance, a big dataset is often needed to get reliable results, and selecting the correct value for K (the number of nearest neighbors to take into account) can significantly impact how well the algorithm performs.

The mathematical expression of the KNN algorithm can be written as

$$\hat{y} = \text{mode}(y\_i) \text{ for } i \in S\_K(x) \tag{6.3}$$

When predicting the class label ŷ for a new input data point x in the K-Nearest Neighbor (KNN) algorithm, the mode function is applied to determine the most common value within a list. The list represents the classes of the K nearest neighbors, which are selected based on their proximity to the input data point x. The variable S_K(x) represents the set of these K nearest neighbors, while y_i denotes the class label of the ith data point in the dataset.

6.5.4 Gradient Boosting

It is a powerful ensemble learning technique that combines multiple weak models into a single robust model. Gradient Boosting is to iteratively add new models that correct the mistakes made by the previous models (Friedman 2001).

The mathematical model of Gradient Boosting can be written as follows:

$$y = F(x) = \Sigma_i f_i(x) \tag{6.4}$$

where

- Y: the dependent variable that we want to predict
- X: the independent variable
- F(x): final prediction, which is the sum of predictions from multiple weak models ($f_i(x)$)
- i: the index of weak models, ranging from 1 to M
- M: total number of weak models.

6.5.5 Extra Trees

This is an ensemble learning technique that shares similarities with Random Forest. However, there is a notable difference in how split points are selected. While Random Forest chooses the best-split point for each feature, Extra Trees takes a different approach by selecting split points randomly. This randomization contributes to a reduction in overfitting and enhances model diversity.

The mathematical model of Extra Trees can be written as follows:

$$y = F(x) = \Sigma_i f_i(x) \tag{6.5}$$

where

- Y: the dependent variable that we want to predict
- x: the independent variable
- F(x): final prediction, which is the sum of predictions from multiple randomized decision trees ($f_i(x)$)
- i: the index of randomized decision trees, ranging from 1 to M
- M: total number of randomized decision trees.

Extra Trees, called Extremely Randomized Trees, is an ensemble learning algorithm combining multiple randomized decision trees to create a robust prediction model. Unlike conventional decision trees that select the best attribute for splitting, Extra Trees takes a different approach. It randomly chooses multiple features and evaluates the best split among them. This randomness plays a crucial role in reducing the model's variance and mitigating the risks of overfitting.

6.5.6 *Naive Bayes*

Naive Bayes uses these probabilities to make predictions in classification tasks. The probability distribution of the classifier is given by

$$P(y|x1, x2, \ldots, xn) = P(y) * P(x1|y) * P(x2|y) * \ldots * P(xn|y)/P(x1, x2, \ldots, xn) \tag{6.6}$$

In the context of Naive Bayes, the class label is denoted as y, while × 1, × 2, ..., xn represent the features associated with the data. P(y) represents the prior probability of the class, indicating the likelihood of encountering a particular class in the dataset. P(xi|y) represents the conditional probability of a feature given the class, signifying the probability of observing a specific feature value given a particular class. Lastly, P(× 1, × 2, ... , xn) represents the probability of the data, which encompasses the joint probability of all the present features.

6.6 Numerical Illustration

In the context of crop recommendation models, a Numerical Illustration refers to the application of mathematical models and to real-world data to predict crop suitability for specific regions or conditions. The Numerical Illustration involves inputting various sets of environmental and soil parameters into the models to assess their performance and accuracy in recommending suitable crops. By conducting a Sensitivity Analysis, the models' responsiveness to changes in input variables can be evaluated, providing valuable insights into the robustness of the recommendations. The observations derived from this process help researchers and agricultural experts understand the strengths and limitations of each model, enabling them to fine-tune the crop recommendation system for better agricultural outcomes and sustainable farming practices.

6.6.1 *Sensitivity Analysis*

Sensitivity, specificity, and accuracy are necessary to evaluate a classification model's performance.

- Sensitivity, also called the true positive rate (TPR), tells us how good the model is at correctly identifying positive cases.
- Specificity, also called the true negative rate (TNR), tells us how well the model identifies negative cases. A higher specificity score indicates that the model effectively detects negative cases and minimizes false positives.

Table 6.1 2 Confusion Matrix

	Actual positive	Actual negative
Predicted positive	TP	FP
Predicted negative	FN	TN

- Accuracy gives an overall measure of how correct the model's predictions are, considering both positive and negative instances.

Sensitivity and specificity assess the model's ability to distinguish between positive and negative instances, while accuracy gives an overall measure of correctness. We can understand how well the model performs in accurately classifying data by considering sensitivity, specificity, and accuracy.

Here's an example table showing how these metrics can be calculated for a binary classification model that predicts whether a person has a disease or not (Table 6.1):

Using this table, we can calculate the sensitivity score, specificity score, and accuracy of the model as follows:

- Sensitivity score = TP/(TP + FN)
- Specificity score = TN/(TN + FP)
- Accuracy score = (TP + TN)/(TP + FP + FN + TN).

The formulas for sensitivity and specificity are still valid for multi-class classification models, but the interpretation and calculation of these metrics have become more complex. In multi-class classification, we can have more than two classes. Let's consider a scenario with three classes: A, B, and C.

- Predicted A True A True Positive (TP) False A as B (FAB) False A as C (FAC)
- Predicted B False B as A (FBA) True B False B as C (FBC)
- Predicted C False C as A (FCA) False C as B (FCB) True C.

In this context, TP represents the count of samples accurately classified as A, B, or C. FAB represents the count of samples that are actually A but wrongly classified as B, and so on. Sensitivity, specificity, and accuracy can be calculated individually for each class. We can use macro-averaging or micro-averaging to calculate the overall metrics for the multi-class classification model. In macro-averaging, we calculate the metrics for each class separately and then take their arithmetic mean. In micro-averaging, we aggregate the results for all the classes and then calculate the metrics. For the multi-class classification model, formulas of accuracy, sensitivity, and specificity are given below.

- Accuracy = (TP_A + TP_B + TP_C)/(TP_A + TP_B + TP_C + FAB + FBA + FAC + FBC + FCA + FCB)
- Sensitivity (macro-average) = (Sensitivity_A + Sensitivity_B + Sensitivity_C)/3
- Specificity (macro-average) = (Specificity_A + Specificity_B + Specificity_C)/3

- Specificity (micro-average) = (TN_A + TN_B + TN_C)/(TN_A + TN_B + TN_C + FP_A + FP_B + FP_C).

In our scenario, the machine learning model focuses not on binary classification but on a multi-class classification problem. The target feature, Crop_label, consists of six different labels: 'Wheat,' 'Soybean,' 'Rice,' 'Watermelon,' and 'Cucumber.' To assess the performance of the model, we can compute the Sensitivity, Specificity, and Accuracy using the following approach:

- Sensitivity_Wheat = TP11/(TP11 + FN1), where FN1 = (FP12 + FP13 + FP14 + FP15)
- Sensitivity_Soyabean = TP22/(TP22 + FN2), where FN2 = (FP21 + FP23 + FP24 + FP25)
- Sensitivity_Rice = TP33/(TP33 + FN3), where FN3 = (FP31 + FP32 + FP34 + FP35)
- Sensitivity_Watermelon = TP44/(TP44 + FN4), where FN4 = (FP41 + FP42 + FP43 + FP45)
- Sensitivity_Cucumber = TP55/(TP55 + FN5), where FN5 = (FP51 + FP52 + FP53 + FP54)
- Specificity_Wheat = (TN22 + TN33 + TN44 + TN55)/(TN22 + TN33 + TN44 + TN55 + FP21 + FP31 + FP41 + FP51)
- Specificity_Soyabean = (TN11 + TN33 + TN44 + TN55)/(TN11 + TN33 + TN44 + TN55 + FP12 + FP32 + FP42 + FP52)
- Specificity_Rice = (TN11 + TN22 + TN44 + TN55)/(TN11 + TN22 + TN44 + TN55 + FP13 + FP23 + FP43 + FP53)
- Specificity_Watermelon = (TN11 + TN22 + TN33 + TN55)/(TN11 + TN22 + TN33 + TN55 + FP14 + FP24 + FP34 + FP54)
- Specificity_Cucumber = (TN11 + TN22 + TN33 + TN44)/(TN11 + TN22 + TN33 + TN44 + FP15 + FP25 + FP35 + FP45).

For example, TP11 represents the number of wheat crops correctly classified as wheat, FP12 represents the number of soybean crops misclassified as wheat, and FP21 represents the number of wheat crops that were misclassified as soybean.

6.6.2 Observation

Exploratory analysis: The first step in exploratory analysis is understanding the data's structure and distributions. We have started by examining the data's summary statistics, including each feature's mean, median, and standard deviation. It gives us a sense of each feature's central tendency and spread and will help identify potential outliers. Another critical aspect of exploratory analysis is to examine the relationships between different features.

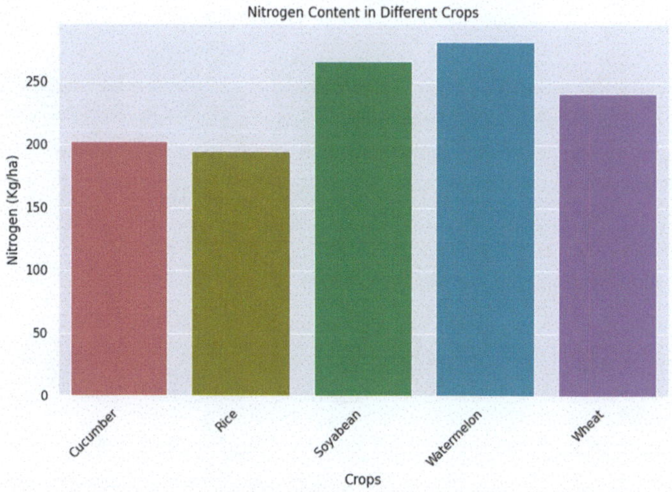

Fig. 6.2 Distribution of crop labels by Nitrogen

Distribution of Crops labeled by Nitrogen: The bar chart shows the distribution of crops labeled by nitrogen, with soybean and watermelon showing higher nitrogen levels while rice has a lower nitrogen level (Fig. 6.2).

This information can help determine which crops may thrive in soil with varying nitrogen levels and in making fertilizer recommendations for each crop. Additionally, this information may help farmers select the appropriate crops to grow in their fields based on the nitrogen levels in their soil.

Distribution of phosphorus for crops: The bar chart depicting the phosphorus distribution among different crops reveals exciting insights. Soybeans and watermelon exhibit higher levels of phosphorus than rice (Fig. 6.3).

This observation suggests that soybean and watermelon crops might require more phosphorus fertilizer for optimal growth and yield. In contrast, rice seems to have lower phosphorus requirements. These findings emphasize the significance of conducting soil tests and tailoring fertilizer applications based on the unique nutrient demands of each crop. Farmers can make informed decisions about fertilizer management practices by understanding the phosphorus needs of specific crops, ensuring efficient resource utilization, and maximizing crop productivity. Adequate phosphorus supply can play a crucial role in meeting the nutrient requirements of soybean and watermelon crops, potentially enhancing their overall performance.

Distribution of Potash for crops: Analyzing the bar chart, it becomes evident that soybean, wheat, and watermelon crops have a higher demand for potash than rice. This discrepancy may stem from several factors. Firstly, the soil where rice is predominantly cultivated naturally may contain higher levels of potash, thereby reducing the need for additional potash-based fertilizers. Conversely, farmers growing soybean,

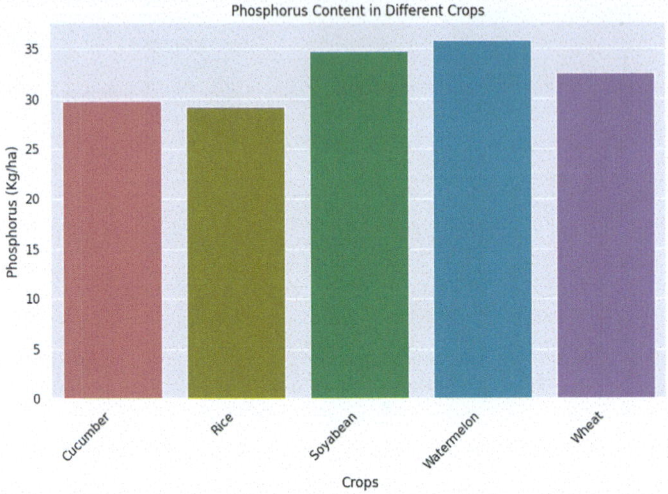

Fig. 6.3 Distribution of phosphorus for crops

wheat, and watermelon may have recognized the importance of supplementing their soil with potash to enhance crop growth and yield (see Fig. 6.4).

Farmers can make informed decisions regarding crop selection and fertilizer application by understanding the varying potash requirements across different crops. This knowledge empowers farmers to optimize fertilizer usage, reduce costs, and maximize crop productivity. Moreover, it highlights the significance of conducting soil tests and tailoring fertilization strategies to meet the specific needs of each crop, ensuring sustainable agricultural practices and improved agrarian outcomes.

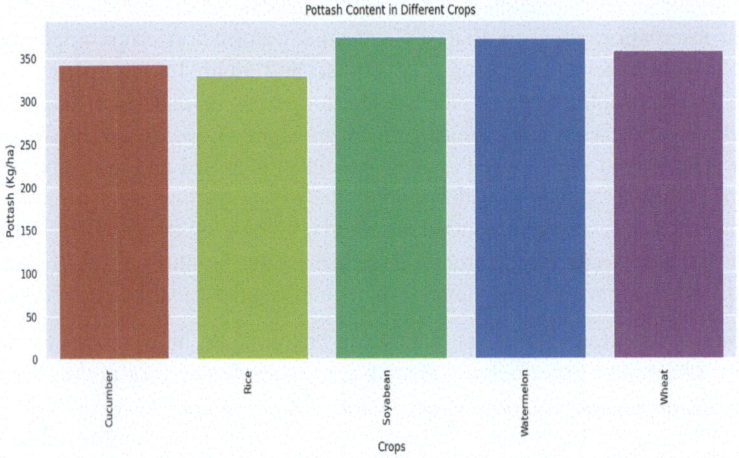

Fig. 6.4 Distribution of Potash for crops

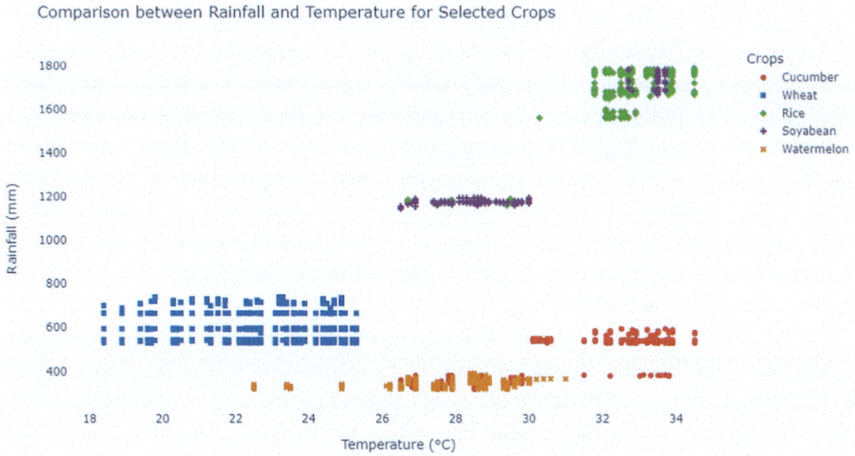

Fig. 6.5 Comparison between rainfall and temperature for each crop

Comparison between rainfall and temperature for each crop: The scatter plot depicting the relationship between rainfall and temperature for various crops in the Madhya Pradesh (MP) region offers valuable insights into the climatic preferences of different crops. By examining the data, we can discern patterns and draw meaningful conclusions regarding the suitability of specific crops for cultivation in different areas within MP. Rice, known for its water-intensive nature, correlates positively with rainfall (Fig. 6.5).

The scatter plot indicates that rice thrives in regions with higher rainfall levels. Additionally, rice cultivation is often associated with warmer climates, as evidenced by the plot's concentration of rice data points at higher temperature values. This suggests that rice is better suited for areas in MP with ample rainfall and higher average temperatures, making it a favorable choice for locations like Bhopal.

On the other hand, cucumber exhibits a different pattern. It requires relatively less rainfall compared to rice but prefers similar temperature levels. The scatter plot reveals a cluster of cucumber data points with moderate rainfall values and comparable temperatures to rice. This indicates that cucumbers may suit regions with less rainfall but similar warm climates. Hence, cucumber cultivation could be a viable option for areas like Anuppur, which may experience lower average rainfall compared to Bhopal.

Meanwhile, crops such as soybean and wheat exhibit intermediate rainfall and temperature preferences. The scatter plot reveals their data points scattered across a broader range of rainfall and temperature values. This implies that soybeans and wheat can adapt to various climatic conditions within the MP region. These crops are less dependent on specific climate requirements, making them versatile options for cultivation in different areas.

The insights derived from the scatter plot have practical implications for farmers and agricultural planners in the MP region. Farmers can make informed decisions regarding crop selection and cultivation strategies by considering the specific climatic preferences of different crops. For instance, focusing on rice cultivation could maximize yields in regions with higher rainfall and temperature. In contrast, areas with lower rainfall but similar temperatures might find cucumber cultivation more feasible. It's important to note that while the scatter plot provides valuable information about the relationship between rainfall, temperature, and crop preferences, other factors such as soil conditions, pests, and market demand should also be considered when making crop cultivation decisions. Nonetheless, the scatter plot is a valuable tool for preliminary crop selection and can guide farmers in optimizing their choices based on local climatic conditions.

The scatter plot comparing rainfall and temperature for different crops in the MP region sheds light on the climate preferences of various crops. The data suggest that rice thrives in areas with higher rainfall and temperatures, while cucumber is more suited for regions with lower rainfall but similar temperatures. Soybean and wheat, on the other hand, exhibit flexibility in adapting to different climatic conditions. By leveraging this information, farmers can make informed decisions about crop selection and cultivation strategies, ultimately enhancing agricultural productivity and sustainability in the MP region.

Hypothesis Analysis

To perform a hypothesis analysis, we must first state our null and alternative hypotheses to check whether a significant relationship exists between temperature and rainfall with crop labels. To test this hypothesis, we used the statistical ANOVA test, and we got this result:

Null hypothesis: no significant relationship exists between temperature and rainfall with crop labels.

Alternative hypothesis: a significant relationship exists between temperature and rainfall with crop labels (Table 6.2).

The ANOVA table presents the results of statistical tests examining the influence of temperature and rainfall on crop labels.

The F-test reveals a highly significant relationship between temperature and crop labels ($F = 11{,}927.78$, $p < 0.0001$). This indicates compelling evidence to reject the null hypothesis, supporting the alternative hypothesis that a significant relationship exists between temperature and crop labels.

Similarly, the F-test for rainfall demonstrates a significant relationship with crop labels ($F = 198.39$, $p = 1.743878e\text{-}44$). Consequently, there is evidence to reject

Table 6.2 ANOVA test result

	Sum_sq	df	F	PR(<F)
Temperature	10,305.55	1.0	11,927.77	0.000000e + 00
Rainfall	171.40	1.0	198.39	1.743878e-44
Residual	6424.68	7436.0	NaN	NaN

the null hypothesis, suggesting a meaningful association between rainfall and crop labels.

The sum of squares (sum_sq) column provides insights into the variation in crop labels that can be explained by temperature and rainfall. The residual column represents the unexplained variation. The degrees of freedom (df) column indicates the number of observations minus the number of parameters in the model.

The ANOVA table furnishes strong evidence supporting the significant relationships between temperature and rainfall with crop labels. These findings emphasize the importance of incorporating temperature and rainfall in predictive models or further analyses.

Machine Learning:

We built several machine learning models to evaluate the best model for crop prediction. We can see the table with the accuracy scores for the given machine learning algorithms.

The Table 6.3 shows the evaluation results for various machine learning models used in the context of crop prediction.

K Neighbors Classifier Model

- CV Mean Score: The Cross-Validation Mean Score is exceptionally high at 0.99770, indicating robust performance during cross-validation.
- Train Score and Test Score: Both the training and test scores are perfect at 1.0, suggesting that the model has learned the training data well without overfitting.
- Accuracy: The overall accuracy on the test set is an impressive 99.77%, emphasizing the model's effectiveness in predicting crop outcomes.

Logistic Regression

- CV Mean Score: The Cross-Validation Mean Score for Logistic Regression is 0.82255, indicating a strong performance during cross-validation but comparatively lower than some other models.
- Train Score and Test Score: Both scores are perfect at 1.0, indicating that the model has learned the training data well without overfitting.
- Accuracy: The accuracy on the test set is 82.25%, suggesting good predictive capabilities but relatively lower than other models.

Table 6.3 Model evaluation result

Model	CV mean score	Train score	Test score	Accuracy
K neighbors classifier model	0.99770	1.0	1.0	99.76953
Logistic regression	0.82255	1.0	1.0	82.25455
random forest classifier model	1.00000	1.0	1.0	100.00
Decision tree classifier model	0.99789	1.0	1.0	99.78866
Extra trees classifier model	0.99789	1.0	1.0	99.78887
AdaBoost classifier model	1.00000	1.0	1.0	100.00000

Random Forest Classifier Model

- CV Mean Score: The model achieves a perfect Cross-Validation Mean Score of 1.0, demonstrating robust performance across different folds during cross-validation.
- Train Score and Test Score: Both scores are perfect at 1.0, indicating that the model has learned the training data well without overfitting.
- Accuracy: The model achieves a flawless accuracy of 100%, suggesting excellent predictive capabilities on the test set.

Decision Tree Classifier Model

- CV Mean Score: The Cross-Validation Mean Score is very high at 0.99789, indicating strong performance during cross-validation.
- Train Score and Test Score: Both scores are perfect at 1.0, indicating that the model has learned the training data well without overfitting.
- Accuracy: The overall accuracy on the test set is 99.79%, emphasizing the model's effectiveness in predicting crop outcomes.

Extra Trees Classifier Model

- CV Mean Score: The model achieves a high Cross-Validation Mean Score of 0.99789, indicating strong performance during cross-validation.
- Train Score and Test Score: Both scores are perfect at 1.0, suggesting that the model has learned the training data well without overfitting.
- Accuracy: The model achieves an accuracy of 99.79%, underlining its effectiveness in predicting crop outcomes.

AdaBoost Classifier Model

- CV Mean Score: The model achieves a perfect Cross-Validation Mean Score of 1.0, demonstrating excellent performance across different folds during cross-validation.
- Train Score and Test Score: Both scores are perfect at 1.0, indicating that the model has learned the training data well without overfitting.
- Accuracy: The model achieves a flawless accuracy of 100%, highlighting its outstanding predictive capabilities on the test set.

The Random Forest, KNN, and Logistic Regression Classification models demonstrate exceptional performance across all evaluation metrics, with perfect scores on training and test sets. These models exhibit a high degree of accuracy, showcasing their suitability for crop prediction tasks. The choice among these models may depend on factors such as interpretability, computational efficiency, and specific requirements of the crop prediction application.

Comparison analysis

The bar chart provides a clear visual comparison of algorithm performance. The choice of the most suitable model depends on the specific requirements of the crop

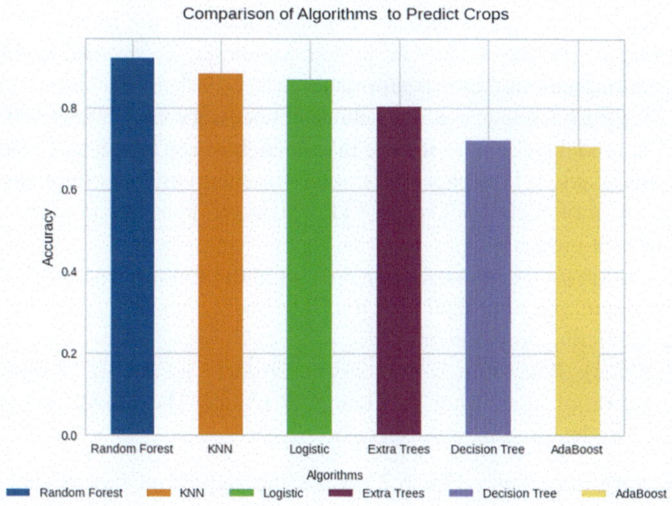

Fig. 6.6 Model performance comparison

prediction application, considering factors such as interpretability, computational efficiency, and the importance of achieving high accuracy (Fig. 6.6).

In this case, random forest, KNN, and logistic regression stand out as top-performing models for predicting crop outcomes.

Feature importance plot: Feature importance plots, often derived from tree-based models like Random Forest or Decision Trees, provide insights into the contribution of different features to the model's predictive performance. In this case, the top important features for crop prediction are listed as follows:

- Season: Seasonality can significantly impact crop growth and yield. Different crops have specific growing seasons, and the model recognizes the importance of this factor in predicting outcomes.
- Rainfall (mm): Rainfall is crucial for providing water to crops. Insufficient or excess rainfall can affect crop growth and health, making it a key predictor in the model.
- Temperature (°C): Temperature directly influences the rate of photosynthesis and other biochemical processes in plants. Different crops have optimal temperature ranges for growth, making this a critical factor.
- Nitrogen (Kg/ha): Nitrogen is an essential nutrient for plant growth, playing a vital role in various physiological processes. The model recognizes its significance in predicting crop outcomes.
- pH: Soil pH affects nutrient availability to plants. Different crops thrive in specific pH ranges, and the model identifies pH as a crucial factor in determining crop success.

- Sulfur (Kg/mg): Sulfur is another essential nutrient for plant development, contributing to processes such as protein synthesis. The model highlights its importance in predicting crop performance.
- Copper (Kg/mg): Copper is a micronutrient necessary for enzyme activation in plants. The model recognizes its role in influencing crop health and productivity.
- Phosphorus (Kg/ha): Phosphorus is critical for energy transfer in plants and is a key component of nucleic acids and ATP. The model identifies phosphorus levels as a significant predictor.
- Organic Carbon (%): Organic carbon content in the soil influences soil structure, water retention, and microbial activity. The model considers it as an important factor in predicting crop outcomes.
- Potash (Kg/ha): Potassium, commonly referred to as potash, is essential for enzyme activation and water regulation in plants. The model recognizes its importance in determining crop success.

These feature importance Fig. 6.7 reveals the significant factors that strongly influence the prediction of the target variable (Cutler et al. 2007). The identified important features align with agricultural science principles, where environmental factors, nutrient levels, and soil characteristics play pivotal roles in crop growth. The model's recognition of these features suggests that it successfully captures the relationships between these agricultural variables and the outcomes of interest. Farmers and agronomists can leverage this information to make informed decisions about crop management practices, ensuring optimal conditions for crop growth and yield. Additionally, these insights may guide further investigations or interventions to improve specific aspects of agricultural practices for enhanced crop production.

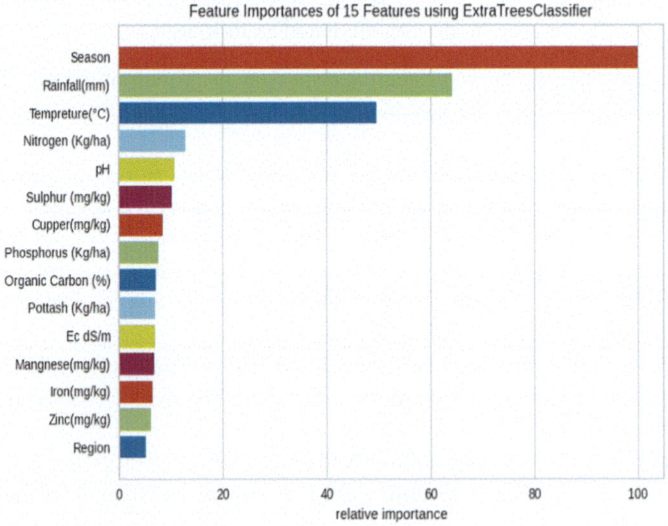

Fig. 6.7 Feature importance plot

Random forest model classification report: The classification report comprehensively assesses the Random Forest model's effectiveness in predicting crop classes using the chosen features. Recall assesses the model's ability to capture the true positive instances out of all instances for a particular class (Powers, 2020). Minimizing false negatives ensures that the model captures as many positive instances as possible (Table 6.4).

Analyzing the results of the tuned Random Forest model provides valuable insights into its performance across different crop classes. The precision and recall scores for each class shed light on the model's ability to make accurate predictions for specific crops. For instance, the model demonstrates exceptional precision and recall scores for class 1 ('Wheat'), indicating accurate predictions for all instances within this class. Similarly, it achieves a high precision score for class 4 ('Watermelon'), suggesting that the majority of predictions made for this class are correct. However, the recall score for class 4 is relatively low, implying that the model missed some instances belonging to this class. The model exhibits a strong recall for class 5 ('Cucumber'), correctly capturing many instances. However, the precision score for this class is comparatively lower, suggesting that some instances predicted as 'Cucumber' may actually belong to other classes. The overall accuracy of the Random Forest model is reported as 1.0, indicating that it correctly predicted 100% of the instances. The average F1-score for all classes is 1.0, suggesting excellent overall model performance across all classes. Additionally, the weighted average F1-score is also reported as 1.0, indicating a robust performance that considers the relative number of instances in each class—a crucial aspect, especially in imbalanced datasets.

We employed hyperparameter tuning methods to enhance the model's performance (Yang and Shami 2020; Hutter et al. 2019). This approach adjusts the F1-score by considering the relative number of instances in each class, resulting in a more comprehensive evaluation. The Random Forest model demonstrates strong predictive capabilities with high precision and recall scores for several classes. The commendable accuracy level and F1-scores underscore its effectiveness in predicting crop classes. However, some limitations are observed, such as a relatively lower recall score for class 4 and a lower precision score for class 5. These insights provide valuable information for further analysis and potential improvements to the model, ensuring its robust performance across various crop categories.

Table 6.4 Classification report of random forest model

Class	Precision	Recall	F1-score	Support
1	1.00	1.00	1.00	330
2	1.00	1.00	1.00	249
3	1.00	1.00	1.00	265
4	1.00	1.00	1.00	193
5	1.00	1.00	1.00	451

Sensitivity analysis result

In a classification problem, sensitivity is a metric that accurately evaluates the model's ability to detect positive instances. It is often referred to as recall or the true positive rate. To determine the sensitivity for each class, one can refer to the 'recall' column in the classification report. This column provides insights into the proportion of true positive instances correctly identified by the model for each class (Table 6.5).

So, the model has perfect sensitivity for classes 1 and 5 but slightly lower sensitivity for classes 2, 3, and 4. It means the model is very good at identifying Wheat and Cucumber but may struggle a bit more with Soybean, Rice, and Watermelon.

ROC Curve

The ROC curve visually represents the binary classification system's performance as its discrimination thresholds are varied. It compares the TP rate versus FP rate at various threshold values, allowing us to evaluate the classifier's capacity to differentiate between positive and negative classes (Fawcett, 2006). They allow us to assess its discriminative capabilities at various threshold settings, enabling us to understand its effectiveness in distinguishing between positive and negative instances. This information is crucial for evaluating and comparing different classifier models and selecting the most suitable one for our specific classification task (Fig. 6.8).

The ROC curve was created for a crop classification system with five classes: Wheat, Soybean, Rice, Watermelon, and Cucumber. The AUC scores for each class were as follows: class 1 (Wheat) had an AUC score of 1.0, class 2 (Soybean) had an AUC score of 1.0, class 3 (Rice) had an AUC score of 1.0, class 4 (Watermelon) had an AUC score of 0.99, and class 5 (Cucumber) had an AUC score of 1.0. The micro-average ROC curve is generated by considering all classes as a single class and calculating the TPR and FPR based on the combined values of true positives, false positives, and negatives. In this specific scenario, the micro-average ROC curve achieved an AUC score of 1.00, indicating the classifier's overall performance when treating all classes as a single entity. This high AUC score suggests that the classifier demonstrated excellent accuracy and efficacy in distinguishing between crop classes.

Similarly, the macro-average ROC curve also attained an AUC score of 1.0, implying that the classifier performed impressively across all individual classes. This result confirms the classifier's ability to classify crops accurately, regardless of the specific class. Overall, the ROC curve and the corresponding AUC scores strongly

Table 6.5 Sensitivity analysis result

Class	Sensitivity (Recall)
1 (Wheat)	1.00
2 (Soybean)	0.80
3 (Rice)	0.89
4 (Watermelon)	0.78
5 (Cucumber)	1.00

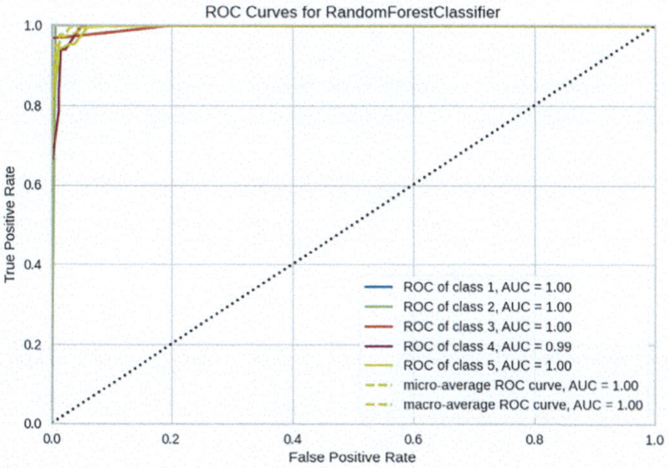

Fig. 6.8 ROC Curve plot

suggest that the crop classification system exhibits high accuracy and effectiveness, showcasing its ability to differentiate effectively between various crop classes.

Precession recall curve

The PR curve is a visual representation that offers an alternative way to evaluate the performance of a binary classifier. The PR curve provides insights into the trade-off between precision and recall for a given classifier (Davis and Goadrich, 2006). By analyzing the PR curve, we can make informed decisions about the classifier's performance and select the optimal threshold setting that best aligns with our specific needs. It helps us understand how changes in the threshold affect the trade-off between recall and precision, allowing for a more nuanced evaluation of the classifier's performance across different operating points (Fig. 6.9).

The PR curve was created for a crop classification system with five classes: Wheat, Soybean, Rice, Watermelon, and Cucumber. The precision scores for each class were as follows: class 1 (Wheat) had a precision score of 1.0, class 2 (Soybean) had a precision score of 0.97, class 3 (Rice) had a precision score of 1.0, class 4 (Watermelon) had a precision score of 0.99, and class 5 (Cucumber) had a precision score of 0.97. The average precision score across all classes was 0.99, indicating that the classifier performed well overall. PR curve and precision scores suggest that the crop classification system is highly accurate and effective at distinguishing between crop classes, particularly for Wheat and Cucumber. However, there may be some trade-off between precision and recall for the other classes, which would need to be evaluated further.

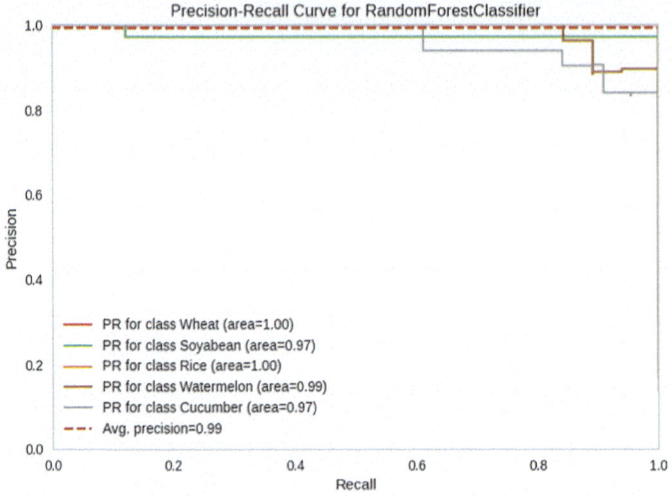

Fig. 6.9 Precession recall curve plot

6.7 Conclusion

In this research study, we comprehensively evaluated different machine learning algorithms to classify soil data and generate optimized crop recommendations for farmers. The data were collected in the Anuppur, Shahdol, and Bhopal regions of Madhya Pradesh, India, and our dataset comprised 7,439 soil samples, each with 17 different soil features. To gather the data, we collaborated with farmers and soil testing experts. Our research process involved several stages: data collection, analysis, and developing, evaluating, and deploying machine learning models. We also fine-tuned and optimized the models to ensure they achieved the highest possible accuracy.

Our analysis focused on using predictive machine learning algorithms to classify soil data and predict the most suitable crop for each soil type. Our findings demonstrated that several machine learning models accurately classified soil data and provided crop recommendations. The results of our research paper underscore the accuracy and effectiveness of machine learning algorithms in classifying soil data and providing optimized crop recommendations for farmers. The analysis of climate preferences for various crops in the MP region, supported by a scatter plot comparing rainfall and temperature, offers valuable insights for farmers. The results suggest that rice thrives in higher rainfall and temperature conditions, while cucumber is better suited for regions with lower rainfall and similar temperatures. Soybean and wheat exhibit adaptability to a range of climatic conditions. By leveraging this information, farmers can make informed decisions about crop selection and cultivation strategies, ultimately enhancing agricultural productivity and sustainability in the MP region. The ANOVA test hypothesis analysis provides robust evidence of significant relationships between temperature, rainfall, and crop labels. The F-tests reveal highly significant relationships, supporting the alternative hypothesis that temperature and rainfall have a meaningful association with crop labels. These findings emphasize the importance of incorporating climate variables in predictive models for crop classification. The evaluation of machine learning models highlights the strengths of various algorithms in predicting crop outcomes. The Random Forest Classifier, Decision Tree Classifier, Extra Trees Classifier, and AdaBoost Classifier models demonstrate exceptional performance with perfect scores on training and test sets. The accuracy, precision, and recall scores provide valuable insights into the models' effectiveness, guiding the choice of algorithms for specific crop prediction tasks. The feature importance plot identifies key factors influencing crop prediction: Season, Rainfall, Temperature, Nitrogen, pH, Sulfur, Copper, Phosphorus, Organic Carbon, and Potash. These features align with agricultural principles, offering valuable information for informed decision-making in crop management practices. Sensitivity analysis reveals variations in the model's ability to detect positive instances for different crop classes.

While classes 1 and 5 show perfect sensitivity, classes 2, 3, and 4 exhibit slightly lower sensitivity. This information helps understand the model's strengths and potential areas for improvement. The ROC curve and AUC scores demonstrate the classifier's excellent accuracy and effectiveness in distinguishing between crop classes. The micro- and macro-average ROC curves reinforce the classifier's overall high performance. The precision-recall curve further confirms the system's accuracy, particularly for Wheat and Cucumber, with an average precision score of 0.99.

The comprehensive analysis of climate preferences, hypothesis testing, machine learning model evaluation, feature importance, and sensitivity analysis provides a thorough understanding of the factors influencing crop prediction. These insights can guide farmers, agronomists, and researchers in making informed decisions for optimal crop management and contribute to the enhancement of agricultural practices in the Madhya Pradesh region. The findings of this research have real-life applications for farmers, soil testing experts, and agricultural researchers. The models developed in this study can be utilized to create more accurate and efficient soil classification and crop recommendation systems. Moreover, this research emphasizes the importance of collecting accurate and relevant data to develop these models. Our research paper comprehensively evaluates various machine learning algorithms for soil data classification and crop recommendation. The findings demonstrate that machine learning algorithms can accurately classify soil data and provide optimized crop recommendations for farmers. These models have significant implications for improving crop yields, minimizing crop failures, and ensuring better utilization of resources. We believe that our research will be valuable for researchers, farmers, and soil testing experts interested in utilizing machine learning algorithms for soil data classification and crop recommendation tasks.

References

Analytical Reports (n.d.) PRS legislative research. https://prsindia.org/policy/analytical-reports/state-agriculture-india

Arooj A, Riaz M, Akram MN (2018) Evaluation of predictive data mining algorithms in soil data classification for optimized crop recommendation. 2018 International conference on advancements in computational sciences (ICACS). https://doi.org/10.1109/icacs.2018.8333275

Breiman L (2001) Random forests. Mach Learn 45(1):5–32. https://doi.org/10.1023/A:1010933404324

Chen W et al (2021b) Machine learning-based fertilizer recommendation system for vegetable fields using the K-Nearest Neighbor (KNN) algorithm. Precision Agric 22(1):112–130

Chen J, Li X, Li Y (2021a) A machine learning-based fertilizer recommendation system for vegetable fields. Comput Electron Agric 184:106022

Cortes C, Vapnik V (1995) Support-vector networks. Mach Learn 20(3):273–297. https://doi.org/10.1007/BF00994018

Cover T, Hart P (1967) Nearest neighbour pattern classification. IEEE Trans Inf Theory 13(1):21–27. https://doi.org/10.1109/TIT.1967.1053964

Cutler DR, Edwards TC, Beard KH, Cutler A, Hess KT, Gibson J, Lawler JJ (2007) Random forests for classification in ecology. Ecology 88(11):2783–2792. https://doi.org/10.1890/07-0539.1

Davis J, Goadrich M (2006) The relationship between Precision-Recall and ROC curves. In: Proceedings of the 23rd international conference on machine learning, pp. 233–240. ACM. https://doi.org/10.1145/1143844.1143874

Doubling of Farmers Income. (2022). Pib.gov.in. https://pib.gov.in/PressReleasePage.aspx?PRID=1883177

Ed-Daoudi R, Alaoui A, Ettaki B, Zerouaoui J (2023) Improving crop yield predictions in morocco using machine learning algorithms. J Ecol Eng 24(6)

Fawcett T (2006) An introduction to ROC analysis. Pattern Recogn Lett 27(8):861–874. https://doi.org/10.1016/j.patrec.2005.10.010

Friedman JH (2001) Greedy function approximation: a gradient boosting machine. Ann Stat, 1189–1232. https://doi.org/10.1214/aos/1013203451

Gopi R, Karthikeyan N (2022) Multimodal Machine Learning-Based Crop Recommendation and Yield Prediction (MMML-CRYP) model for the Indian agricultural context. Comput Electron Agric 193:106264

Hosmer Jr, DW, Lemeshow S, Sturdivant RX (2013) Applied logistic regression (Vol 398). Wiley

Hutter F, Kotthoff L, Vanschoren J (2019) The springer series on challenges in machine learning. https://library.oapen.org/bitstream/handle/20.500.12657/23012/1/1007149.pdf#page=15

Iniyan G et al (2023) Machine learning methods for predicting crop yields:aA comprehensive exploration

Kuradusenge T et al (2023) Impact of climate change on crop production: a data mining approach for Irish potatoes and maize in Musanze. Rwanda. Comput Electron Agric 193:106247

Land Degradation and Desertification in India (n.d.) Drishti IAS. Retrieved August 6, 2023, from https://www.drishtiias.com/daily-news-analysis/land-degradation-and-desertification-in-india

Li X et al (2019b) Machine learning-based fertilizer recommendation system for tea plantations using the Random Forest algorithm. J Agric Sci Technol 21(5):1123–1135

Li X et al (2021b) Machine learning-based fertilizer recommendation system for rice fields using the Artificial Neural Network (ANN) technique. Comput Electron Agric 186:106042

Li J, Zhang J, Li Y (2019a) A machine learning-based fertilizer recommendation system for tea plantations. Comput Electron Agric 157:1–7

Li X, Zhang J, Li Y (2021a) A machine learning-based fertilizer recommendation system for rice fields. Comput Electron Agric 183:106019

Liaw A, Wiener M (2002) Classification and regression by randomForest. R News 2(3):18–22. Retrieved from https://cran.r-project.org/doc/Rnews/Rnews_2002-3.pdf

Liu Z et al (2021b) Machine learning-based fertilizer recommendation system for cotton fields using the Decision Tree algorithm. Front Plant Sci 12:654751

Liu Y, Zhang J, Li Y (2021a) A machine learning-based fertilizer recommendation system for cotton fields. Comput Electron Agric 184:106023

Mamatha R, Kavitha K (2023) Hydroponics farming within a smart greenhouse environment: an automated solution using the K-nearest neighbors (KNN) algorithm. Measurement. J Int Measur Confederation 192:108892

Nehra R, Anand S, Kumari P (n.d.) Fertilizer and crop recommendation using the Internet of Underground Things (IoUT) and machine learning

Oikonomidis A, Catal C, Kassahun Y (2022) Application of deep learning for crop yield prediction: a systematic literature review. Comput Electron Agric 195:106429

Pasquel D, Roux S, Richetti J, Cammarano D, Tisseyre B, Taylor JA (2022) A review of methods to evaluate crop model performance at multiple and changing spatial scales. Precision Agric 23(4):1489–1513. https://doi.org/10.1007/s11119-022-09885-4

Peraka S et al. (2020) Smart irrigation based on crops using IoT. 2020 IEEE 15th international conference on industrial and information systems (ICIIS), IEEE, 611–616

Powers DMW (2020) Evaluation: from precision, recall and F-measure to ROC, informedness, markedness and correlation. [Cs, Stat]. https://arxiv.org/abs/2010.16061

Rashid M, Bari BS, Yusup Y, Kamaruddin MA, Khan N (2021) A comprehensive review of crop yield prediction using machine learning approaches with special emphasis on palm oil yield prediction. IEEE Access 9:63406–63439. https://doi.org/10.1109/access.2021.3075159

Reddy, Anantha D et al. (2019) Crop recommendation system to maximize crop yield in Ramtek region using machine learning. Int J Scientific Res Sci Technol Technoscience Academy, 485–489. https://doi.org/10.32628/ijsrst196172

Senapaty G et al. (2023) IoT-enabled soil nutrient analysis and crop recommendation (IoTSNA-CR) model using multi-class Support Vector Machine (MSVM) with directed acyclic graph (DAG) structure and fruit fly optimization method (FFO). *Computers, Vol-ume(Issue), Page range

Yang L, Shami A (2020) On hyperparameter optimization of machine learning algorithms: theory and practice. Neurocomputing 415:295–316. https://doi.org/10.1016/j.neucom.2020.07.061

Zhang Y et al (2020b) Machine learning-based fertilizer recommendation system for apple orchards using the Gradient Boosting method. Agric for Meteorol 280:107781

Zhang Y et al (2021b) Machine learning-based fertilizer recommendation system for corn fields using the Support Vector Machine (SVM) algorithm. Comput Electron Agric 186:106041

Zhang J, Li X, Li Y (2020a) A machine learning-based fertilizer recommendation system for apple orchards. Comput Electron Agric 178:105758

Zhang J, Li X, Li Y (2021a) A machine learning-based fertilizer recommendation system for corn fields. Comput Electron Agric 184:106021

Chapter 7
Application of Neural Network-Based Techniques to Network Intrusion Detection

Ashalata Panigrahi and Manas Ranjan Patra

Abstract Intrusion detection has been a challenge to safeguard computer networks. Several approaches have been proposed to make intrusion detection systems more effective in detecting intrusive attempts. In this work, we analyze the performance of four Neural Network-based techniques, namely Radial Basis Function Network (RBFN), Self-Organizing Map (SOM), Sequential Minimal Optimization (SMO), and LVQ 3 on intrusion data. Further, wrapper subset evaluator feature selection and standardization have also been considered for performance enhancement. The efficacy of the detection models has been evaluated using different parameters such as false alarm rate, negative predictive value, specificity, Matthew's correlation coefficient, kappa statistic, and geometric mean.

Keywords Artificial neural network · Self-organizing map · Confusion matrix · Wrapper approach · Standardization · Matthew's correlation coefficient

7.1 Introduction

Rapid development in information and communication technology has raised several challenges to ensure reliable computer networks which are being threatened by issues like confidentiality and integrity. Some of the biggest challenges today are dealing with intruder activities such as guessing and cracking passwords, defacing websites, and viewing sensitive data without authorization. With the emergence of new attacks, intrusion detection systems (IDS) have a crucial role to play in identifying new unknown attacks and responding suitably. IDS is a system which monitors

A. Panigrahi (✉) · M. R. Patra
Department of Computer Science & Engineering, NIST University, Institute Park, Pallur Hills, Berhampur, India
e-mail: ashalata.panigrahi@nist.edu

M. R. Patra
e-mail: mrpatra@nist.edu

D. K. Sharma et al. (eds.), *Machine Learning for Real World Applications*, Transactions on Computer Systems and Networks, https://doi.org/10.1007/978-981-97-1900-6_7

and analyzes malicious activities to detect possible network attacks. The major challenge in designing anomaly-based intrusion detection is to maximize the percentage of detection and minimize the percentage of false alarms. Essentially, one can adopt either a signature-based or anomaly (heuristic)-based approach to build an IDS. While the signature-based IDS tries to match with previously known attack patterns, anomaly-based intrusion detection classifies intrusions into three categories: good or benign, suspicious, or unknown (Pfleeger and Pfleeger 2008). Anomaly-based IDS are capable of identifying zero-day attacks (Alazab et al. n.d.).

7.2 Related Work

The authors have designed a Semi-supervised AdaBoost technique which integrated three AdaBoost algorithms, namely discrete, real, and gentle AdaBoost to constitute a tri-training method (Yuan et al. 2016). Using the KDD Cup 1999 dataset, a detection of 89.02%, false alarm of 1.38%, and precision of 98.63% were achieved. In Azad and Jha (2017), the authors have developed an IDS by applying a genetic algorithm and decision tree. The authors used the KDD Cup 1999 dataset for their experiment and reported an accuracy of 99.99% and an error of only 0.01% which outperforms the C4.5 decision tree and Naive Bayes. The authors in Almseidin et al. (2017) proposed an IDS that combines random forest and decision tree techniques. With the KDD Cup 1999 dataset, it yielded an accuracy of 93.77% for random forest and 92.44% for decision tree. In Khammassi and Krichen (2017), a wrapper technique based on a genetic algorithm has been applied as a search strategy for minimizing the number of features of the KDD Cup 1999 dataset. On applying logistic regression as a learning algorithm for classification, they could obtain higher accuracy. Researchers in Aljawarneh et al. (2018) proposed a hybrid intrusion detection model based on Random tree, J48, Meta Bagging, AdaBoost M1, Decision Tree, Naive Bayes, and Decision Stump. They have reported a detection accuracy of 99.81%. In Pham et al. (2018), a hybrid model has been proposed with a gain ratio for feature selection and bagging for combining tree-based classifiers. Performance was enhanced with the bagging model when used along with the classifier J48 on a feature subset of 35 attributes taken from the NSL-KDD dataset. The authors of Hota and Shrivas (2014) proposed an IDS based on decision tree methods with different feature selection techniques such as correlation, IG, Relief, and symmetrical uncertainty on the NSL-KDD dataset. Empirical results show that C4.5 along with IG and symmetrical uncertainty yielded an accuracy of 99.68% and 99.64% with 17 and 11 selected features, respectively.

7.3 ANN-Based Techniques

In this work, we have tried to apply some of the neural network-based techniques to develop an effective intrusion detection system to deal with a variety of network threats.

7.3.1 *Radial Basis Function Network*

The basic RBFN (Haykin 1999) is depicted in Fig. 7.1.

Non-linear transformation is used at the hidden layer, and linear transformation between the hidden and the output layers. The RBFN learns its parameters during training in two steps (Karray and Silva 2009):

Step 1: Training of the RBFN using unsupervised training.

Step 2: Adaptation of weights of the output layer by applying some supervised learning algorithm.

An RBF structure with s units in the hidden layers and t output units is represented as

$$o_y(X) = \sum_{y=1}^{s} wyz\, g_y(X) \quad z = 1, \ldots, t \tag{7.1}$$

where wyz is the connection weight between the y-th receptive field unit and the z-th output, and g_y is the y-th receptive field unit. Some commonly used radial basis functions are

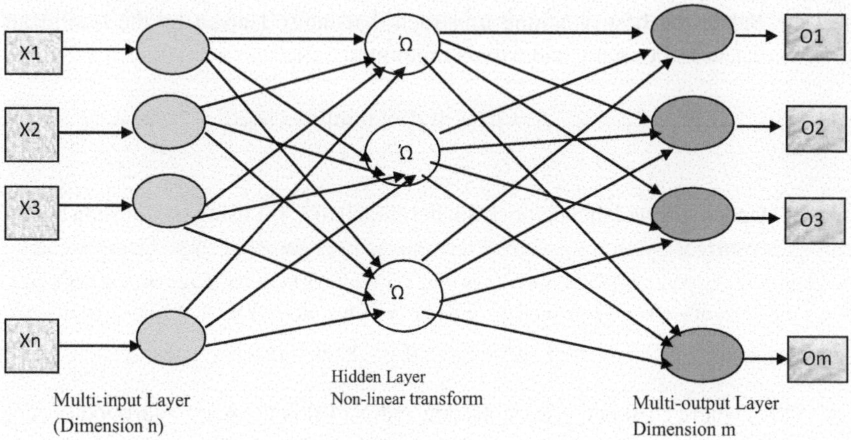

Fig. 7.1 RBFN structure

(a) *Gaussian Function:* $g(x) = exp\left(\frac{-(x-v)^2}{2\sigma^2}\right)\sigma > 0, x, v \in R$

(b) *Multi-quadratics:* $\phi(r) = \left(r^2 + e^2\right)^{\frac{1}{2}}$, *e is a positive constant and* $r \in R$.

Inverse Multi-quadratics: $\phi(r) = \frac{1}{(r^2+e^2)^{\frac{1}{2}}}$, $e > 0;\ r, e \in R$.

(c) The most commonly used Gaussian kernel function has been applied:

$$g(x) = \exp\left(\frac{-(a-b)^2}{2\sigma^2}\right)\sigma > 0, x, v \in R \tag{7.2}$$

where a is the input vector and b is the centre of the receptive field unit g with σ as its unit width parameter.

7.3.2 Self-Organizing Map (SOM)

SOM is a NN-based unsupervised algorithm which works on the principle of winner-takes-all all. The feature of SOM is the existence of a topological structure among the cluster units (Karray and Silva 2009). The algorithm tries to cluster input data into smaller groups of data with similar features.

Given a SOM with r input units $y_1, y_2, y_3, \ldots\ldots, y_r$ and t output units $p_1, p_2, \ldots p_t$, it is to be trained with a set of i number of patterns $s_1, s_2, \ldots, s_i$.

Algorithm:

Step1: Initialize weight matrix, i.e., wuv for all u and v, topological neighbourhood Nd, and learning rate ς.

Step 2: Select the input pattern y from the dataset.

Step 3: Select the best matching unit such that index J given by the Euclidean distance between y and wuv is minimum:

$$J = ||y - w_a|| = \min_{uv} ||y - w_{uv}|| \tag{7.3}$$

Step 4: Update the weight vectors from iteration h to h + 1 using the formula below given:

$$\begin{aligned} w_{uv}(h+1) &= w_{uv}(h) + \varsigma(h)\,(y - w_{uv}(h)) \quad, if\ (u,\ v) \in N_d(h) \\ &= w_{uv}(h) \qquad , if\ (u,\ v) \notin N_d(h) \end{aligned}$$

where ς (h) is the learning rate and N_d (h) is the neighbourhood of unit d at iteration h.

Step 5: If required reduce Nd and ς.

Step 6: Test for the stopping condition.

7.3.3 Sequential Minimal Optimization (SMO)

SMO technique applies an iterative procedure to quickly find the solution to SVM QP problems. SMO decomposes the big QP problem into possible smallest QP sub-problems. Each sub-problem then solves analytically until convergence. The process starts with the initialization of $\beta_1, \beta_2, \beta_3, \beta_4, \ldots, \beta_m$.

There are two components of SMO, namely, (i) an analytic method for solving the two Lagrange multipliers, and (ii) a heuristic that helps in choosing the multipliers to be optimized.

Step 1: Pick exactly two β 's from total m variable betas. For large datasets, it is very useful because there are m (m–1) possible choices for the pairs. The most promising pairs of β 's are found by using certain heuristic methods.

Step 2: Optimize two betas for example β1 and β4 at a time in order to find the optimal values for these two multipliers. Hence re-optimize β1 and β4, and keep all other betas fixed.

Step 3: Repeat Step 1 and Step 2 until the algorithm converges. The convergence criteria for SMO are taken to be the Karush–Kuhn–Tucker (KKT) conditions. We select threshold b such that KKT conditions are satisfied. That is SMO converges when all the Lagrange multipliers (i.e., all β1, β2, β3, β4, ………., βm) satisfy the criteria.

7.3.4 LVQ 3

LVQ 3 is a supervised learning algorithm (Kim and Oommen 2003) in which the two closest vectors V_e and V_f, represent the two nearest neighbours of z.

Let C(z) be the class label associated with z.

Let $d_e = ||Z - V_e||$ be the Euclidean distance of the input vector from the e-th weight vector.

Let φ be the width of the "window" and let $\zeta = (1 - \varphi)/(1 + \varphi)$.

The input vector z is considered to be within the window, and the below given updates are computed if the relative distance satisfies the below condition:

$$\min\left(\frac{de}{df}, \frac{df}{de}\right) > \zeta$$

Case—1:

If $C_f = C(z)\, and C_f \neq C_e$, *then*

$V_e(s+1) = V_e(s) - \delta(s)\,(z(s) - V_e(s))$ *and*

$$V_f(s+1) = V_f(s) + \delta(s)\left(z(s) - V_f(s)\right)$$

Case—2.

$C_f = C(z) = C_e$, *then*

$$V_k(s+1) = V_k(s) + \varepsilon\,\delta(s)\,(z(s) - V_k(s)),\; k \in [e, f]$$

where s: Discretized time index
δ(s): Learning Rate.

7.4 The Proposed Design Model

The proposed model has three phases as shown in Fig. 7.2.

Phase-1: Pre-processing (select relevant features using wrapper subset evaluator).
Phase-2: Post-processing (standardization of the dataset).
Phase-3: Classification of dataset using Neural Network-based classifiers using tenfold cross-validation.

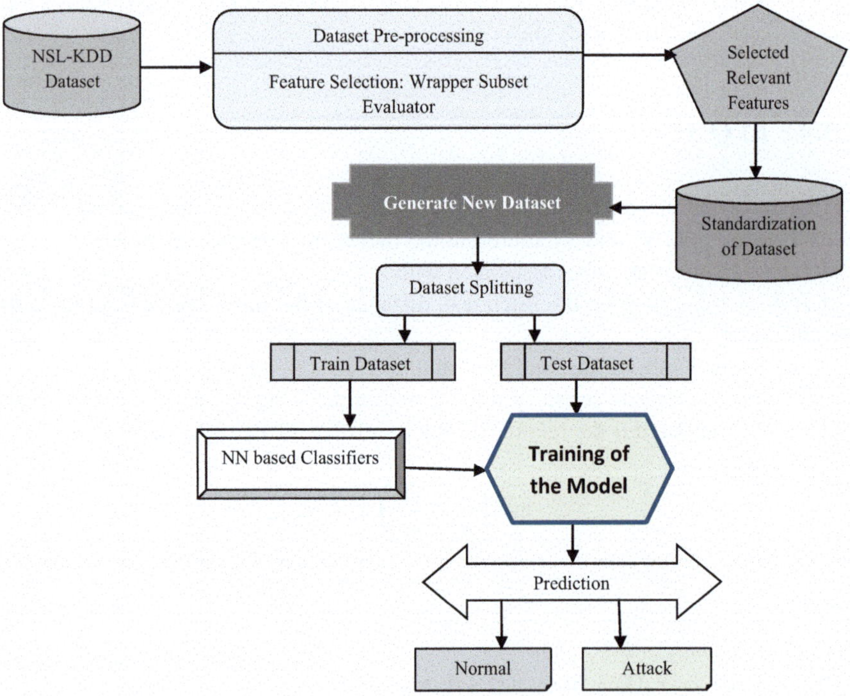

Fig. 7.2 Proposed Design Model

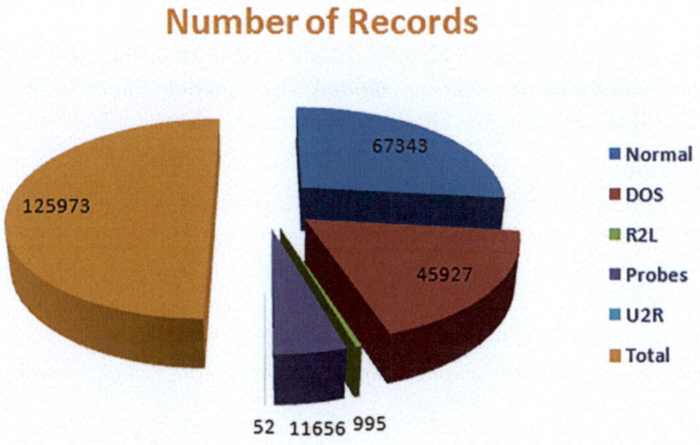

Fig. 7.3 Distribution of records

7.5 Experimentation

7.5.1 Dataset

The NSL-KDD dataset proposed by Tavallaee et al. (2009) has been used to build the model. The dataset has 41 features with one class label which identifies whether a particular instance is harmful or normal. The dataset is imbalanced and it consists of 125,973 data points. The data points are distributed unequally. The number of instances of normal (67,343), DoS (45,927), and Probes (11,656) are much larger than the number of instances of U2R (52) and R2L (995) as evident from Fig. 7.3.

7.5.2 Pre-processing of Data

The first step before proceeding with the analysis of intrusion data is pre-processing which consists of two components—feature selection and standardization.

Feature Selection: The feature selection process does not create any new features but selects the important existing features. Three main categories of feature selection algorithms are wrapper, filter, and embedded (Guyon and Elisseeff 2003). The wrapper algorithm tries to figure out the most relevant features. The filter procedure tries to find features without applying any learning algorithms. The embedded methods perform feature selection during the learning time (Tang et al. 2014). Wrapper-based feature selection method selects an optimal subset of the original features by removing repetitive and unrelated features. The wrapper algorithm in Liu and Yu (2005) tries to search through the feature space using an appropriate algorithm and evaluate each subset by running a model on the subset.

In this work, the potential features are selected using a wrapper-based subset evaluator. The potential features can identify different types of attacks more accurately. Various combinations of wrappers are applied which include BayesNet + Best First Search, J48 + Rank Search, and OneR + Genetic Search.

Standardization: The raw values of features degrade the performance of the algorithm. Therefore, standardization is necessary to transform the values into scaled values. After feature selection, standardization is applied to the selected features of the dataset which is rescaled to ensure the mean and standard deviation (S.D.) are 0 and 1, respectively.

Standardization is defined as in Han and Kamber (2006):

$$Z = \frac{(x - \mu)}{\sigma} \tag{7.4}$$

Z = Generated value.

μ = Mean.

σ = Standard Deviation (S. D.).

7.5.3 *Confusion Matrix*

Confusion matrix is a S × S table which depicts four aspects of a classifier, namely A_{TP} (True Positive), D_{FP} (False Positive), B_{FN} (False Negative), and C_{TN} (True Negative). A_{TP} indicates the attacks that are successfully detected by the model, D_{FP} indicates those wrongly detected as an attack but actually the instances are normal. B_{FN} indicates that the model failed to detect an attack, and C_{TN} indicates the model correctly detected normal instances (Sokolova et al. 2006) (Table 7.1).

The performance of the model can be evaluated by calculating the percentage of accuracy, precision, sensitivity or detection rate, false alarm rate, F-value, specificity, negative predictive value (NPV), false negative rate (FNR), Matthews correlation coefficient (MCC), Kappa statistic (KS), geometric mean (GM), Youden's index (YI), balanced classification rate (BCR), false discovery rate (FDR), and error rate which are expressed as

Table 7.1 Confusion matrix

		Predicted class	
		Normal	Attack
Actual class	Normal	C_{TN}	D_{FP}
	Attack	B_{FN}	A_{TP}

$$\text{Accuracy} = \frac{(A_{TP} + C_{TN})}{(A_{TP} + C_{TN} + D_{FP} + B_{FN})} \tag{7.5}$$

$$\text{Precision} = \frac{(A_{TP})}{(A_{TP} + D_{FP})} \tag{7.6}$$

$$\text{Recall or Sensitivity or Detection Rate} = \frac{(A_{TP})}{(A_{TP} + B_{FN})} \tag{7.7}$$

$$\text{False Alarm Rate } (FAR) = \frac{(D_{FP})}{(C_{TN} + D_{FP})} \tag{7.8}$$

$$F - \text{Value} = 2 \times \frac{(\text{Precision} \times \text{Recall})}{(\text{Precision} + \text{Recall})} \tag{7.9}$$

$$\text{Specificity or True Negative Rate } (TNR) = \frac{C_{TN}}{(C_{TN} + D_{FP})} \tag{7.10}$$

$$\text{Negative Predictive Value } (NPV) = \frac{C_{TN}}{(C_{TN} + B_{FN})} \tag{7.11}$$

$$\text{False Negative Rate } (FNR) = \frac{B_{FN}}{(A_{TP} + B_{FN})} \tag{7.12}$$

$$\text{Matthews Correlation Coefficient } (MCC) = \frac{(A_{TP} \times C_{TN}) - (D_{FP} \times B_{FN})}{\sqrt{(A_{TP} + D_{FP}) \times (A_{TP} + B_{FN})} \times (C_{TN} + D_{FP}) \times (C_{TN} + B_{FN})} \tag{7.13}$$

$$\text{Kappa Statistic } (KS) = \frac{(\text{Total Accuracy } (T_A) - \text{Random Accuracy } (R_A))}{(1 - \text{Random Accuracy } (R_A))} \tag{7.14}$$

where Total Accuracy $= \frac{(A_{TP}+C_{TN})}{(A_{TP}+C_{TN}+D_{FP}+B_{FN})}$

$$\text{Random Accuracy} = \frac{((C_{TN} + D_{FP})(C_{TN} + B_{FN}) + (A_{TP} + B_{FN})(A_{TP} + D_{FP}))}{(A_{TP} + C_{TN} + D_{FP} + B_{FN})^2}$$

$$\text{Geometric Mean } (GM) = \sqrt[2]{(A_{TP}/(A_{TP} + B_{FN})) \times (C_{TN}/(C_{TN} + D_{FP}))} \tag{7.15}$$

$$\text{Youden's Index } (YI) = (A_{TP}/(A_{TP} + B_{FN})) + (C_{TN}/(C_{TN} + D_{FP})) - 1 \tag{7.16}$$

$$\text{Balanced Classification Rate } (BCR) = \frac{(\text{Sensitivity} + \text{Specificity})}{2} \tag{7.17}$$

$$\text{False Discovery Rate } (FDR) = \frac{D_{FP}}{(D_{FP} + A_{TP})} \tag{7.18}$$

$$\text{Error Rate} = \frac{(D_{FP} + B_{FN})}{(A_{TP} + C_{TN} + D_{FP} + B_{FN})} \tag{7.19}$$

7.5.4 *Cross-Validation*

In cross-validation (CV), the entire dataset is divided into two subsets—training and testing. The training subset is used to build the model and the test subset is used to validate the model. A tenfold cross-validation strategy has been applied on the NSL-KDD dataset wherein the entire dataset is randomly divided into 10 folds of approximately equal size, of which nine folds are used for training and the remaining one fold is used as the test/validation set. To evaluate the effectiveness of the model, error estimation is computed as the average of all 10 trials.

7.6 Result Analysis

Quantitative IDS performance measurement results are necessary to compare different classification techniques. Different combinations of four neural network-based classifiers with wrapper subset evaluator were applied to the NSL-KDD dataset and their performance was measured using different metrics such as detection rate, false alarm rate, specificity, kappa statistic, and geometric mean. The false alarm rate and undetected attack rates are the two parameters that define the cost function of IDS. Table 7.2 depicts the performance of four neural network-based classifiers in terms of accuracy, precision, detection rate, false alarm rate, and F-value. We observed that in Table 7.2 the performance results are close to each other and acceptable results are obtained. The results show that SMO with J48 and Rank Search gives the best result (highest accuracy of 0.9735, highest detection rate of 0.9625, and lowest false alarm rate of 0.017). Figures 7.4, 7.5, 7.6, 7.7, and 7.8 provide a comparative view of the performance in terms of accuracy, FAR, precision, detection rate, and F-value of four classifiers.

A high specificity value and low false negative rate (FNR) indicate that the proposed model correctly predicted normal records. MCC calculates the relation between actual and predicted values. When the performance values in Table 7.3 are compared, it is observed that all the classifiers have approximately the same values for specificity, MCC, and Kappa statistics. The results show that SMO with (J48 and Rank Search) wrapper approach gives the best result (highest specificity value of 0.983, highest NPV of 0.9678, and lowest FNR value of 0.0375). MCC value of 0.9468 indicates that our proposed model is good in classifying the attacks in IDS. Figures 7.9, 7.10, 7.11, 7.12, and 7.13 provide a comparative analysis of the performance of specificity, NPV, FNR, MCC, and Kappa Statistic of four classifiers, respectively.

Error rate refers to the proportion of the samples that the classifier incorrectly predicts in all samples. Minimizing the error is very important in IDS where the best-case error rate is 0.0 and the worst-case value is 1.0. A higher value of YI indicates a very low false positive and false negative value. When the performance values in Table 7.4 are compared, it is observed that SMO with (J48 and Rank Search) wrapper

Table 7.2 Comparison of accuracy, precision, detection rate, false alarm rate, and F-value of four NN-based classifiers

Wrapper Subset Evaluator	Classifier techniques	Evaluation criteria				
		Accuracy	Precision	Detection rate	False alarm rate	F-Value
Bayesnet + Best First Search	RBFN	0.9455	0.9685	0.9125	**0.0258**	0.9397
	SOM	0.8986	0.8839	0.9004	0.1029	0.8921
	SMO	**0.9646**	0.9659	**0.9578**	0.0294	**0.9618**
	LVQ3	0.9023	0.9551	0.8291	0.0339	0.8876
J48 + Rank Search	RBFN	0.9285	0.9435	0.9003	0.0469	0.9214
	SOM	0.9327	0.9335	0.9211	0.0571	0.9273
	SMO	**0.9735**	**0.9802**	**0.9625**	**0.017**	**0.9712**
	LVQ3	0.8434	0.9352	0.7131	0.043	0.8092
OneR + Genetic Search	RBFN	0.8845	0.9225	0.8208	0.06	0.8687
	SOM	0.8895	0.9234	**0.8315**	0.06	0.875
	SMO	**0.8926**	**0.9311**	0.831	**0.0537**	**0.8782**
	LVQ3	0.8607	0.9079	0.7798	0.0689	0.839

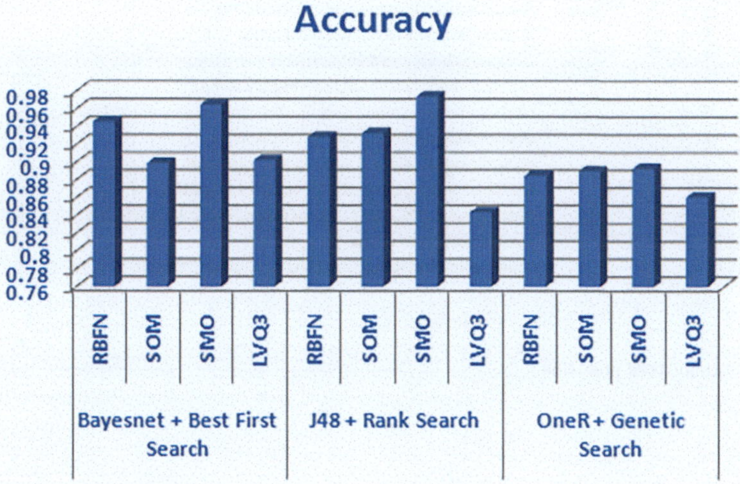

Fig. 7.4 Analysis of accuracy

approach gives the best result (highest GM value of 0.9727, YI value of 0.9455, low FDR of 0.0198, and low error rate of 0.0265). Figures 7.14, 7.15, 7.16, and 7.17 provide a comparative analysis of the performance of GM, YI, BCR, and FDR of four classifiers, respectively.

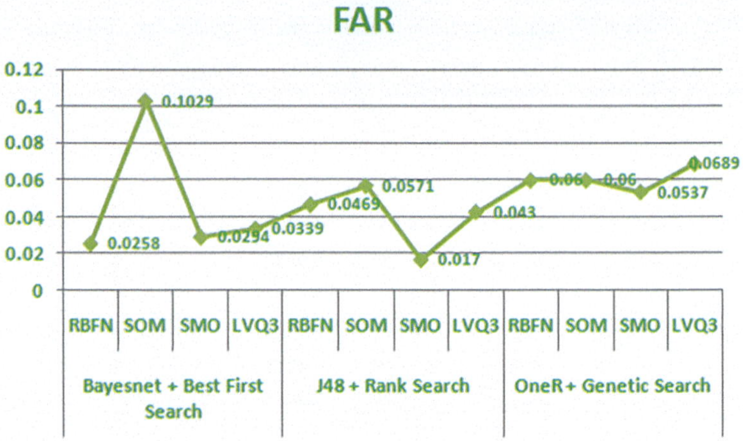

Fig. 7.5 Analysis of false alarm rate

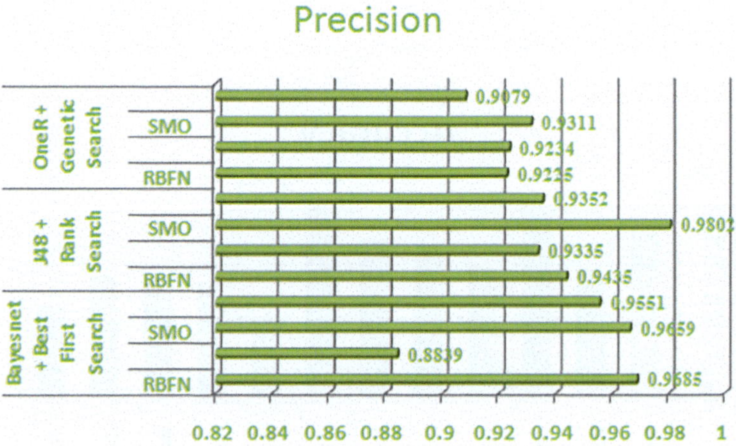

Fig. 7.6 Analysis of precision

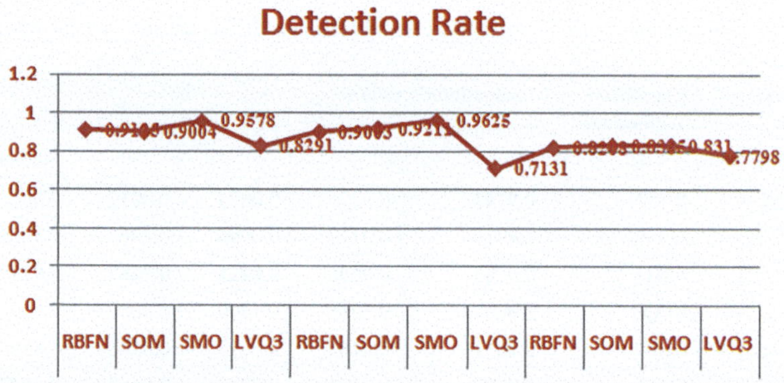

Fig. 7.7 Analysis of detection rate

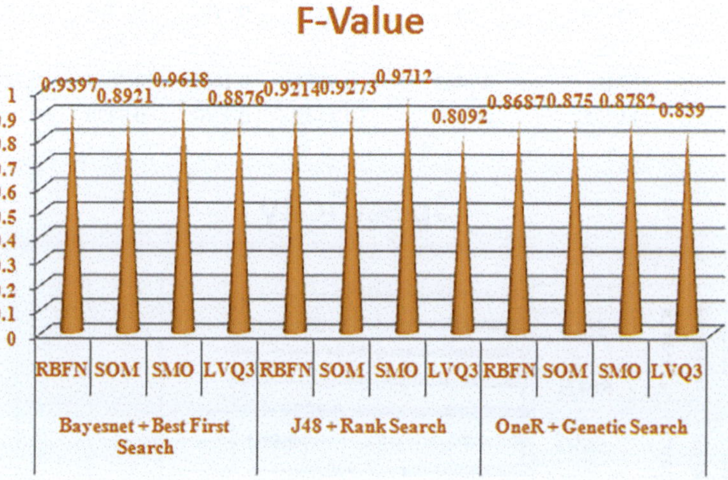

Fig. 7.8 Analysis of F-value

Table 7.3 Comparison of specificity, NPV, FNR, MCC, and Kappa statistic of Four NN-based classifiers

Wrapper Subset Evaluator	Classifier Techniques	Evaluation Criteria				
		Specificity	NPV	FNR	MCC	Kappa Statistic
Bayesnet + Best First Search	RBFN	**0.9742**	0.9275	0.0874	0.8913	0.89
	SOM	0.897	0.9119	0.0996	0.7966	0.7965
	SMO	0.9705	**0.9636**	**0.0421**	**0.9289**	**0.9289**
	LVQ3	0.966	0.8665	0.1709	0.8082	0.802
J48 + Rank Search	RBFN	0.9531	0.9165	0.0997	0.8568	0.856
	SOM	0.9428	0.9321	0.0788	0.8648	0.8647
	SMO	**0.983**	**0.9678**	**0.0375**	**0.9468**	**0.9466**
	LVQ3	0.957	0.793	0.2869	0.6985	0.6805
OneR + Genetic Search	RBFN	0.94	0.8576	0.1792	0.7704	0.7662
	SOM	0.934	0.865	0.1685	0.7799	0.7765
	SMO	**0.9463**	**0.8651**	**0.1689**	**0.7867**	**0.7827**
	LVQ3	0.9311	0.8292	0.2202	0.7329	0.7174

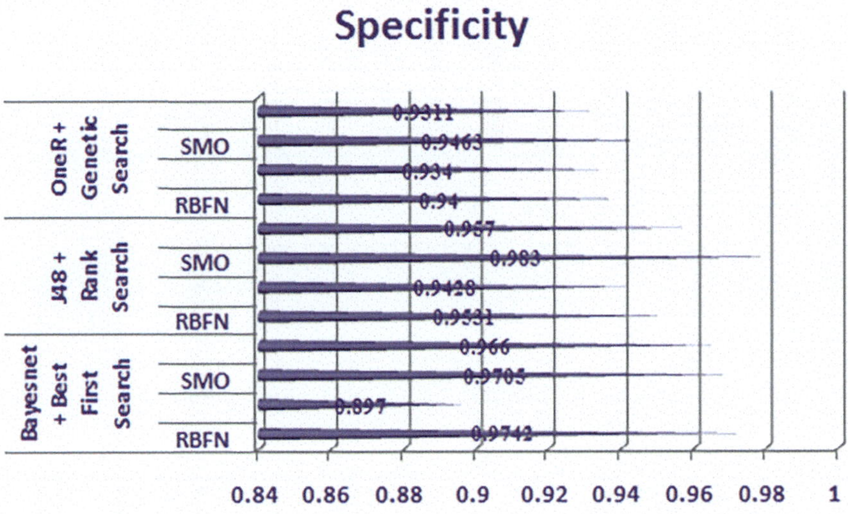

Fig. 7.9 Analysis of specificity

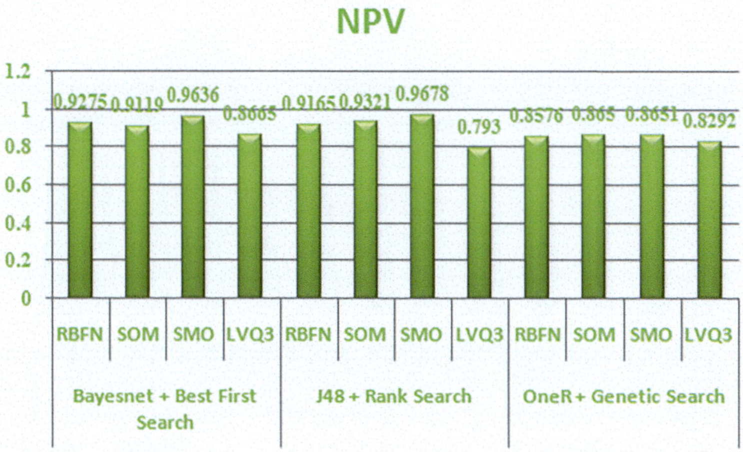

Fig. 7.10 Analysis of NPV

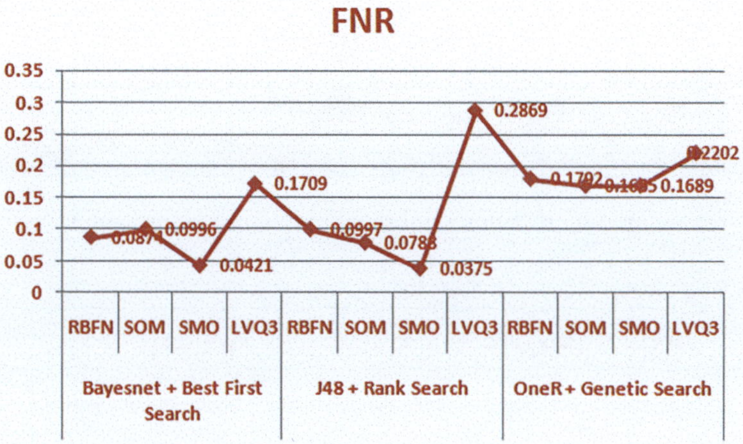

Fig. 7.11 Analysis of FNR for four classifiers

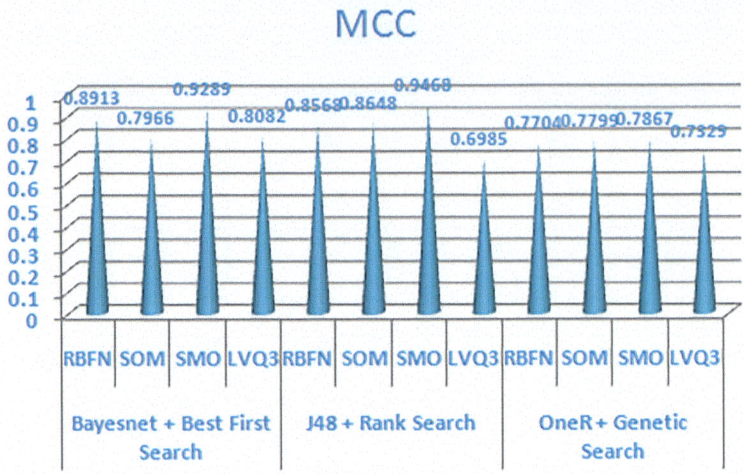

Fig. 7.12 Analysis of MCC for four classifiers

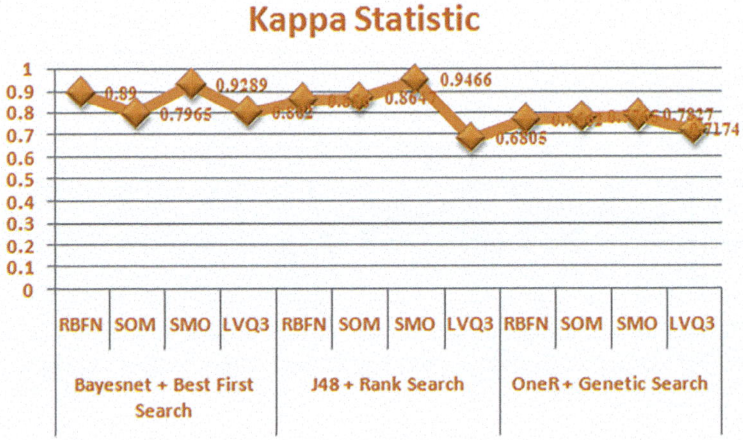

Fig. 7.13 Analysis of Kappa statistic

Table 7.4 Comparison of GM, YI, BCR, FDR, and error rate of four NN-based classifiers

Wrapper subset evaluator	Classifier techniques	Evaluation criteria				
		GM	YI	BCR	FDR	Error rate
Bayesnet + Best First Search	RBFN	0.9428	0.8867	0.9433	**0.0315**	0.0545
	SOM	0.8987	0.7975	0.8987	0.1161	0.1014
	SMO	**0.9642**	**0.9284**	**0.9641**	0.0341	**0.0354**
	LVQ3	0.8949	0.7951	0.8975	0.045	0.0977
J48 + Rank Search	RBFN	0.9263	0.8534	0.9265	0.0565	0.0715
	SOM	0.9319	0.864	0.9319	0.0665	0.0673
	SMO	**0.9727**	**0.9455**	**0.9727**	**0.0198**	**0.0265**
	LVQ3	0.8261	0.67	0.835	0.0648	0.1566
OneR + Genetic Search	RBFN	0.8784	0.7608	0.8804	0.0775	0.1155
	SOM	0.8841	0.7715	0.8857	0.0766	0.1105
	SMO	**0.8868**	**0.7774**	**0.8886**	**0.0689**	**0.1074**
	LVQ3	0.8521	0.7109	0.8554	0.0921	0.1393

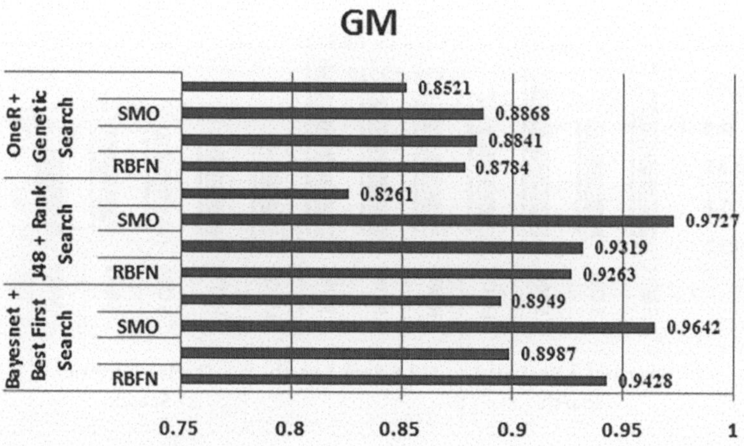

Fig. 7.14 Analysis of geometric mean

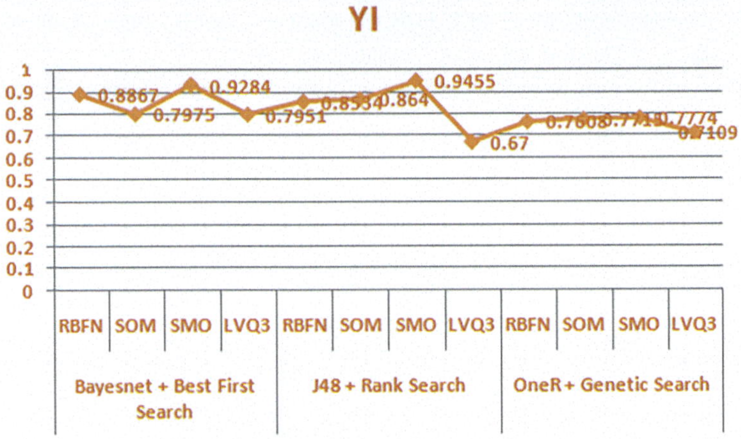

Fig. 7.15 Analysis of Youden's Index among four classifiers

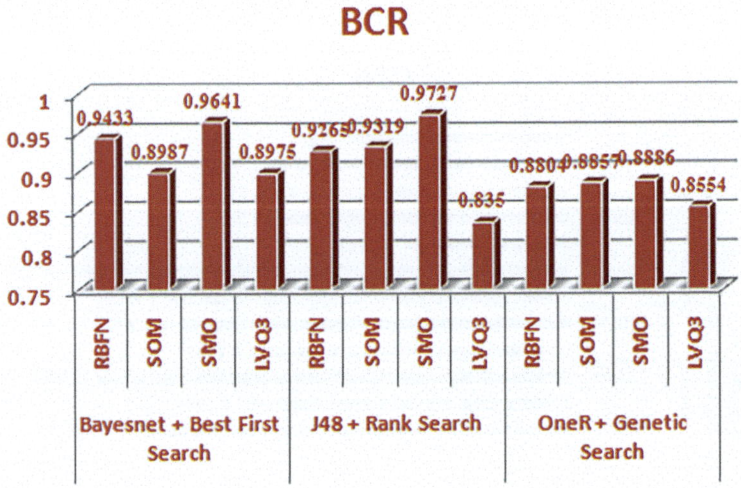

Fig. 7.16 Analysis of BCR

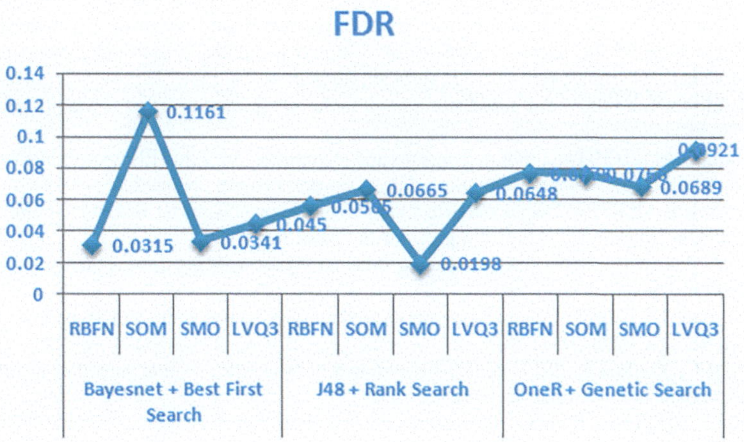

Fig. 7.17 Analysis of FDR

7.7 Conclusion

In this work, some neural network-based techniques were used to develop an effective intrusion detection system. While analyzing the data, it is found that some of the features are irrelevant to the intrusion detection process and thus increase the time of analysis and decrease the accuracy of detection. Therefore, with the help of the wrapper-based feature selection method potential features were first selected which were then standardized to enhance the performance of classification results. Next, the reduced dataset was subjected to four neural network-based classifiers, viz., RBFN, SOM, SMO, and LVQ3. The performance of different combinations of feature selection and ANN-based classifiers was measured in terms of different parameters. It was observed that Sequential Minimal Optimization (SMO) provided the highest accuracy, highest detection rate, low false alarm rate, and low FNR value irrespective of the feature selection method used. In our future work, we shall continue to explore various other classifiers and feature selection techniques and measure their performance with a view to building efficient intrusion detection systems.

References

Aljawarneh S, Aldwairi M, Yassein MB (2018) Anomaly-based intrusion detection system through feature selection analysis and building hybrid efficient model. J Comput Sci 25:152–160

Alazab A, Hobbs M, Alzab M, Using feature selection for intrusion detection system. International symposium on communications and information technologies (ISCIT), pp 296–301, IEEE, Gold Coast, QLD, Australia

Almseidin M, Alzubi M, Kovacs S, Alkasassbeh M (2017) Evaluation of machine learning algorithms for intrusion detection system. In: 2017 IEEE 15th international symposium on intelligent systems and informatics (SISY), pp 277–282. IEEE

Azad C, Jha VK (2017) Decision tree and genetic algorithm based intrusion detection system. In: Proceeding of the second international conference on microelectronics, computing & communication systems, pp 141–152, Springer, Singapore
Guyon I, Elisseeff A (2003) An introduction to variable and feature selection. J Mach Learn Res 3:1157–1182
Han J, Kamber M (2000) Data mining concepts and techniques. 3rd Edn., Elsevier Publishers
Haykin S (1999) Neural networks. Prentice Hall, A Comprehensive Foundation
Hota HS, Shrivas AK (2014) Decision tree techniques applied on NSL-KDD data and its comparison with various feature selection techniques. Adv Comput Netw Inf
Karray FO, Silva CD (2009) Soft computing and intelligent systems design theory, tools and applications, pearson
Kim S, Oommen BJ (2003) Enhancing prototype reduction schemes with LVQ3-type algorithms. Pattern Recogn 36(5):1083–1093
Khammassi C, Krichen S (2017) A GA-LR wrapper approach for feature selection in network intrusion detection. Comput Secur 70:255–277
Liu H, Yu L (2005) Towards integrating feature selection algorithms for classification and clustering. IEEE Trans Knowl Data Eng 17(4):491–502
Pham NT, Foo E, Suriadi S, Jeffrey H, Lahza HFM (2018) Improving performance of intrusion detection system using ensemble methods and feature selection. In: Proceedings of the Australasian computer science week multiconference, pp1–6, Association for Computing Machinery Brisbane, Australia
Pfleeger CP, Pfleeger SL (2008) Security in computing, 4th edn, Pearson Education
Platt JC (1999) Fast training of support vector machines using sequential minimal optimization. In: Advances in Kernel methods—support vector learning. MIT Press
Sokolova M, Japkowicz N, Szpakowicz S (2006) Beyond accuracy, F-score and ROC: a family of discriminant measures for performance evaluation. Australasian joint conference on artificial intelligence, Springer, pp 1015–1021
Tavallaee M, Bagheri E, Lu W, Ghorbani A (2009) A detailed analysis of the KDD CUP 99 data set, 2nd IEEE symposium on computational intelligence for security and defense applications. IEEE, Ottawa, ON, Canada
Tang J, Alelyani S, Liu H (2014) Feature selection for classification: a review. Data classification: algorithms and applications, pp 37–64. CRC Press
Yuan Y, Kaklamanos G, Hogrefe D (2016) A novel semi-supervised adaboost technique for network anomaly detection. In: Proceedings of the 19th ACM international conference on modeling, analysis and simulation of wireless and mobile systems, pp 111–114. Association for Computing Machinery, New York, NY, United States

Chapter 8
Smart Algorithm as a Machine Learning Application for Elderly People Safety Using Mobile Platforms

Aaron Rasheed Rababaah, Yousef Al-Tabtabaei, Yousef Hamadah, Lubna Amin, and Hisham Al-Ruwaih

Abstract This chapter introduces a smart algorithm in Machine Learning (ML) applications that aims to enhance elderly people safety in daily life. The main focus was the idea of creating a fall detecting mobile application for people who are paralyzed from waist down (Paraplegic) or have injured their legs. The algorithm uses the training data collected from real-world scenarios to closely learn the best calibration for system parameters. The smart phone fall detection system detects falls using sensors available in the phone itself. The chapter starts off by explaining the acronym of the application PFAST, then discusses the functions, necessity for building such an app. It also discusses the problems that prior applications have faced and how exactly will it work, which algorithm and sensors from Android devices will be used. Using various UML diagrams, it displays how the user will interact with the system. We address the issue "sudden falls" which is very common in elder people and has led to serious health injuries.

Keywords Machine intelligence · Machine learning · Smart algorithms · Physical safety · Elderly fall detection · Mobile platforms

A. R. Rababaah (✉) · Y. Al-Tabtabaei · Y. Hamadah · L. Amin · H. Al-Ruwaih
American University of Kuwait, Salmiya, Kuwait
e-mail: haroun01@gmail.com

Y. Al-Tabtabaei
e-mail: S00027751@auk.edu.kw

Y. Hamadah
e-mail: S00012249@auk.edu.kw

L. Amin
e-mail: S00031531@auk.edu.kw

H. Al-Ruwaih
e-mail: S00001212@auk.edu.kw

D. K. Sharma et al. (eds.), *Machine Learning for Real World Applications*, Transactions on Computer Systems and Networks, https://doi.org/10.1007/978-981-97-1900-6_8

8.1 Introduction

Machine learning applications have transformed diverse sectors, enhancing efficiency and decision-making. In healthcare, algorithms predict diseases and personalize treatments, while finance benefits from risk assessment and fraud detection. Personal assistants and recommendation systems leverage machine learning for tailored user experiences. Autonomous vehicles rely on advanced algorithms for navigation and decision-making, shaping the future of transportation. Manufacturing benefits from predictive maintenance, minimizing downtime. In agriculture, machine learning aids crop monitoring, disease detection, and yield prediction. These applications showcase the versatility and innovation of machine learning across industries, revolutionizing our daily lives.

Machine learning excels in person fall detection through the analysis of sensor data. By training algorithms on patterns associated with falls, such as sudden accelerations or changes in orientation, systems can accurately identify and alert caregivers or emergency services in real-time, ensuring prompt assistance for individuals at risk of injury.

Our Application name is "PFast" which stands for "Paraplegic Fall Assist System Technology". Paraplegic which is sometimes called "Partial Disability" is the medical term for patients who are disabled from their lower half and below. Many patients who suffer from this condition have perfectly functional legs, but the problem lies with the brain or the spinal cord which are unable to send signals to the lower body parts. As the name suggests, it will be an application that assists people who are paraplegic with their falls. Thankfully, with the rapidly evolving technology this can be achieved. For instance, many smart phones and tablets applications that depend on movement rely heavily on the combination gyroscope and the accelerometer sensors. Using those two sensors in a creative way allows for many interesting ideas in the application development area. In our case, we will be using those sensors to determine if a person fell. This will be done using mathematical and physics models to develop an algorithm programmed within the application to determine that a fall occurred and alert a contact which can either be a relative or a friend depending on the patient's preference.

PFast application is not just limited to paraplegia patients. It can be used by anyone who has a disability from the waist down whether it is temporary such as a broken leg or people who are unfortunate enough to have to rely on crutches or wheelchairs permanently due to an accident or an injury. It can also be used by elderly who may have trouble walking and a sudden fall occurred with no one around to help.

As previously mentioned, the applications area is flooded with applications that make use of the user's movement such as counting steps or how fast a person is running or detection of vehicles as is (Viangteeravat et al. 2007). Those applications obviously cater to people with a fully functional movement parts, but they offer little to no help to people who are handicapped from the waist down. Taking advantage of the potential of the technologies within the smartphones, such as sensors, can enable developers to create amazing tools to make the life of paraplegia patients safer and more convenient. However, there is little to no applications that are catered to people who are paraplegic. This unfortunate because there is a huge potential in this area that has not been fully explored. The main purpose behind the idea was how to make those patients feel safer with just a smartphone in their pocket or hand. In addition to that, our purpose is not only just to be known as a group who created an app that helps paraplegic patients but also a group that made an impact on the physically disabled community in general. Moreover, expanding upon the idea in the future and updating it constantly with more features for handicapped people, such as people with back problems, is also one of our purposes for the future. Hopefully, if the project takes off and becomes successful, this can inspire others to help people who are handicapped and allow them to make their lives even more convenient and safer.

As previously mentioned, our application relies heavily on the accelerometer and the gyroscope of the smart phones. Once a combination of those two sensor readings reach or exceed a certain threshold, the application will immediately notify a contact. The app will also make use of the sound input and output to minimize false flags which is through a voice assistant. A smart watch can also be used however, it is not mandatory and can act as a boost to accuracy which has been proven in several studies (Cao et al. 2012). After the testing phase of our project, we can decide if the current design is robust enough or it needs more resources to boost reliability and confidence in its accuracy such as a smart watch aforementioned. This further investigation is planned to be pursued in future work of this project.

The goal of our applications is to provide aid to elderly or people who are paralyzed from waist down (paraplegic) by accurately detecting a fall and calling for help instantly. The application will be very beneficial as it would avoid the delay in getting help. It will also increase the safety level as more people will be comfortable in living independently and it would eliminate the cost of hiring a nurse.

According various statistics, for example, from World Health Organization 28% to 35% of people over the age of 64 fall at least once a year (Casilari and Oviedo-Jiménez 2015). Furthermore, in Canada 62% of injury-related hospitalization for older people was due to fall the immediate treatment after a fall is very crucial for a patient's health. The cost of medical treatment affiliated with the falls crossed $20 billion in 2010, by 2020 it is predicted to reach $67.7 billion (Casilari and Oviedo-Jiménez 2015). Helping the patient as soon as he/she fall can decrease the morbidity and mortality of falls. Hence, the research being done regarding Fall Detecting Systems have increased over the last decade. Making an application that works the same as any other fall detecting system would be very efficient and convenient. Given the use of smart phones and other technological devices the older generation isn't lagging behind in the world of technology.

8.2 Literature Review

The population of elderly people in our society has been increasing, which means the healthcare to assist them is also increasing. In USA more than one third of the population over the age of 65 falls at least once a year. Usually older people who live alone cannot get help right after they fall (Cao et al. 2012). These accidental falls can cause serious injuries to the person if no assistance is provided in time. Some stats show the aftermath of these accidents is the delay in getting help not the actual fall itself. These delays can be very dangerous for the medical condition of older people if aid is not provided in time (Jia 2009). Falling of elderly people is one of the main causes of getting home nurse and hospitalization, which can be very costly. Due to these factors a lot of research has been done in developing sensors and devices to detect and assist as soon as a fall occurs.

One of the sensors that is mostly used to detect fall is accelerometers, they measure the high acceleration proceeding ground impact, and they are very inexpensive and lightweight. More often than not they are accurate. However, they have their drawback as well, other activities also result in high acceleration, and not all fall result in high acceleration (fall Sensors). There many factors that should be taken when classifying an activity as a fall. Example lying on the bed or ground may seem as a fall for elderly. The main challenge in this field is using sensors and other devices to accurately sense a fall, because other events may seem like a fall but, they are not. Such situations should be avoided to prevent false alarms. Another cost-effective way of implementing fall detection is using android devices, using the sensors and computing capabilities of an android phone we can detect falls.

The figure below demonstrates the basic architecture for most fall detections systems the phone or smart watch connected via Bluetooth, detects if the phone has fallen and then places a call to the emergency contact.

Majority of the older people are reluctant to use phones which could be based on many reasons example lack of interest in technology, vision impairment or economical restraints. Due to these factors the most or only feature of a cell phone that is used by them are "emergency calls/alarms" (Casilàri et al. 2015). The way they look at their phones and use them is very constrained, for them it's a safety device rather than a platform for social media activities.

As far as we know, the usage of smart phones by older people is less compared to other age groups in our society. However, this is starting to change in the western countries by the year 2020, it is predicted the technological gap within generation will decrease. There has been an increase of 10% in the usage of mobile phone in British people over the age of 55 (Casilari et al. 2015). In less than 10 years it is expected that older generation will use mobile devices on a daily basis (Casilari et al. 2015). In an opinion poll "82% of the older people who owned a smart phone described their phone as liberating experience" (Casilari et al. 2015). In spite of these technological obstacles among older people this opinion poll displays that our fall detection application will be appealing and feasible to many elder people.

Technology has helped out people with special needs at different ages from children with disabilities right up to the elderly in our society. A lot of progress has been made in the field of creating assistive technology for senior citizen, as there is a great deal of demand of such products/technologies in the healthcare industry.

More significantly these advancements are supported by governments in many developed countries. Such as, "Singapore government held a conference and exhibition in Jan 2008 SiCEX, which promoted concepts, product and technologies related to healthcare for elderly and patient" (Approaches).

A fall alert system is not a new concept and it has been around for a while. However, each system differs based on the method it uses to detect the fall, how it alerts the emergency contact and the device itself. The Life Alert system offers a pendant that has GPS and can either be worn as a necklace or on the wrist as a wristband. This has been around for years however it requires that the user press a button on the pendant if they require medical attention and it connects them with the monitoring centre of the company who can then notify the emergency services.

Life Alert doesn't have international monitoring centres, so it will only work in the USA. Furthermore, the user will have to pay a monthly fee which is based on whether they have a landline ($50) or if they require a cellular line ($60). It also requires that the user commit to their service for 36 months and should they decide to cancel before then they have to pay for the remainder of the contract.

Life Station is another FDS that has no commitment requirements and like Life Alert it only works in the USA. It does have a cheaper monthly subscription compared to Life Alert as they only charge $29.95 per month and if the customer commits to a 1-year contract they can get it lowered to $25.95. Life Station also has a built-in sensor that detects if the user falls which negates the need for them to press the button in case of a fall.

All the previous systems require the use of a separate hardware system specifically designed for the purpose of providing people with ease of access to medical emergency personnel. While they may be specialized systems their functions are limited. However, most smart phones in the market today are sufficient enough to be able to detect if the user falls and with the various features these smart phones have, such as voice command, GPS and WIFI building an app that makes use of these features is efficient and the correct algorithm can make it better than any of those systems while only requiring that the user carry a smart phone with them.

One such application is the Fade: fall detector which was developed by the Institute of Technology and Renewable Energies based in Spain. It detects falls using the sensors in Android phones such as the accelerometer. Once it detects a fall it sends a message to the emergency contact specified when first setting the application. This message contains the time and GPS location of the user and can either be sent as an SMS or an email. It can also make phone calls to the emergency contact and allow them to check up on the user.

Another similar application is Emergency Fall Detector which detects when a fall occurs and then waits for a bit to reduce the chances of false alerts, then it sends an SMS to the predefined emergency contact informing them that the user had a fall and it attaches the users' current location.

Fall Detection Systems (FDSs) can be categorized into three general typologies: vision based, ambient based and wearable device-based approaches (WDA), each with its own advantages and disadvantages.

Vision based relies on cameras that are placed around a room and they detect if a user has a fall. The problem with this type of system is that the user needs to be confined to the area covered by the camera and this will reduce their ability to move around freely. If they move out of the cameras line of sight and they happen to fall the system will not detect this and thus it will be like the system doesn't exist. Invasion of privacy is another issue as the camera will constantly have to monitor the user throughout the day and this might make the user uncomfortable. A hacker may also gain access to the system and scout the house of the user for a robbery.

Similarly, ambient based FDS are conventionally founded by analyzing audiovisual signals together with other specific information (such as floor vibrational data or microphone signals) captured by environmental sensors. The downside of this is that it requires equipment to be installed around the house, and under the floors for better safety measures. This equipment can become expensive especially in larger houses.

Wearable based FDS depend on the information they receive from sensors which are embedded in clothes or other wearable devices such as smart watches. This tends to be cheaper and more efficient since it's always with the user wherever they go. Android apps fall under this category and with the high market penetration of smart phones today it's very unlikely to find someone without one.

Fade: fall detector and Emergency Fall Detector are both outdated and had some bugs that needed to be fixed but they haven't been updated in years and they are most likely discontinued projects and a newer app is needed to be compatible with the latest versions of Android OS.

8.3 Technical Background

The project will heavily rely on the movement of the user that is holding the phone. It also needs to take into account the acceleration of the phone, the altitude and the gyroscopic values. Some physics will have to be applied in the coding process. The reason for that is the phones do not have a collision sensor so we will need a certain combination of values of acceleration and altitude which determines that the phone really fell. For example, a phone moving at a speed of 10 k/h but the altitude did not drop by a reasonable amount then that could mean the phone was thrown or a similar scenario happened. However, if the acceleration suddenly increased and there was a sharp drop in altitude then that is a flag for a phone that has fallen. A person posture after a fall needs to be taken into consideration as well since someone who fell will most likely lay flat on the ground and that's where the gyroscope will come into use.

The method mentioned previously is called "The Confidence System". The activities and the movement of the phone are monitored after the fall with the Confidence System (Luštrek and Gjoreski 2015). For example, if after a potential fall there was no movement or activity whatsoever then that may indicate an alarm that the user fell. Another similar method is called "The Basic Threshold Monitoring" which also uses some algorithms to detect a sharp increase in the body acceleration.

The main technologies that will be used are the built-in phone sensors with little outside equipment. The Accelerometer will be used to decide the increasing speed of the phone during a fall. Moreover, the Gyroscope will determine the values of the x, y and z axis during a fall which will help determine if a fall happened or not (Casilari et al. 2015).

For the future development of this project we propose to use a smart watch as a complement to the phone app which also takes in the movement and the posture of the person the smart watch will use Bluetooth to send messages to the application in case of a fall. In addition to that, the remote alerting will not be initiated in the fall was from a single device in order to eliminate false positives. Several studies have been done on this subject and according to them, using a smart watch definitely help produce more accurate results and less false alarms. The smart watch that runs on Android system would be recommend since the application is built for Android phone and it will run parallel with the development tools of the project (Cao et al. 2012).

Outside of the physical aspects, Android Studios is going to be used for coding purposes and a Smartphone or a Virtual Machine will be used for testing and debugging. A MySQL database will be used to register the records of falls such as the time of the fall or any another important information. Finally, Fluid UI is an online service that offers the ability to create prototypes of phone applications interface which we will incorporate in our project.

Android is a Linux based operating system used for mobile devices, the programmer can use many major hardware and software components such as accelerometer, and create an interactive user interfaces for mobile applications. Which in this case is very helpful as the target audience are older people? As of 2015 nearly 260 million android devices were shipped worldwide making it the most relevant OS in the market of mobile devices (Casilari et al. 2015, Luštrek and Gjoreski 2015, Yu 2008). As a result, Android is especially used by programmers to develop medical and social network applications. While doing my research most articles used Android Operating System for smart phone fall detection systems.

Apple IOS has a very closed approach also, iPhones can be costly especially considering the fact elder people use mobile phone for limited number of applications mainly to get assistance not, for social media communications or other functionalities that Apple provide. That is why we opted to use Android operating system than Apple IOS.

The application is deep when it comes to technicality and algorithms implemented but the user interface and navigating through it is going to be as simple as possible. The user is greeted with a menu where they can choose their emergency contacts numbers. Once that is done, the user then has the ability to toggle how the contacts will be alerted. It can be a phone call, text message or sharing their location through a link. The user can also use the voice assistance settings to set an easy phrase to call the specified contact instead of spelling out the number or the whole name which may cause complications with the voice recognition of the device. Another option is the records management. They can look at previous records of falls such as location and time which can be used for medical purposes.

The application will be running in the background and once it detects a fall, the voice assistance service will immediately ask if there was a fall or it's just a false alarm. The user can respond with Yes or No. If it's a yes then the user can ask to alert the contact, which will be named according to the user. How the contact will be alerted is determined by the permissions of the user (SMS Message, Phone Call or Send Location). In Cases where there is no response, the application will notify the contact with the most permission with a text message and a location. If there is more than one contact then it will be sent to the others as well.

The smart watch will be an optional accessory with the application available for download in the play store. The application will automatically detect that a smart watch is being used and a message will be displayed to notify the user and instruct them how to link it to the phone app. Linking it with the mobile phone is an easy process since all that is required is having the Bluetooth turned on in both sides. The smart watch can greatly enhance the accuracy and reduce false flags.

The sensors that will be used to implement the algorithms are tri axial accelerometer, tri axial gyroscope and magnetometer which are available in most android devices. Android also provides linear acceleration and orientation of the device which assist in detecting a fall (Figueiredo et al. 2016).

Tri Axial Accelerometer—It calculates the acceleration force applied to the phone along each axis in m/s^2, it also includes the force of gravity (Figueiredo et al. 2016).

Tri Axial Gyroscope—It senses the rate of rotation along each axis, it calculates the angular velocity in radian per second rad/s (Figueiredo et al. 2016).

Magnetometer—It sense the magnetic field in microtesla (μT), which can calculate the azimuth (Figueiredo et al. 2016).

Linear Acceleration—It is acceleration without gravity, it can be calculated by projecting an acceleration into a fixed coordinate system, removing the gravity vector and using orientation information (Figueiredo et al. 2016).

Orientation of the device—It provides the rotation angles for roll, pitch and azimuth which are then used by gyroscope, accelerometer and magnetometer sensors as shown in the figure below (Figueiredo et al. 2016) (Fig. 8.2).

Comparisons of Fall Detection Algorithms

Name	Algorithm	Pros	Cons
Basic threshold monitoring	$\mathrm{SMV}_i = \sqrt{\lvert A_{xi}\rvert^2 + \lvert A_{yi}\rvert^2 + \lvert A_{zi}\rvert^2}\ \mathrm{m/s^2}$	Easy to implement as it is not that complex	Can detect false positives
Fall index	$\mathrm{FI}_i = \sqrt{\sum k = x, y, z \sum_{i-19}^{i} (A_k - A_{ki-1})^2}$	Prevents false positive	Slow falls may not be noticed

(continued)

(continued)

Name	Algorithm	Pros	Cons
Two-phase detection	$\lvert A_{vi}\rvert =$ $\lvert A_{xi}\sin\theta_{zi} + A_{yi}\sin\theta_{yi} - A_{zi}\cos\theta_{yi}\cos\theta_{zi}\rvert$	Most accurate out of the fours, least number of false positives	Difficult to implement as it is too complex
iFall	No equation just checks the SMV value within a range	Easy to implement	Works accurately only when a person is standing

Basic Threshold Monitoring—Fall occurs when the body acceleration reaches a sudden peak. In this method if the value of SMV (Single Magnitude Vector) goes beyond a certain threshold a fall has happened. A_{xi}, A_{yi} and A_{zi} are direction values of x, y and z-axis from the accelerometer of the smart phone or smart watch (Casilari and Oviedo-Jiménez 2015).

Fall Index—This algorithms compares the value against a certain Fall Index threshold FITh, which is calculated by adding the gradual variation of the last 20 measured values from the accelerometer (Casilari and Oviedo-Jiménez 2015).

Two-Phase Detection—This method uses both SMV and absolute value of vertical direction, θ_{yi} and θ_{zi} are angles values for pitch and roll as shown in Fig. 8.1. It uses two phases free fall and impact to detect a fall. In the first phase the algorithm always check if the absolute maximum difference of the SMV values in a short time frame is outside the threshold value (SMV FF), then it initiates the impact phase. In which the mobile again calculates the difference among minimum and maximum values of SMV, in the second time frame. If that value exceeds the decision threshold (SMV IP), which means impact against the floor has happened and a fall is detected. If the two phase and two detection values meet the threshold a fall is identified (Casilari and Oviedo-Jiménez 2015).

iFall—When a fall happens it causes an abrupt change in acceleration, this technique works by checking if the value of SMVi exceeds lower SMVi and upper threshold SMVu, in a certain time interval. It only works when a person goes from a upright posture to a horizontal position, other times the fall is neglected (Casilari and Oviedo-Jiménez 2015).

The algorithm we are going to use for our application is Basic Threshold Monitoring, because it is easier to implement and the results are more accurate than iFall and Fall Index. The problem with iFall is fall is only detec (Igual and Medrano 2013) ted accurately when the person is standing, and with Fall Index slow falls are neglected. Although the results of Two-Phase detection are the most precise, due to time constraints and the complexity of the algorithm we decided to opt for Basic Threshold Monitoring.

Fall detection system faces many challenges in real life application. Few of them are mentioned below;

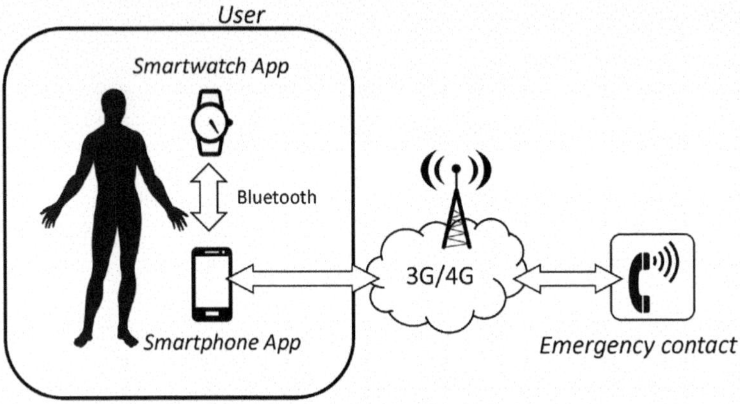

Fig. 8.1 Basic architecture for fall detection systems (FDS)

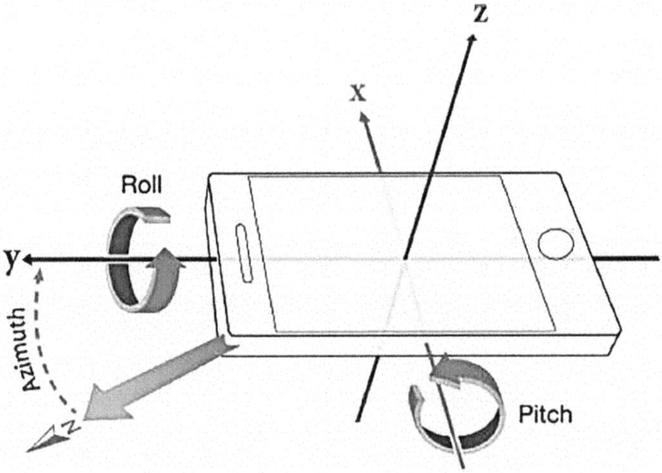

Fig. 8.2 Smart phone reference frame and orientation angles

- In some cases, fall might be initiated by another person or an accident, let's take this scenario as an example; the user lives with a pet that causes the mobile to fall, thus creating a false alarm to the emergency contacts.
- Smart-phone fall detection is an appealing idea, but we have to consider the way people carry their phones on a daily basis. For example: in pockets or handbags, this could affect the readings collected by the sensors (Igual and Medrano 2013).
- Smart-phones were not created for fall detection, there might be technical problems when implementing the application in real-time operation. Such as stability of the accelerometer and specific features of the operating system, also the same fall detector might behave slightly differently depending on the smart phone model in which it is installed (Igual and Medrano 2013).

- With the continuous sensing, the performance of the device will be affected. For example it would reduce the battery power which may cause concern for some people (Igual and Medrano 2013).

8.4 Methodology

8.4.1 Fishbone Diagram

Fishbone diagram (Also called Cause and Effect diagram) is a tool to visualize a problem and what the causes of the problem may be. In our case, the main problem is the device not registering a fall or not alerting a contact regarding a fall. It can be due to four main sections which are Hardware, Software, Network or from the Users themselves.

In the case of a network, the fall could have occurred but didn't get reported due to some potential causes. It could be due to the location service of the phone is not working properly. Another cause could be that the phone network service is down or facing some issues. Being in a location where the phone signal may be disturbed could cause the alert to not be sent.

The software can also play a big role. The application version may be too out of date to function or several bugs have been fixed since then. The data of the application may get corrupted whether its during the installation process or the corruption happened later. The device operating system itself could be outdated and may be incompatible or not run well with the newer version of the application which has been development with a newer operating system in mind. Background apps that are competing for the same resources may the disturb PFast while its running.

The user misuse can also cause the application to malfunction. The calibration process provided by the application could have not been done properly. The user also can refuse access to some permissions which may cause the application to partially or completely stop performing its function.

Finally, a damaged hardware interior can interfere with the functions of the phone which will in turn affect the application. Faulty device sensors where they are not functional enough to provide the service will cause some problems (Fig. 8.3).

8.4.2 Requirements Modelling

Requirement modelling displays the requirements of a system in text and diagrammatic form, which is easier to understand. Modelling diagrams are very helpful when developing mobile applications or any other software of any size. They describe the needs of the users and stakeholders, reduce irregularities in the requirements, and map out which functionalities/features should be developed first. There are

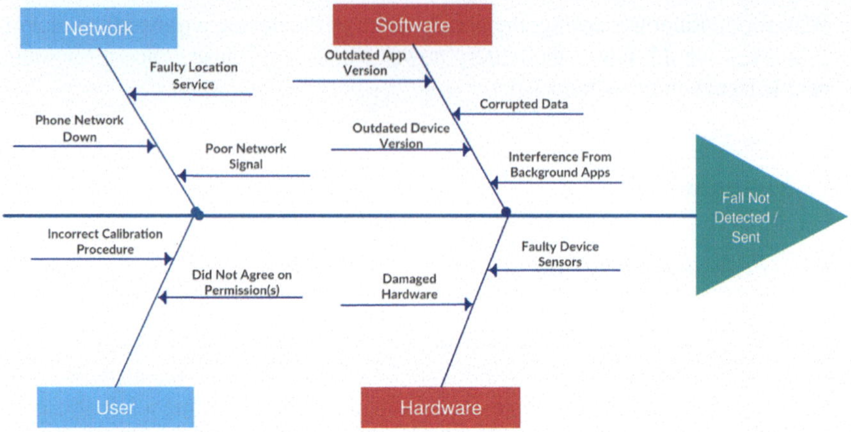

Fig. 8.3 Fishbone diagram

many different diagrams you could use to demonstrate various functionality of your software/application (Model user requirements 2015).

The table below shows all the functional requirements of our application input, output, the name of the process and the description. The highlight grey area is a requirement for future work which can use in conjunction with the application to increase its accuracy.

ID	Process	Input	Output	Description	Control
1	Emergency contact	• Adding phone number • Selecting the priority	Contact number and priority/order is added	User adds the phone number and selects the order/priority first, second, so on, in which the person should be called after a fall	User
2	Detecting fall	• Falling phone	Waits for a while for any voice assistance commands. If no response, then calls or send messages to emergency contacts	Using certain algorithms and the sensors in the phone to calculate if the phone has fallen or not	None
3	Calling emergency contacts	• Detecting a fall	Calling the contact	After detecting the fall has occurred, calling the emergency contacts based on their priority	None
4	Calling second contact	• A fall was detected • First contact did not answer the call	Calling the second contact	After detecting the fall has occurred and the first person has not answered the call, calling the second emergency contact	None

(continued)

(continued)

ID	Process	Input	Output	Description	Control
5	Text messaging	• Fall has been detected	Sends text message to the emergency contact	Sending texts message to emergency contact after detecting a fall	None
6	Privacy settings	• User selects the settings for the contact	Updates the contact settings	This allows the user to control which contact should be called and send text message, also which contact can access their location	User
7	Location	• Fall has been detected	Sends the location link via text message	Based on the privacy settings the location will send to those contact which are selected by the user. Sends the link of the location which can be viewed in Google Maps	None
8	Smart watch	• Go to the Bluetooth settings of the mobile • Connect the mobile phone to smart watch	Device and watch are connected	Using Bluetooth connecting the device and the smart watch to increase the accuracy of predicting a fall	User

8.4.3 Performance

Product Functions These requirements include all the functions that the product should do. Reliability For system's reliability the target accuracy is at least 90%. As the contact has to know if a fall has occurred or not, false positives could make frustrate both the user and the contact. Several tests will be conducted before submitting the final version of the application to make sure that our app calculates the accurate results Extensibility/Maintainability. Our application can be modified in the future to add more functions Bluetooth connectivity through smart watch, notifications for predicting a fall. Also, the application can be made to form it function for mountain climbers, hikers or cyclers. Portability The application uses Java and oracle, which will enhance the portability of the system since java is a portable language.

Performance The system should be fast in responding to the user's gesture input, such that the checkboxes and buttons selection would not take more than a few seconds to be recognized. User Friendliness Our application target users are going to be mostly older people who are not tech savvy so, the UIs and layouts of each interface has to be very user friendly such as bigger fonts or using buttons or check boxes instead of text boxes. Organizational Functions these include requirements which are a consequence of managerial standards. Delivery Requirements Team members have to abide by predefined deadlines for accomplishing different planned phases such as writing drafts making diagram for the report, making the power point for the presentations and showing up on time in every meeting with the course coordinator.

System Development The system will be developed by the Enterprise Architecture, Microsoft project, Microsoft Visio and android studios.

External Functions—These requirements consist of functions that are outside the framework of the project.

Privacy The system will be secure and won't share the location information with the contact if the user doesn't want too. The user can control which contact can access their information about the fall.

8.4.4 System Flowchart

System Flow Chart is a method used to show how data moves within the project and the decisions that determine which step the data should take next. In our case, it shows the scenario that starts with the data that represent the axis and acceleration values which determines whether it's a fall or not. Depending on whether the voice assistant is turned on or not, it can either directly check for contacts to alert or proceed to wait for an input from the user through the voice assist. In the case of no response or incorrect input within 5 s, it will stop the voice assist path and move to the checking contacts process. During this process, each contact will be checked for permissions for the application to determine how they will be alerted. The application job will be done once it reaches the "End" part (Fig. 8.4).

8.4.5 Use Case Diagram

Unified Modelling Language UML use case is a common tool to draw software Models. Use case is used to identify, clarify and organize a systems requirements, an actor represents the roles played by users or admin of the system. The system boundary determines what is external or internal to the system (Create a UML use case diagram n.d.). In our system there will be two actors Patient and caregiver (emergency contact) the processes the actors interact with, in the system is shown in the figure (Figs. 8.5, 8.6 and 8.7).

Machine Learning Model:

In person fall detection using machine learning, data collection involves gathering sensor data, including accelerometers and gyroscopes, capturing movements. Relevant features such as sudden velocity changes are extracted from this data. The

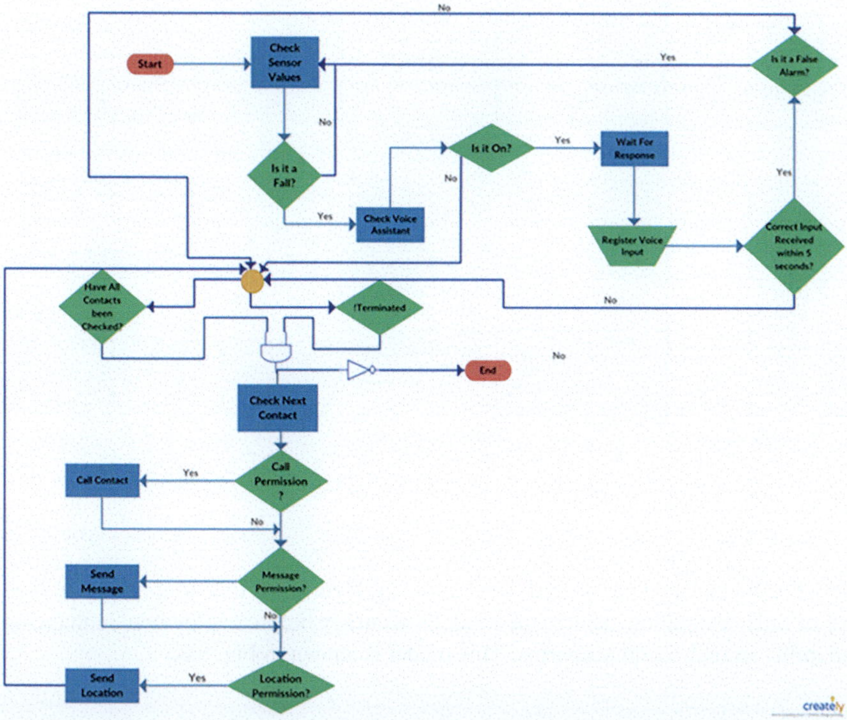

Fig. 8.4 System flowchart

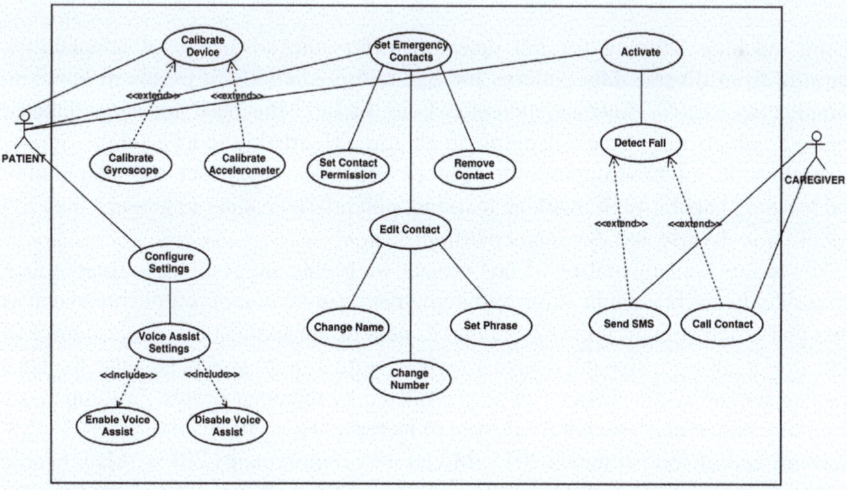

Fig. 8.5 Use case diagram

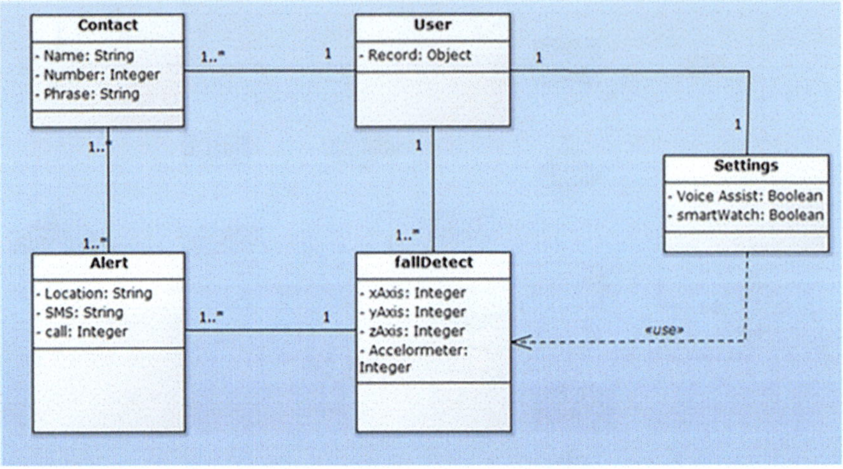

Fig. 8.6 Class diagram

model is then trained on labeled datasets, using algorithms to learn patterns indicative of falls. Validation ensures the model's accuracy before deployment, enhancing reliability in real-world scenarios. This model is shown in Fig. 8.8.

8.5 Conclusions

Using machine learning for fall detection offers the advantage of adaptability, learning from diverse data patterns for accurate predictions. It excels in handling complex scenarios. However, potential cons include the need for large labeled datasets, which may be challenging to acquire. Heuristic models, while simpler to implement, might struggle with nuanced situations. They lack the adaptability and learning capability of machine learning, potentially leading to less accurate fall detection in diverse and dynamic environments.

For future implementation of this project we highly suggest to use smart watch alongside the mobile application, and implement a more complex algorithm such as Two-phase detection to improve the preciseness of the application. The accuracy of detecting a false positive fall decreases when using a smart watch, because the false positive sensed by one device can be neutralized by the other device. Combined use of mobile phone and watch have proven to increase the accuracy of a fall by 5–15% (Casilari and Oviedo-Jiménez 2015; Model user requirements 2015). Also, people wear watches almost all the time they don't take their phone everywhere they go. Compared to other generations elder people do not have the habit of carrying their phone 24/7.

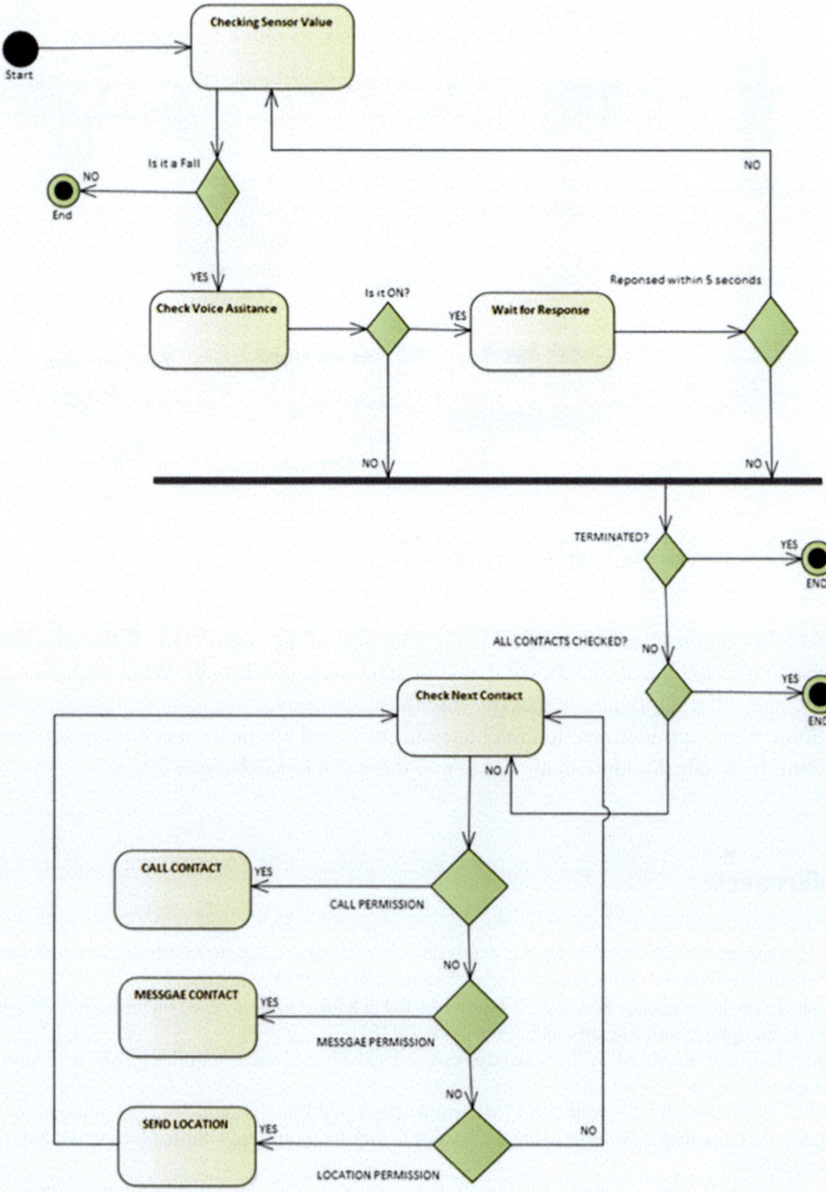

Fig. 8.7 Activity diagram

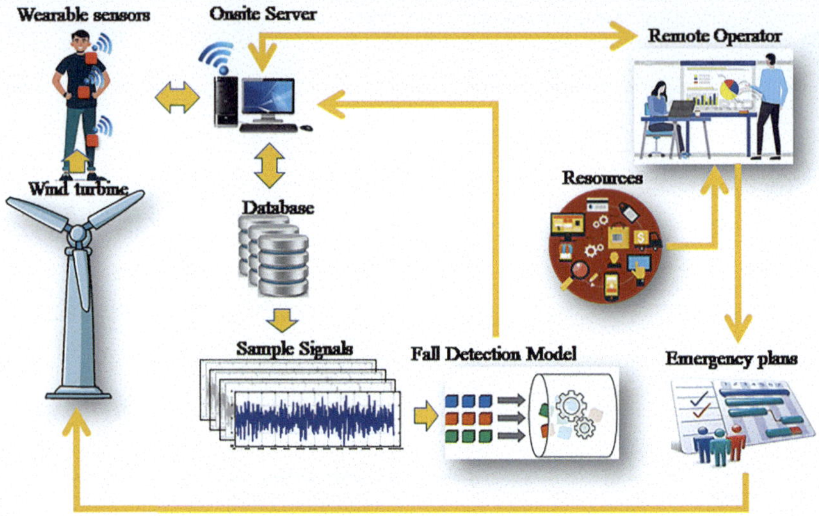

Fig. 8.8 Machine learning model for fall detection

Another feature that could be added is, placing call through VOIP. If the user has access to internet the application can call or send message through Whatsapp, Viber, Skype and other applications that provide these services.

Some other applications that may use our proposed application is human activity labeling is in (Shirkhodaie et al. 2011) and the work in (Rababaah 2023).

References

Cao Y, Yang Y, Liu W (2012) E-FallD: A fall detection system using android-based smartphone. In: 9th International conference on fuzzy systems and knowledge discovery

Casilari E, Oviedo-Jiménez MA (2015) Automatic fall detection system based on the combined use of a smartphone and a smartwatch. Plos ONE 10(10)

Casilari E, Luque R, Mo M-J (2015) Analysis of android device-based solutions for fall detection. Sensors

Create a UML use case diagram (n.d.) (Microsoft) Retrieved December 1, 2017, from https://support.office.com/en-us/article/Create-a-UML-use-case-diagram-92cc948d-fc74-466c-9457-e82d62ee1298

How long does it take to make an app? (n.d.). Retrieved November 29, 2017, from https://savvyapps.com/blog/how-long-does-it-take-to-make-an-app

Figueiredo N, Leal C, Pinto L, Bolito J, Lemos A (2016) Exploring smartphone sensors for fall detection. J Mob User Exp

Igual R, Medrano C (2013) Challenges, issues and trends in fall detection systems. NCBI

Jia N (2009) Fall detection application by using 3-axis accelerometer ADXL345. Analog Devices, 43

Kroputaponchai T (n.d.) Vision-based fall detection and alert system suitable for the elderly and disabled peoples. Research Gate

List of Risks (n.d.). Retrieved November 29, 2017, from http://groups.engin.umd.umich.edu/CIS/course.des/cis375/projects/risktable/risks.htm

Luštrek M, Gjoreski H (2015) Fall detection using location sensors and accelerometers. Prevaise Comput

Medical alert systems with fall detection (n.d.). Retrieved November 20, 2017, from https://medical-alert-systems.bestreviews.net/fall-detection/

Model user requirements (2015). Retrieved November 20, 2017, from https://msdn.microsoft.com/en-us/library/dd409376.aspx

Rababaah A (2023) Machine learning comparative study for human posture classification using wearable sensors. Int J Comput Sci Math 18(1):54–69. https://doi.org/10.1504/IJCSM.2023.133518

Shirkhodaie A, Elangovan V, Rababaah A (2011) Acoustic semantic labeling and fusion of human-vehicle interactions. In: Signal processing, sensor fusion, and target recognition XX, vol 8050, 80500Q. Event: SPIE Defense, Security, and Sensing, Orlando, Florida, United States. https://doi.org/10.1117/12.883544

Viangteeravat T, Shirkhodaie A, Rababaah A (2007) Multiple target vehicles detection and classification based on low-rank decomposition. In: Automatic target recognition XVII, vol 6566, 65660R. Event: Defense and Security Symposium, Orlando, Florida, United States. https://doi.org/10.1117/12.718770

Who are we—ITER (n.d.). Retrieved November 20, 2017, from http://www.iter.es/el-iter/quienes-somos/?lang=en

Yu X (2008) Approcahes and principles of fall detection for elderly and patient. Institute for Infocomm Research

Chapter 9
Machine Learning Based Parkinson's Disease Detection Using Voice and Handwriting Analysis

Sauransh Singh, Ruchir Kumar Kadwey, Shrey Srivatsava, Shikhar Singh, and M. R. Shrisha

Abstract This paper is based on detection of Parkinson disease using machine learning models accurately. In this paper two major symptoms of Parkinson are taken into consideration one is shimmering of voice and another is tremor in hands. They are applied with sufficient datasets two obtain accurate results. For Shimmering in voice we visualized the data and divided the dataset in 23 unique features and then we applied Extreme Gradient Boosting algorithm (XGBoost) to train the model and obtained an accuracy of 92.3%, and for tremors in hands we collected dataset of patients with drawing of spirals and waves, then we passed it through feature processing and used Histogram of oriented gradient, to obtain features and then attached them with tags of Parkinson and Non Parkinson, then passed it to random forest classifier algorithm to train and obtain an accuracy of 82%. We also obtained results for real time detection and data set management, using Praat software and able to extract features of voice on real time.

Keywords Parkinson's disease · Machine learning · Extreme gradient boosting · Histogram of oriented gradient · Praat

9.1 Introduction

Parkinson's disease is the most common neuro-degenerative disease after Alzheimer's disease. More than 1 crore people worldwide suffer from Parkinson's disease. It is estimated that Parkinson's disease is common among 1.8% of people aged 65 and over. Dopamine acts as a messenger molecule in the brain that allows certain nerve cells to communicate with one another. Dopamine has been shown to

S. Singh · R. K. Kadwey · S. Srivatsava · S. Singh
Students of BMS College of Engineering, Bangalore, India

M. R. Shrisha (✉)
BMS College of Engineering, Bangalore, India
e-mail: shrishamr.ece@bmsce.ac.in

D. K. Sharma et al. (eds.), *Machine Learning for Real World Applications*, Transactions on Computer Systems and Networks, https://doi.org/10.1007/978-981-97-1900-6_9

be important for central nervous system functions such as mood, motivation, pleasure, movement (Rao et al. 2020). There is no way for the general purpose clinical diagnosis. It is thought that Parkinson disease can only be diagnosed with certainty when a corpse is found, which also highlights the complexity of the diagnosis.

In the United States alone, about ten lakh people are diagnosed with Parkinson's disease each year with an economic burden of $14.4 billion or more. There is currently no cure for Parkinson Disease, although studies show that certain therapies are effective in improving the quality of life for patients (Sakar et al. 2013), if used early, thus reducing the estimated cost of pathology (Rajani Kumari et al. 2021).

During early days before the onset of tremors or motor symptoms, dopaminergic neurons begin to change. Some of the early symptoms of Parkinson's disease (early) include decreased sense of smell, disturbed sleep by Rapid Eye Movement, small handwriting, difficulty moving, etc. (Al-Islam and Akter 2020). Early diagnosis of Parkinson's disease leads to effective control and avoidance of unnecessary tests, treatments, costs, risks etc.

Parkinson's is quickly detected using brain scans such as PET, SPECT, MRI, fMRI etc. Currently, in almost every clinic, these images are translated by medical professionals and have increased risk of human errors. The study reported that the combined clinical accuracy of Parkinson's disease was only 80.6%.

9.2 Literature Survey

Medical data analysis is one of the most important applications of data mining and important in the use of medical data analysis is the diagnosis of Parkinson's disease. The affected patients with Parkinson's have damaged brain cells are emotional unstable due to the disease. Symptoms of the disease start mildly and increase over time which includes hand tremors, limited leg movements, the ability to think is increasingly impaired, and so on. Currently the Parkinson's disease Diagnosis is done using PET scans and MRI scans or through minimally invasive tests (Politis et al. 2010). The proposed model is a predictive tool that can use a combination of voice databases and a set of whirlwind writing data and can provide the intensity of Parkinson's disease to affected individuals.

Parkinson's disease is the most common neurological disorder in people over 45 years of age. The condition causes swallowing problems, muscle numbness, back pain, trembling hands, bad breath, speech impairment (Wu et al. 2017), hearing loss (Sadat 2016) etc. Parkinson's disease should be diagnosed as soon as possible. The results of clinical trials, takes hours to diagnose and can cost more money and time ("Parkinson's Disease", National Institute on Aging 2020). An automated model for the diagnosis of Parkinson's disease in a more accurate person is proposed and illustrated in this paper. Although different models of Parkinson's disease have been developed, the strategies are less reliable and more accurate. The model is developed using an advanced decision tree, which not only helps to accurately predict Parkinson's disease in a person but also to predict it quickly.

The model prediction feature set consists of 22 voice signal parameters, a data set, provided by the XG Boost separator. The feature model predicts Parkinson's disease with 96.6% accuracy, 95.6% sensitivity, F-Score 97.7%.

Parkinson's disease is one of the most important neurological disorders in the human nervous system. Parkinson's disease is not contagious, as early detection of Parkinson's disease can prevent trauma in patients with the disease. However, it has been noted that the presence of Parkinson's disease in humans is related to the size of the manuscript and the cleanliness of the handwritten articles (Binder et al. 2014).This view has led to a number of researchers proposed to diagnose Parkinson's disease in hand-drawn pictures of patients diagnosed with Parkinson's disease (Xu et al. 2012).

9.3 Methods for Analysis

In this study, it was recommended that the Parkinson's disease prediction method be developed from hand pulses and wind images using machine learning methods (O'Sullivan et al. 2008) such as Decision Tree, Gradient Boosting, K-Nearest Neighbor, Random Forest, and other split algorithms with a HOG feature that defines the algorithm. The proposed strategy with Gradient Boosting and K's closest Neighbors achieved the best performance with accuracy, sensitivity, specificity and flexibility in system design (Rumman et al. 2019). The Gradient Boosting algorithm scored 86.67% accuracy, 93.33% sensitivity, and 80.33% clarity and KNN scored 89.33% accuracy and 91.67% sensitivity.

There are two major symptoms that have been used in this paper:

(1) Shimmering of voice
(2) Tremor in hands

9.3.1 Data Set Analysis for Voice

For analysis of voice (Giuliano et al. 2019) the data set of 193 participants was considered of which 147 were of Parkinson patients and the rest 48 were of non-Parkinson patients. The dataset was created by University of Oxford Max little with collaboration of national Centre for voice (Bocklet et al. 2011) and speech, Denver, Colorado, as they recorded the speech signals of 23 different features have been used for the voice extraction and analysis (Fig. 9.1).

For visual analysis, firstly a box plot of the data set is made. In general statistics, a box plot also known as box and whisker plot is a typical chart used in exploratory data analysis. Box plots show the distribution of numerical data and skewness through showing the data percentile and averages scores (Fig. 9.2).

By mapping the relation and frequency of each feature that has been extracted is obtained.

ABBREVIATIONS	FEATURE DESCRIPTION	ABBREVIATIONS	FEATURE DESCRIPTION
MDVP:FO	Average vocal fundamental frequency	MDVP:Shimmer	MDVP local shimmer
MDVP:Fhi	Maximum vocal fundamental frequency	MDVP:Shimmer(dB)	MDVP local shimmer in dB
MDVP:Flo	Minimum vocal fundamental frequency	Shimmer:APQ3	Three-point amplitude perturbation quotient
MDVP:Jitter(%)	MDVP jitter in percentage	Shimmer:APQ5	Five-point amplitude perturbation quotient
MDVP:Jitter(Abs)	MDVP absolute jitter in ms	MDVP : APQ11	MDVP 11-point amplitude perturbation quotient
MDVP:RAP	MDVP relative amplitude perturbation	Shimmer:DDA	Average absolute differences between the amplitude
MDVP:PPQ	MDVP five-point period perturbation quotient	NHR	Noise to harmonics ratio
Jitter:DDP	Average absolute difference of differences between jitter cycles	HNR	Harmonics to noise ratio

Fig. 9.1 Features of voice extraction

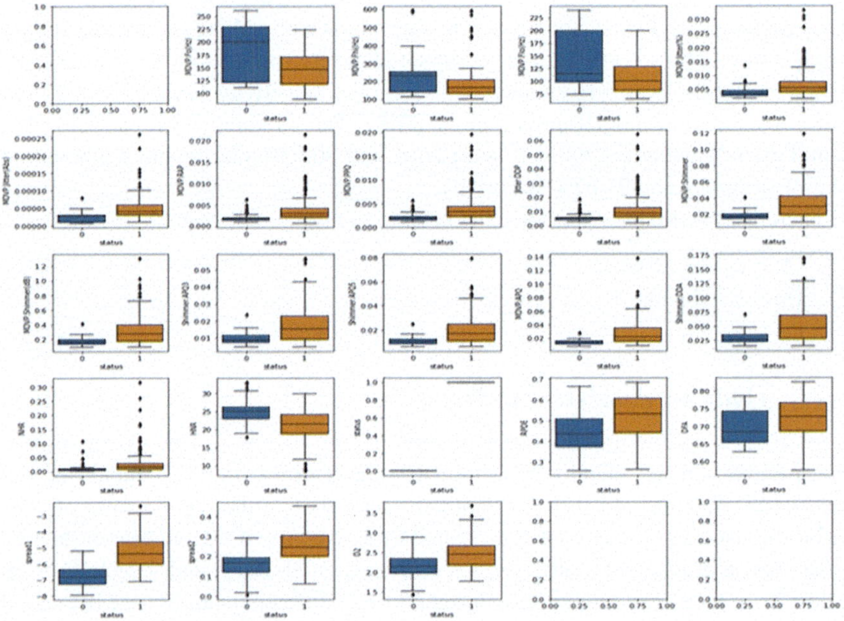

Fig. 9.2 Box plot of various features

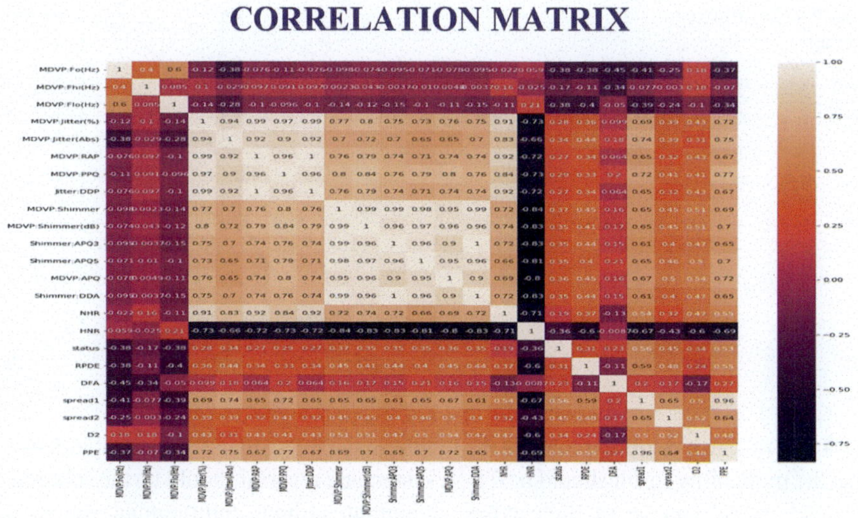

Fig. 9.3 Correlation matrix for relativity among features

There are 23 features, so there is high possibility that not every feature would be as important as other for detection, so to understand the relation between all the features it is visualized it in form of heatmap or correlation matrix (Fig. 9.3).

As in this score of 1 signifies that there is very high correlation and score of -0.75 signifies that there is very low correlation between two features. If there is very less correlation then, features are assigned according to their priorities, for detection of Parkinson.

9.4 Algorithm Selection

9.4.1 Logistic Regression

Moving backwards defines the process of modelling the probability of a different outcome when given a different input. Typical retrospective models have a binary effect which can take two values as true or false, yes or no and more. Multiplication of signals may indicate situations where there are more than two possible effects. Moving backwards is a useful analysis method for classification problems, in which it is determined that a new sample fits better into a phase. While using this model the two values are kept a 0 for healthy and 1 for Parkinson's.

Reversal of objects is a retrospective model with phase-dependent variability. It is a machine learning method for binary separation problems. Mainly due to the

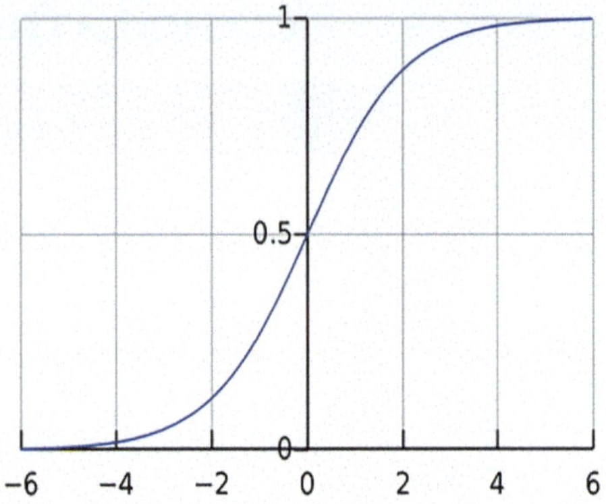

Fig. 9.4 Logistic regression model

problem of linear regression where the output may be less than 0 and greater than 1 but the probability result should be 0 to 1, systematic reversal should be the solution.

This model can be of three types—binomial, ordinal, multinomial Binomial logistic regression always provides two displayed outputs.

In a 0 or 1 way, most deals in situations where three or 18 exits are expected to be more than three but the exits are not ordered and ordinal deals have three or more order outcomes.

Moving backwards is represented by Fig. 9.4.

This algorithm has not been used as it creates two possibilities of 0 or 1, but this would not be the case in all cases, as some values may be among them which may be included in the false category.

Accuracy achieved was 89.23%.

9.4.2 K-Nearest Neighbor

K-Nearest Neighbor is the simplest Machine Learning algorithm based on Supervised Learning technique (Celik and Omurca 2019). The KNN algorithm assumes the similarity between the new data and available cases from the data set and puts the new case into the category that is most similar to the available categories. K-Nearest Neighbor algorithm takes all the available data and classifies a new data point based on the similarity observed. This signifies when new data comes then it can be easily classified into a well suited and appropriate category by using K-Nearest Neighbor algorithm. K-Nearest Neighbor algorithm can be used for Classification as well as for regression but mostly it is used for the Classification problems.

K-Nearest Neighbor is a non-parametric algorithm, as it does not make any assumptions on the underlying data set. It is also referred to as the lazy learner algorithm because it does not learn from the training set immediately but it stores the data set and performs an action on the dataset at the time of classification. At the initial training phase the K-Nearest Neighbor algorithm just stores the data set and when it gets new data from sources, it tries to classify that data into a category that is much similar to the new data (Figs. 9.5 and 9.6).

Fig. 9.5 KNN Model

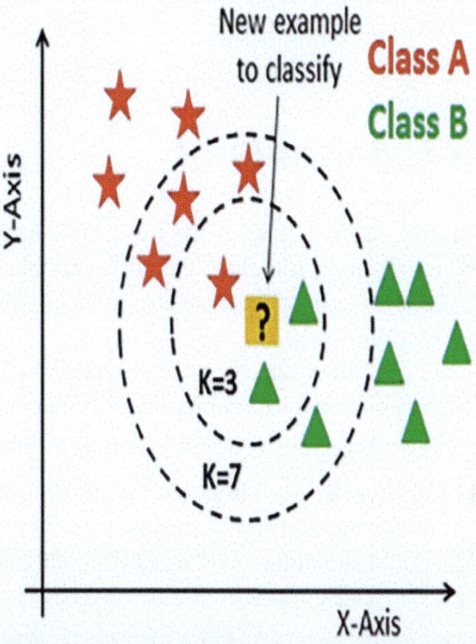

Fig. 9.6 KNN error rate graph

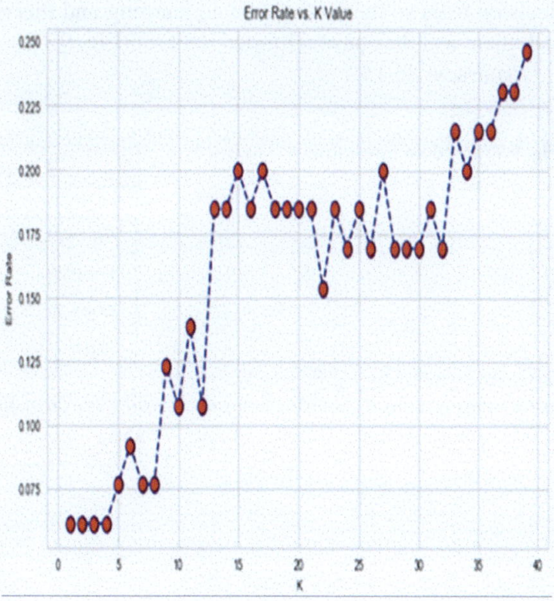

As and when this model was used, the K value that was obtained was 3. This has some limitations like a very high computational head depending on the value of K. Also as it assumes and gives equal importance to all the features which are also not required in our problem statement.

Accuracy: 90.7%

9.4.3 Random Forest

Random forest is one of the machine learning methods used for both classification and regression problems. This is a variant of the ensemble method in which groups of weak models become strong models when combined. Multiple trees are created in a random forest.

For classification, each tree must provide a classification and vote for that class. The forest chooses the classification that gets the most votes.

Random forest Selection Prediction Pseudo Code is as follows.

(1) Build a decision tree that predicts outcomes using features from the test set and stores the prediction results
(2) Count the number of votes for each prediction result
(3) The prediction with the most votes is considered the final prediction

The data was provided with a label of Parkinson and Not Parkinson, which gave this as the output. This algorithm was not used as this model provides independent decision trees so there would be no learning and this is not suitable for the problem statement as it has have a limited data set (Fig. 9.7).

Accuracy: 86.15%

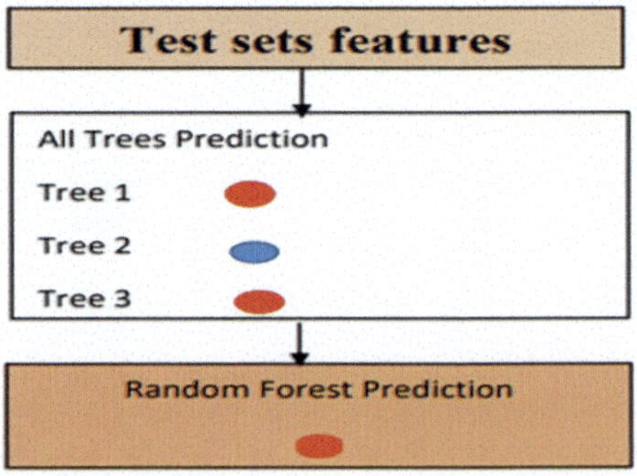

Fig. 9.7 Random forest flow

9.4.4 XGBoost

An extreme gradient gain is a very flexible and universal tool that exceeds the calculation of tri-algorithms and engineering, which is an abbreviated semester, which is especially used for all the memory and hardware resources used. Tianqi was first used in the command-line interface but it can be used after downloading in a variety of programming languages such as Python, C + + , R, Julia, Scala and Java. It can work successfully through most regression issues, classifications, and ranking and user-built objective features, quickly adds setting and normalization parameters, and reliably supports other and excellent normalization parameters than other libraries. SziRd Pafka, created by the landmark, shows how XGBoost (Rustempasic 2013) does not damage the speed or accuracy, and exceeds other similar algorithms.

The working of eXtreme Gradient Boosting:

(1) First, boosting technique is done by generating a strong classifier from a weak classifier by uninterruptedly adding models one by one so that the errors of the previous model are corrected in the next model until the training model is fully fixed or the best ones are reproduced first
(2) Next is the boosting part of the gradient. The same process for adding a model is followed but in its place of assigning weights to the classifier after each step, the new model is fitted to the rest of the old model. It is therefore called gradient boosting because it uses gradient descent to update the previous model. Both regression and classification problems are supported
(3) eXtreme Gradient Boosting implements the same decision tree boosting with added computational speed and precision

Improve the accuracy of the model by combining multiple weak learners (accuracy = 92.307%). Fewer computation steps are required to train the model (Fig. 9.8).

Tree Pruning: Nodes that do not contribute significantly to leaf node classification are removed.

Parallelism: Divides data into blocks allocated in parallel across multiple threads, optimized for CPU usage.

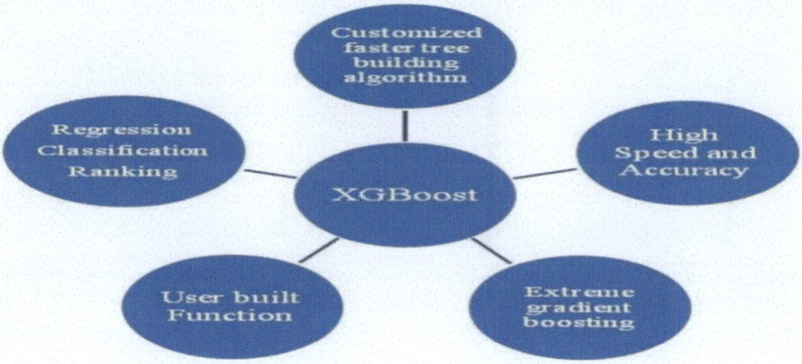

Fig. 9.8 Features of XGBoost Algorithm

Fig. 9.9 Cross validation score for various models

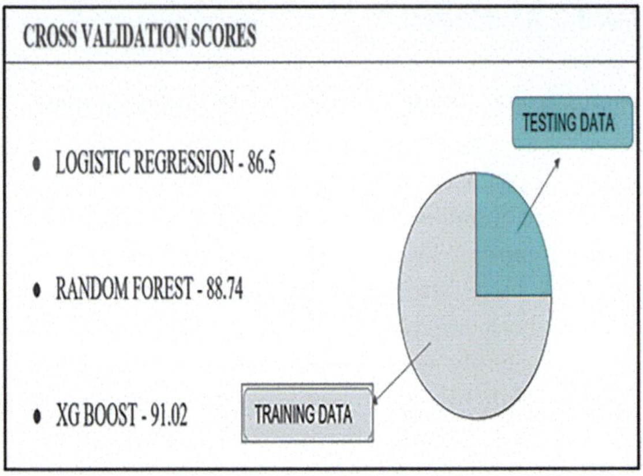

The optimal algorithm was obtained using cross validation scores.

Cross-validation is a statistical method that evaluates and compares learning algorithms by dividing data into two segments. One is used to train or train the model and the other is used to test the model. In a typical cross, the learning and validation kits must cross in a continuous round, so each data point may be confirmed.

As XGBoost has maximum score so it was decided to be accurate (Fig. 9.9).

9.5 Flow Chart

See Fig. 9.10.

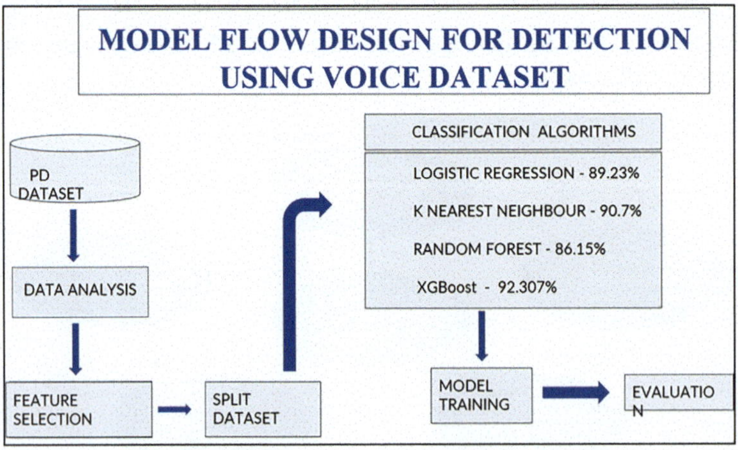

Fig. 9.10 Flow of voice data model for prediction

9.5.1 *Handwriting Detection*

For detection using tremor of hands, handwriting or drawing of specific figures is a very decent approach for differentiating the patients. In this patients would be drawing spiral and wave, and that would pass through the feature extraction, then through specific algorithm to obtain the required results.

Data Set Analysis

The data set was obtained from an open source and was segregated into testing data and training data. As for spiral testing 102 data sets from which testing 30 entries were used out of which 15 were labelled as healthy and 15 were labelled as Parkinson. And for training 72 entries were used out of which 36 were labelled as healthy and 36 were labelled as Parkinson.

Similarly, as for wave dataset, testing 102 data sets from which testing 30 entries were used out of which 15 were labelled as healthy and 15 were labelled as Parkinson. And for training 72 entries were used out of which 36 were labelled as healthy and 36 were labelled as Parkinson (Fig. 9.11).

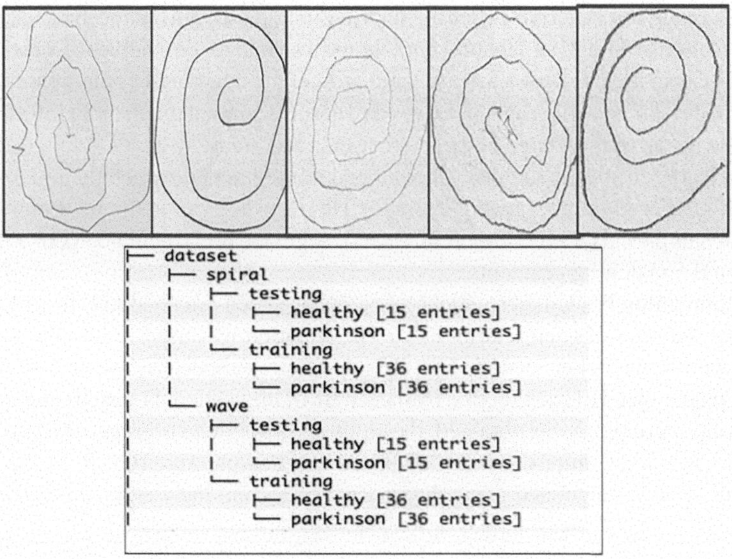

Fig. 9.11 Data set for handwriting detection

Fig. 9.12 HOG gradient calculation

$$d_i = \sum_{n=0}^{m_i} \sqrt{(x_{n+1} - x_n)^2 + (y_{n+1} - y_n)^2} \quad (1)$$

$$T_i = \frac{m_i}{133} \quad (2)$$

$$s_i = \frac{d_i}{T_i} \quad (3)$$

$$\overline{S}_w = \frac{\sum_{i=1}^{N} d_i s_i}{\sum_{i=1}^{N} d_i} \quad (4)$$

$$\overline{pr}_w = \frac{\sum_{i=1}^{N} d_i pr_i}{\sum_{i=1}^{N} d_i} \quad (5)$$

$$\overline{I}_{spr} = \overline{S}_w * \overline{pr}_w \quad (6)$$

9.6 Methodology

9.6.1 Histogram of Oriented Gradient

The HOG function descriptor uses the distribution (histogram) of the gradient direction (direction gradient) as a feature. Gradients in an image (derivatives of x and y) are useful because the gradient sizes are large around the edges and corners (areas with sharp changes in intensity) and the edges and corners contain much more information about the shape of the image objects rather than flat areas.

Data has been split to define each wait and the segment between them. Pendown is identified based on $pr > 0$ and has index label i. where m_i is the total number of segment samples. The total length of each d_i segment was calculated (Eq. 9.1) and less than 0.5 matching segments.

The remaining N segments and parameters have been renamed to i (from 1 to N). The total duration of each section is T_i. The equation was used to find the average speed in millimeters per second. The weighted average speed S_w was calculated by weighting the speed of each section by the length of the section. For each segment, the average pressure (pr_w) was obtained, the equivalent weighted average was calculated using the average Composite Index of Speed and Pen pressure (CISP) of the spiral plot, and I_{spr} was calculated using the product of S_w and the weighted pressure pr_w (Fig. 9.12).

9.7 Results

9.7.1 For Handwriting Detection

When random forest was used, it gave an accuracy of 82% (Figs. 9.13 and 9.14).

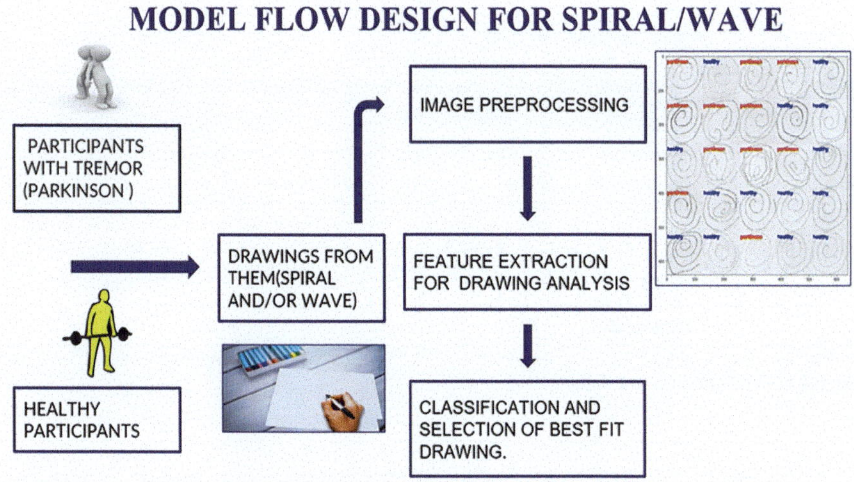

Fig. 9.13 Flow of handwriting detection for prediction

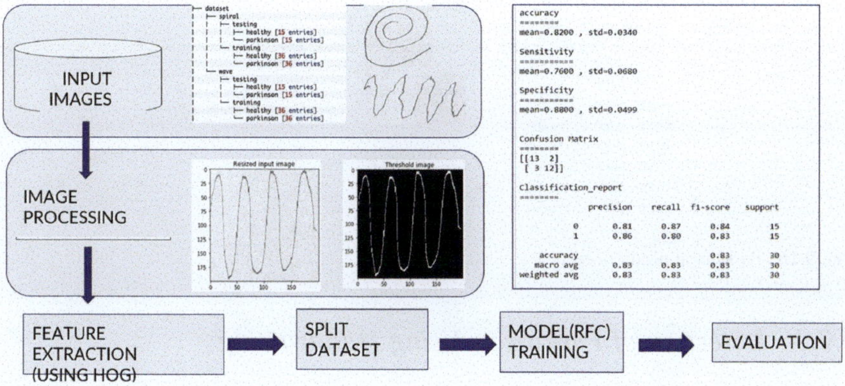

Fig. 9.14 HOG flow for detection

For voice detection using XGBoost model: Obtained 92.307% accuracy.

9.7.2 *Detection of Parkinson Using Voice*

See Figs. 9.15 and 9.16.

```
[65]:  pred1 = LR.predict(x_test)

126]:  #########prediction on test data


[66]:  pred1

:[66]: array([1, 1, 0, 0, 1, 1, 1, 0, 0, 1, 1, 1, 1, 1, 1, 1, 1, 1, 1, 1, 0, 0,
              1, 1, 1, 0, 1, 0, 1, 0, 1, 1, 0, 1, 1, 1, 1, 1, 1, 1, 1, 0, 1, 1,
              1, 1, 0, 1, 1, 1, 1, 0, 0, 1, 1, 1, 1, 1, 1, 1, 1, 1, 1, 1, 1],
             dtype=int64)
```

Fig. 9.15 Prediction on test data

```
n [121]:  pred1 = LR.predict(dfsamplexxxx)

n [122]:  pred1

ut[122]:  array([0, 0, 0, 0, 0, 0], dtype=int64)

n [125]:  for t in pred1:
              if t==0:
                  print(' not parkinson')

           not parkinson
           not parkinson
           not parkinson
           not parkinson
           not parkinson
           not parkinson
```

Fig. 9.16 Parkinson prediction

9.7.3 Detection of Parkinson Using Handwriting

See Fig. 9.17.

9.8 Future Work

As there's not much data available for processing, another way is to create a data set, for that open source software PRAAT has been used, which helps in extracting features of voice signals. As this was incorporated into python, the library that was used was parselmouth. In this paper 13 features out of 23 were obtained. For testing purpose the rest of features are considered as mean so that there is not much deviation of result.

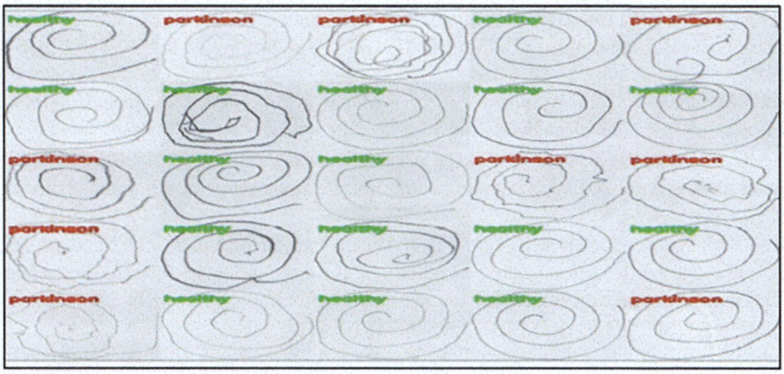

Fig. 9.17 Result of handwriting prediction

9.9 Conclusion

In order to reduce the effects of Parkinson's disease, it is of greatest importance that the patient is diagnosed with Parkinson's disease early. In this article, a machine learning predictive model to improve accuracy over humans and provide early detection capabilities is introduced. With the proposed predictive model, we aim to make it easier and more accurate for doctors to detect Parkinson disease compared to traditional detection techniques. In the proposed method, the accuracy of speech is 92.307% and the accuracy of handwriting is 82%. Contributions of this study are:

i. This model is very effective in assisting clinicians because it uses only two functions
ii. The sample size used is sufficient and very accurate to make this model efficient
iii. Software such as PRAAT can aid real-time detection with the required precision and accuracy.

References

Al-Islam F, Akter L (2020) Early identification of Parkinson's disease from hand-drawn images using histogram of oriented gradients and machine learning techniques. Emerging technology in computing, communication and electronics (ETCCE)

Bocklet T, Noth E, Stemmer G, Ruzickova H, Rusz J (2011) Detection of persons with Parkinson's disease by acoustic, vocal, and prosodic analysis. 2011 IEEE workshop on automatic speech recognition & understanding. https://doi.org/10.1109/asru.2011.6163978

Bonfiglio M, Chiaramonte R, Speech acoustics in Parkinson's disease: a systematic study of vocal impairment and meta-analysis of studies. https://doi.org/10.33588/rn.7011.2019414 PMID: 32436206

Binder H, Gefeller O, Schmid M, Mayr A (2014) The evolution of boosting algorithms. Methods Inf Med 53(06):419–427

Celik E, Omurca S (2019) Improving Parkinson's disease diagnosis with machine learning methods. Electr Electron Biomedical Eng Comput Sc (EBBT), 1–4. https://doi.org/10.1109/EBBT.2019.8742057
Goberman AM, Coelho C (2002) Acoustic analysis of parkinsonian speech i: Speech characteristics and l-dopa therapy. NeuroRehabilitation 17:237–246
Giuliano MG et al (2019) Selection of voice parameters for Parkinson´s disease prediction from collected mobile data. 2019 XXII Symposium on image, signal processing and artificial vision (STSIVA), pp 1–3
Mohan Rao KM, Naveen Reddy MS, Teja VR, Krishnan P, John Aravindhar D, Sambath M (2020) Parkinson's disease detection using voice and spiral drawing dataset. Proceedings of the third international conference on smart systems and inventive technology (ICSSIT 2020) IEEE Xplore Part Number: CFP20P17-ART; ISBN: 978-1-7281-5821-1
O'Sullivan SS, Williams DR, Gallagher DA, Massey LA, Silveira-Moriyama L, Lees AJ (2008) Nonmotor symptoms as presenting complaints in Parkinson's disease: a clinicopathological study. Mov Disord 23:101–106
Politis M, Wu K, Molloy S, Bain PG, Ray Chaudhuri K, Piccini P (2010) Parkinson's disease symptoms: the Patient's perspective, movement disorders. Movement Disorder Soc 25(11):1646–1651
"Parkinson's Disease", National Institute on Aging, 2020. [Online; Accessed: 01- Oct- 2020]
Rajani Kumari LV, Aatif Jaffery M, Saketh Sai Nigam K, Manaswi G, Tharangini P (2021) Detection of Parkinson's disease using extreme gradient boosting proceedings of the fifth international conference on trends in electronics and informatics (ICOEI). IEEE Xplore Part Number: CFP21J32-ART; ISBN: 978-1-6654-1571-2
Rustempasic F (2013) Mehmet Can: diagnosis of Parkinson's disease using principal component analysis and boosting committee machines. Southeast Europe J Soft Comput 2(1)
Rumman M, Tasneem A, Pavel M, Farzana S, Alam Md. A (2019) Early detection of Parkinson's disease using image processing and artificial neural network. https://doi.org/10.1109/ICIEV.2018.8641081
Sakar B, Isenkul M, Sakar CO, Sertbaş A, Gurgen F, Delil S, Apaydin H, Kursun O (2013) Collection and analysis of a parkinson speech dataset with multiple types of sound recordings. Biomed Health Inform IEEE J 17:828–834. https://doi.org/10.1109/JBHI.2013.2245674
Sadat J (2016) Preliminary results of a cross-linguistic comparison on dysarthria in Parkinson's disease movement disorders 31:S133–S134
Wu Y, Chen P, Yao Y, Ye X, Xiao Y, Liao L, Wu M, Chen J (2017) Computational and statistical approaches of medicine by Hindawi Volume 2017, dysphonic speech pattern study in Parkinson's disease patients using risk feature selection with minimum interclass probability and bagging ensemble learning method, 4201984 post, 11 p. https://doi.org/10.1155/2017/4201984
Xu B, Ye Y, Nie L (2012) An improved random forest classifier for image classification. 2012 IEEE international conference on information and automation, Shenyang, pp 795–800

Chapter 10
CCTV AI-Based System with Custom Object Detection and Video Upscaling Model

Talal Al-Failakawi, Yousef Landoulsi, and Shereef Abu Al-Maati

Abstract This chapter presents an affordable and secure CCTV system that integrates custom object detection and video upscaling models. The system aims to enhance the security capabilities of CCTV systems while remaining cost-effective and accessible. It incorporates an alert mechanism to promptly notify security personnel via popups, sound, and email notifications. The custom object detection model, based on Yolov5 architecture, accurately detects potential threats by leveraging convolutional neural networks and pooling layers. The video upscaling model, utilizing Real-ESRGAN architecture, improves the resolution and quality of surveillance videos. Custom datasets were curated for both models, including high-resolution videos with motion and firearm instances. The system's user interfaces were developed using Kivy and Tkinter, offering intuitive interactions. With features such as object detection, alerts, and an "Enhance" option, the system provides an effective security solution. The technical analysis of the models, dataset evaluation, and modifications made to ensure secure communication are discussed. This affordable and secure CCTV system with custom object detection and video upscaling models offers a valuable contribution to video surveillance, benefiting users with limited budgets while maintaining a high level of security.

Keywords Object detector · Super resolution · Alert System · Threat detection · Generative adversarial networks · Convolutional neural networks

T. Al-Failakawi (✉) · Y. Landoulsi · S. A. Al-Maati
American University of Kuwait, Salmiya, Kuwait
e-mail: S00047597@auk.edu.kw

Y. Landoulsi
e-mail: S00035712@auk.edu.kw

S. A. Al-Maati
e-mail: smaati@auk.edu.kw

D. K. Sharma et al. (eds.), *Machine Learning for Real World Applications*, Transactions on Computer Systems and Networks, https://doi.org/10.1007/978-981-97-1900-6_10

10.1 Introduction

Video Surveillance plays a crucial role in modern society in ensuring the security of various infrastructures from modern threats. However, due to the rising costs of these security systems we have curated a way in which these CCTV methods are more accessible and affordable to a wider range of users due to being more cost effective. In this chapter we will demonstrate an extra layer of security to these CCTV systems while remaining cost effective, affordable, and usable.

The primary purpose of our chapter is to enhance the security capabilities of CCTV systems through the integration of custom object detection and video upscaling models. By leveraging these models, we aim to detect and identify potential threats accurately while minimizing false positives. Our objective is to create an effective security solution that is accessible to organizations and individuals with budget constraints.

To achieve our goals, we employed a series of methodologies. Initially, we conducted a thorough analysis of potential threats to identify the key areas where security enhancements were required. Based on our findings, we implemented an alert system that promptly notifies security personnel on the infrastructure. In scenarios where officers may not be actively monitoring the screen, our AI object detector freezes the frame and plays a sound to draw their attention. Additionally, an email notification is sent to all personnel on the infrastructure, ensuring widespread awareness of potential security incidents.

The main features of our system include object detection, which enables accurate identification of threats, as well as alerts through popups, sound, and email notifications. We have also incorporated an "Enhance" option to allow users to double-check potential threats and minimize false positives.

For the implementation of our system, we utilized several architectural frameworks and libraries. Notably, we leveraged Yolov5, a powerful object detection architecture, to develop our custom object detection AI model. Additionally, we employed Real-ESRGAN which is made by xinntao (xinntao/Real-ESRGAN n.d.), a video upscaler architecture, to enhance the quality and resolution of surveillance videos. The graphical user interfaces of our system were developed using Kivy and Tkinter, providing intuitive and user-friendly interactions.

In terms of data collection and processing, we curated custom datasets for both the video upscaling and object detection models. For the video upscaling model, we collected high-resolution (3840 × 2160) and low-resolution (960 × 540) videos with significant motion and instances of firearms from reliable sources. We leveraged ffmpeg to extract frames from the videos and implemented automated image rotation and flipping using YAML scripts during the training process, with the assistance of Real-ESRGAN. Similarly, for the object detection model, we expanded our existing dataset to include additional classes such as guns, knives, and people. The dataset was labeled and published on Roboflow, facilitating easy export and maintenance for training the Yolov5 model. To ensure a balanced distribution of data, we utilized

Roboflow's capabilities to allocate the dataset among validation, training, and testing subdirectories.

In this Chapter, we provide a technical analysis of the Yolov5 object detection model, highlighting its utilization of convolutional neural networks and pooling layers to detect object features and reduce feature map sizes. Furthermore, we discuss the evaluation of accuracy using the Intersection over Union metric, which compares predicted bounding boxes with ground truth annotations. Additionally, we present an analysis of our dataset, showcasing example images and the training process, while highlighting the modifications we made to Yolov5 to include secure email communication.

In the context of video upscaling, we delve into the technical analysis of Real-ESRGAN, which employs a generative adversarial network (GAN) to generate high-resolution data. We elaborate on our dataset creation process, involving the extraction of frames from high- and low-resolution videos using ffmpeg and the pairing of metadata using custom scripts. We will also discussed the improvements in training time that were achieved by installing new hardware which has more VRAM and thus allows for larger batch sizes.

10.2 Review of Related Literature

10.2.1 Theoretical Background

Object detection is a computer vision method used to find objects found in images or instances in a video. Object detection algorithms use machine learning or deep learning in their algorithms to supply more exact results. Humans can recognize objects in images and videos very easily; object detection simulates this ability to the best possible result. Object detection is a key technology used in image classification that retrieves images and puts them into their respective classes. Objection detection is also a key technology in other applications such as video surveillance or driver aid systems. Object detection can be done using a variety of techniques. The most popular deep-learning technique uses convolutional neural networks (CNNs), such as R-CNN and YOLO v2 that automatically learn to detect objects within images. There are two key approaches to object detection using deep learning, either to create and train our own object detector or use an already pretrained object detector. Our approach is to create our own custom object detector while using an online dataset. We must create a network architecture to learn the characteristics for the items of interest if we want to train a custom object detector from scratch. To train the CNN (convolutional neural networks), we also need to assemble a sizable quantity of labeled data. A custom object detector can produce amazing results. Nevertheless, we must manually configure the CNN's layers and weights, which takes a lot of effort and training data. Image and video upscaling is a process done by an algorithm that takes a low-quality image and improves its quality with the help of an AI. First, the AI needs to be trained to know where and how to draw the pixels. This process requires datasets holding thousands of images and hundreds of hour-long videos. The process of training the AI can take around an entire week. Once training is complete, we obtain various models that can be used for different purposes. These can be for images and videos, 3d graphics and real-life images. The more suitable the model is the more correct the result will be. Let's assume that the AI was fed an image of a building, the AI will separate all the unique shapes from the original image and compare them to other images and videos in an attempt of finding similar pixel patterns. The AI then uses these pixel patterns to draw more pixels in certain areas which increases the number of pixels in the image resulting in a clearer higher quality image.

10.2.2 Related Literature

Literature Review on Object Detection with Backbone Network by the International Journal for Research in Applied Science and Engineering Technology (iJRASET) (Rakshitha and Revathi 2020) dives into the development of object detection using CNNs like what we have done and other methods. It explains the main methods of object detection, one where the image is segmented from the background of an image or a video and from there the object is detected from the background. Describing how the process can be challenging given that these objects can have various shapes and colors that can be like the background. Another literature review on object detection this time by the International Research Journal of Engineering and Technology (IRJET) (Muskan et al. 2022). In this article a different algorithm is used to detect objects, a method by the name of "You Only Look Once" (YOLO). This algorithm uses a technique called image localization, where the location of one or multiple objects is identified using bounding boxes. This process is not to be confused with image classification, which is a method that identifies to which class a certain image or object belongs. An example of YOLO can be seen in the image below (Fig. 10.1).

10.2.3 Related Systems

Physical hardware such as a LiDAR sensor can be used to detect objects. However, implementing a system of LiDAR sensors throughout campus can be costly. LiDAR sensor is a popular object detector that uses a pulsed laser technology to detect and categorize objects. LiDAR sensors produce precise, three-dimensional data on the surface properties and shape of nearby objects. The system builds a 3D model of the inspected area using laser beams that are safe for human eyes to see. While surveying, map navigation, atmospheric physics, and archeology are some of the typical uses

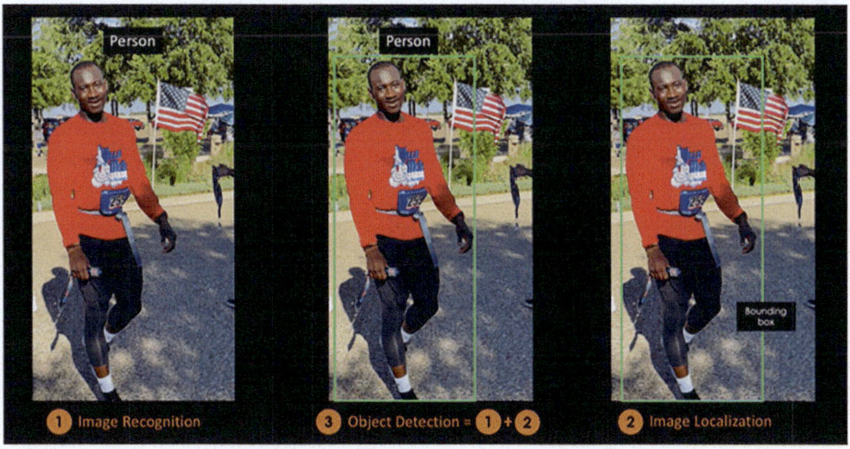

Fig. 10.1 YOLO application demonstration (YOLO Object Detection Explained n.d.)

of LiDAR sensors, they are also employed in many applications today, including robots, smart cities, and autonomous cars.

Software such as Topaz Vide AI will analyze your video input, upscale and enhance your video to 4 times its native video resolution however it costs $299 and requires a yearly subscription to get regular updates to the AI upscaling model that Topaz Labs have made (Topaz Video n.d.).

10.2.4 Issues with Prior Systems

Topaz Video AI requires a $299 initial license fee and then requires a yearly subscription for regular updates to the AI model (Topaz Video n.d.). Topaz Video AI also adds chromatic aberration onto the upscaled video which is not ideal. In comparison to Topaz Video AI, our solution is a lot more cost effective as it handles realistic video compressions and as of now it is completely open source and free. LiDAR sensors are considered better than a camera software-based detector. However, for simpler cases using a camera would be the better choice given how expensive a LiDAR sensor is.

10.3 Technical Background

10.3.1 Object Detector

See Fig. 10.2.

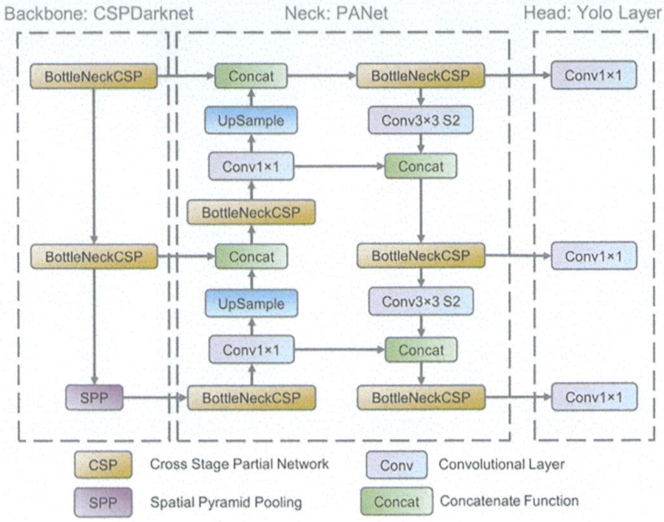

Fig. 10.2 YOLO network model (YOLO v5 model architecture [Explained] n.d.)

In Yolov5's network architecture, there are three components to it: CSPDarknet for the backbone, PANet for the neck, and Yolo Layer for the head. Before being sent to PANet for feature fusion, the data are first sent to CSPDarknet for feature extraction. The Yolo Layer then outputs the results of the detection (class, score, position, and size).

After the models are trained and due to the requirement of CUDA they can be then used on a lower end system such as a laptop that has an NVIDIA GPU. That laptop will be connected to multiple cameras then MonitorAI pings every camera and creates an array of buttons that are then displayed via the tkinter python library, after selecting which camera we plan to use it will pass an instruction onto YOLO to use said camera with our custom trained AI object detection model that was trained via the dataset we published onto Roboflow to classify people, guns and knives, each object was given an ID and was defined through a bounding box, Roboflow balanced the training, testing and validation datasets and it exported a text file for us that declared the ID and position of the bounding box onto the matrix of each image. We have set up YOLO to train for 500 epochs however it stopped at epoch 347 as the AI model stabilized at that training state and further epochs from that point will not increase the accuracy of our model. After the object gets classified, a simple instruction that we wrote in the YOLO source code calls a python class to send the alert via pop-up, email and play a sound. For post-processing, the Real-ESRGAN model we trained will be used to upscale the video and enhance the quality of the feed so that further research can be done on whether the alert was a false positive or not (Fig. 10.3).

Convolution Neural Networks work by having at least one convolution layer and typically consist of convolution layers, pooling layers and dense layers. To put simply,

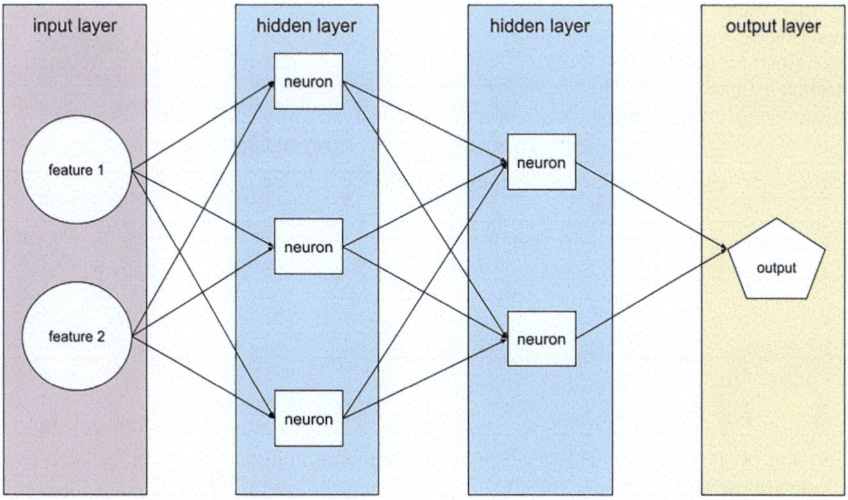

Fig. 10.3 Neural network (Machine Learning Glossary | Google for Developers n.d.)

convolution layers work by passing a convolutional filter to a deep neural network that consists of multiple hidden layers along the input matrix in order to extract specific features from the input. Pooling layers then use the convolutional layer output and reduces it to a smaller matrix by taking either the average or the maximum values across the pooled area and it does so through the use of strides. Finally dense layers define the complete connection between all these nodes (Machine Learning Glossary | Google for Developers n.d.).

Yolov5 work by utilizing a Convolution Neural Network technique titled Intersection Over Union. Which is used to measure the accuracy of the object detector model. First, we need to measure the Area of Overlap and to do that we need to find the intersection between the ground-truth bounding box which is what we manually labeled in our dataset and the predicted bounding box which is what the AI model predicts. Then we need to measure the Area of Union by finding the union of the ground-truth bounding box and the predicted bounding box. Afterwards we divide the Area of Overlap over the Area of Union to get the accurate measurement of the IoU and thus check the accuracy of our AI model (Introduction to YOLO Algorithm for Object Detection n.d.) (Fig. 10.4).

We looked for datasets that consisted of guns, knives and people however to our surprise none were found online so we collected three different datasets and merged them into our Roboflow repository so that we can easy balance and maintain said dataset. Two datasets were sourced from Kaggle one of which is a Knife dataset that was made by Shashank Shekhar (Knife Dataset n.d.), the other dataset is a Guns dataset that was made by Sai Sasank (Guns Object Detection n.d.). The final dataset we sourced was from Roboflow and it is a dataset that consisted of people however we cannot find the name of the author of said dataset so we will credit their work by listing their dataset in the references page at the end (People Detection Object

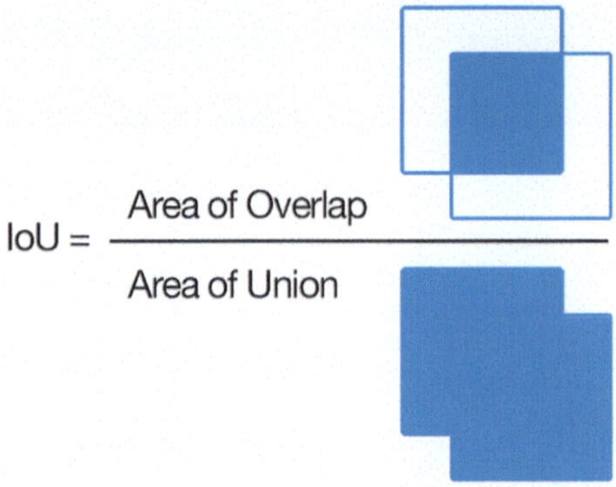

Fig. 10.4 Intersection over union (Intersection over Union (IoU) for object detection n.d.)

Detection Dataset n.d.). After we sourced all of these datasets we then combined them and posted them onto Roboflow for further maintenance however we also had to redo the bounding boxes on all the objects which lead to some loss in terms of accuracy. Our Roboflow repository now has a dataset large enough to detect potential threats (Monitor n.d.).

The following image represents the training log accuracy of our AI model at epoch 20 (Fig. 10.5).

The following image represents the training log accuracy of our AI model at epoch 347 (Fig. 10.6).

We notice that at epoch 347 our model stabilizes at an accuracy of 74% (Fig. 10.7).

We collected this YAML script by exporting our dataset from Roboflow. It defines the number of classes and the names of each object with the location of each dataset directory.

Each directory has two folders inside of it, one for the images the other is for labels.

epoch,	train/box_loss,	train/obj_loss,	train/cls_loss,	metrics/precision,	metrics/recall,
0,	0.086495,	0.029652,	0.034154,	0.063874,	0.25848,
1,	0.063529,	0.027095,	0.014287,	0.24615,	0.26264,
2,	0.060553,	0.025381,	0.0099757,	0.12644,	0.23978,
3,	0.05946,	0.024613,	0.01001,	0.18204,	0.1389,
4,	0.05912,	0.02521,	0.0093446,	0.15987,	0.25346,
5,	0.057715,	0.025297,	0.0091277,	0.12811,	0.18352,
6,	0.058364,	0.025021,	0.0094669,	0.20795,	0.14011,
7,	0.057264,	0.025694,	0.0094655,	0.10982,	0.069824,
8,	0.055829,	0.026035,	0.0088407,	0.37724,	0.25494,
9,	0.054487,	0.024762,	0.0066468,	0.40149,	0.27674,
10,	0.051777,	0.024391,	0.005509,	0.35138,	0.39485,
11,	0.050714,	0.023861,	0.0049518,	0.31002,	0.36018,
12,	0.051827,	0.024266,	0.005548,	0.25197,	0.2846,
13,	0.050251,	0.024634,	0.0056431,	0.34073,	0.33817,
14,	0.0488,	0.023786,	0.0040507,	0.57054,	0.24494,
15,	0.048389,	0.023123,	0.0038902,	0.43179,	0.42916,
16,	0.048031,	0.022875,	0.0035409,	0.51545,	0.25452,
17,	0.04719,	0.023511,	0.0044041,	0.31615,	0.37844,
18,	0.048402,	0.022121,	0.0040817,	0.38842,	0.32552,
19,	0.046769,	0.022147,	0.0037194,	0.45316,	0.36725,
20,	0.045535,	0.021958,	0.0042095,	0.47923,	0.36996,

Fig. 10.5 Yolov5 training log

335,	0.020533,	0.010658,	0.00037881,	0.69499,	0.5739,
336,	0.020655,	0.010496,	0.00022463,	0.69211,	0.5725,
337,	0.020051,	0.010536,	0.00035875,	0.68233,	0.59207,
338,	0.019395,	0.01025,	0.00025417,	0.7417,	0.56462,
339,	0.019634,	0.010327,	0.00016957,	0.72178,	0.59061,
340,	0.020517,	0.01031,	0.00020867,	0.69504,	0.60471,
341,	0.019416,	0.010212,	0.00025684,	0.71058,	0.60857,
342,	0.020023,	0.010321,	0.00018002,	0.69328,	0.59704,
343,	0.019951,	0.010147,	0.00015654,	0.68883,	0.59634,
344,	0.019803,	0.010271,	0.00027001,	0.69794,	0.59207,
345,	0.019617,	0.010193,	0.0006093,	0.73403,	0.56313,
346,	0.019394,	0.0099952,	0.00044611,	0.73135,	0.58205,
347,	0.01985,	0.0099526,	0.00036544,	0.74376,	0.56674,

Fig. 10.6 Yolov5 training log continued

```yaml
train: ../train/images
val: ../valid/images
test: ../test/images

nc: 3
names: ['Person', 'gun', 'knife']

roboflow:
  workspace: auk-lsntl
  project: monitorai
  version: 6
  license: CC BY 4.0
  url: https://universe.roboflow.com/auk-lsntl/monitorai/dataset/6
```

Fig 10.7 data.yaml file from MonitorAI object recognition dataset

The images directory is filled with all of the images in the respected subset however the labels directory is filled with text files with the same exact names as its sibling image directory and it maps the bounding box and index number of an object onto an image from the sibling image with the same file name.

10.3.2 Changes to Yolov5's Source Code

The following adjustments were made to the detect.py source code from Yolov5. It checks if the name an object and if it is a threat it sounds an warning through an mp3 file afterwards it calls the detectedGun.py script to send an email to the security personnel then final it shows a Message box window with the alert message (Fig. 10.8).

detectedGun.py represents the email alert class and it is called when a gun gets detected. It classifies the sender and receiver, and it sends the email via an smtp protocol cameraAdapter.py pings every single camera that is connected onto your computer and it creates an array of buttons that pass on a parameter to the detect.py class from Yolov5 and it send the camera ID as a source for Yolov5 to use (Figs. 10.9 and 10.10).

```
#-----------------------------------------------------------------------------------------
                    if names[c] == 'knife':
                        print('Knife Detected!')
                        audio_file = os.path.dirname(__file__) + '\sounds\warning.mp3'
                        playsound(audio_file)
                        subprocess.call('python detectedKnife.py')
                        ctypes.windll.user32.MessageBoxW(0, "Knife DETECTED!", "Alert", 1)
                    elif names[c] == 'gun':
                        print('Gun Detected!')
                        audio_file = os.path.dirname(__file__) + '\sounds\warning.mp3'
                        playsound(audio_file)
                        subprocess.call('python detectedGun.py')
                        ctypes.windll.user32.MessageBoxW(0, "Gun DETECTED!", "Alert", 1)
                    else:
                        print('No Threats :D')
#-----------------------------------------------------------------------------------------
```

Fig. 10.8 Modified Detect.py file from Yolov5s source code

```
import os
from email.message import EmailMessage
import ssl
import smtplib

email_sender = 'aukmonitorai@gmail.com'
email_password = 'HIDDEN FOR SECURITY'
email_reciever = 's000xxxxx@auk.edu.kw'

subject = 'ALERT!'
body = """
A gun has been detected!
"""

em = EmailMessage()
em['From'] = email_sender
em['To'] = email_reciever
em['Subject'] = subject
em.set_content(body)

context = ssl.create_default_context()

with smtplib.SMTP_SSL('smtp.gmail.com', 465, context=context) as smtp:
    smtp.login(email_sender, email_password)
    smtp.sendmail(email_sender, email_reciever, em.as_string())
```

Fig. 10.9 detectedGun.py file for email alerts via smtp

10.3.3 Video Upscaler

We used Real-ESRGAN for our upscaler as it is a Generative Adversarial Network meaning that it creates new data through a generator and a discriminator determines whether that created data is valid or invalid. A generator is a subsystem in the GAN and it creates new examples, it takes noise as an input and it uses that noise to produce a new output then the discriminator classifies the generators output and calculates the loss from the discriminator classification then it backpropagates through both the discriminator and generator to obtain gradients and finally it uses those gradients to change only the generator weights (Generator and | Machine Learning | Google for Developers n.d.).

To train our REAL-ESRGAN model for video upscaling, we opted to use a 3840 $\times$ 2160p gun video and a 3840 $\times$ 2160p video of someone running across the beach

```
import os
import subprocess

#cameraAdapterStart
import cv2
index = 0
arr = []
while True:
    cap = cv2.VideoCapture(index)
    try:
        if cap.getBackendName()=="MSMF":
            arr.append(index)
    except:
        break
    cap.release()
    index += 1

from tkinter import *
import tkinter as tk

# create an instance of tkinter
win = tk.Tk()

#Define the size of the window
win.geometry("400x200")

#Name the title of the window
win.title("Select Camera")

# number of buttons
n=len(arr)

my_str=tk.StringVar()
l1=tk.Label(win, textvariable=my_str, width=60)
l1.grid(row=1, column=60, columnspan=60)
#Defining the row and column
i=3

#Iterating over the numbers till n and
#creating the button
def yolov5(k):
    subprocess.call('python detect.py --source ' + str(k))
for j in range(n):
    e = Button(win, text='Camera ID ' + str(j) , command=lambda k = j:yolov5(k))
    e.grid(row=i, column=j)

# Keep the window open
win.mainloop()

#CameraAdapterEnd
```

Fig. 10.10 cameraAdapter.py file used for camera detection

from pexels.com. The reason we picked those two videos is because enhancing a gun video frame takes a high priority while the video of someone running across the beach was used as it had a lot of motion so that the video upscaler trains to handle motion better.

For the REAL-ESRGAN dataset. We used Hand Brake to import those videos and downscale to 960×540p and compress each video with a realistic bitmap compression level. ffmpeg was then used to turn these high and low resolution videos into a sequence of images. MagickUtils was used to check for the divisibility of both the height and width of each image in the dataset by making sure that both dimensions are divisible by 16. MagickUtils was then used to further compress the low-resolution images. Our training script for Real-ESRGAN specified that it will randomly flip and rotate each image during training. The high-resolution video frames were named HR which is what we use to refer to a high-resolution dataset and the low-resolution video frames were named LR is what we use to refer to a Low-Resolution dataset. Since images in the LR dataset are 540p and since images from the HR dataset are 2160p then if we multiplied 540p by 4 it will give us 2160p meaning that the model will be trained to upscale videos to 4 times their native resolution. After prepping both the HR and LR datasets, we used another function from REAL-ESRGAN to pair both HR and LR datasets with a metadata configuration file. Since we are using an RTX 4090, we have 24 GB of VRAM which is plenty enough for our dataset configuration. A few additional adjustments have been made to the training script, we chose Adam as our optimizer, and picked NVIDIA Collective Communications Library which is NCCL as our backend. For our Visual Geometry Group, we picked vgg19 as that is the deepest level we could reach with our current system. We picked 68,000 iterations to be our limit as it leads us to 1440 epochs which is plenty enough for the video upscaling model. Finally for our data loader we configured it to shuffle the dataset and set the number of workers per GPU to 6 and it sets the batch size per GPU to 12.

We notably picked Adam due to its extreme efficiency as it is an algorithm for first order gradient-based optimization of stochastic objective functions, (Diederik and Jimmy 2015) based on adaptive estimates of lower-order moments and it is an industry standard as it has been used on IMDB sentiment analysis datasets and CIFAR-10 image recognition dataset (Gentle Introduction to the Adam Optimization Algorithm for Deep Learning n.d.) (Figs. 10.11 and 10.12).

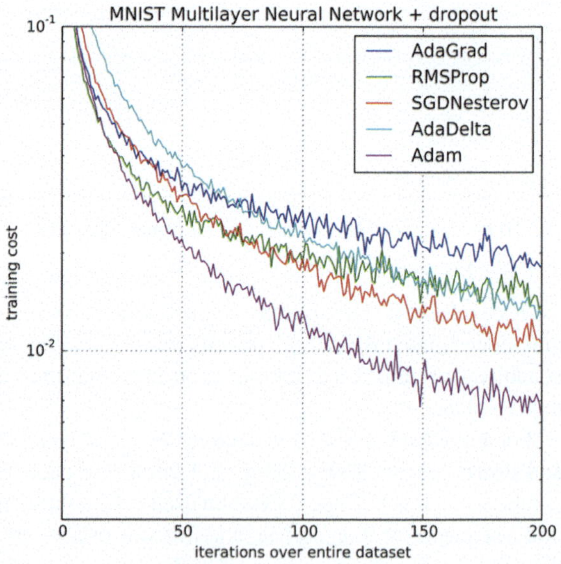

Fig. 10.11 Adam optimizer's efficiency demonstration over the MNIST dataset (Diederik and Jimmy 2015)

```
2023-05-23 22:58:38,165 INFO: Saving models and training states.
2023-05-23 22:59:49,503 INFO: [Video..][epoch:1429, iter:  67,100, lr:(1.000e-04,)] [eta: 2 days, 17:11:28, time (data): 0.592 (0.001)]
l_g_pix: 1.7764e-02 l_g_percep: 4.3320e+00 l_g_gan: 3.3328e-01 l_d_real: 2.4321e-01 out_d_real: 5.9538e+00 l_d_fake: 1.3134e-01
out_d_fake: -3.2013e+00
2023-05-23 23:01:23,053 INFO: [Video..][epoch:1431, iter:  67,200, lr:(1.000e-04,)] [eta: 2 days, 17:18:11, time (data): 0.878 (0.178)]
l_g_pix: 1.9673e-02 l_g_percep: 4.1200e+00 l_g_gan: 2.4621e-01 l_d_real: 4.8752e-01 out_d_real: 2.1399e+00 l_d_fake: 2.4945e-01
out_d_fake: -2.2125e+00
2023-05-23 23:02:48,364 INFO: [Video..][epoch:1432, iter:  67,300, lr:(1.000e-04,)] [eta: 2 days, 17:21:59, time (data): 1.162 (0.549)]
l_g_pix: 1.5577e-02 l_g_percep: 3.9976e+00 l_g_gan: 1.8148e-01 l_d_real: 1.1938e-01 out_d_real: 5.3767e+00 l_d_fake: 4.3717e-01
out_d_fake: -1.3777e+00
2023-05-23 23:03:58,298 INFO: [Video..][epoch:1433, iter:  67,400, lr:(1.000e-04,)] [eta: 2 days, 17:20:33, time (data): 0.773 (0.176)]
l_g_pix: 1.4277e-02 l_g_percep: 3.4809e+00 l_g_gan: 2.4797e-01 l_d_real: 2.2792e-01 out_d_real: 4.8214e+00 l_d_fake: 2.2398e-01
out_d_fake: -2.2555e+00
2023-05-23 23:05:07,695 INFO: [Video..][epoch:1434, iter:  67,500, lr:(1.000e-04,)] [eta: 2 days, 17:18:55, time (data): 1.168 (0.577)]
l_g_pix: 1.1229e-02 l_g_percep: 3.4440e+00 l_g_gan: 2.7229e-01 l_d_real: 2.2123e-01 out_d_real: 4.9685e+00 l_d_fake: 1.6219e-01
out_d_fake: -2.5606e+00
2023-05-23 23:06:20,450 INFO: [Video..][epoch:1435, iter:  67,600, lr:(1.000e-04,)] [eta: 2 days, 17:18:25, time (data): 0.795 (0.184)]
l_g_pix: 1.5381e-02 l_g_percep: 3.5003e+00 l_g_gan: 3.8103e-01 l_d_real: 2.9667e-01 out_d_real: 4.9833e+00 l_d_fake: 9.0150e-02
out_d_fake: -3.7200e+00
2023-05-23 23:07:31,838 INFO: [Video..][epoch:1436, iter:  67,700, lr:(1.000e-04,)] [eta: 2 days, 17:17:28, time (data): 1.220 (0.609)]
l_g_pix: 1.4660e-02 l_g_percep: 3.5766e+00 l_g_gan: 2.9004e-01 l_d_real: 3.4346e-01 out_d_real: 4.2859e+00 l_d_fake: 1.8190e-01
out_d_fake: -2.7183e+00
2023-05-23 23:08:41,172 INFO: [Video..][epoch:1437, iter:  67,800, lr:(1.000e-04,)] [eta: 2 days, 17:15:49, time (data): 0.770 (0.178)]
l_g_pix: 1.2739e-02 l_g_percep: 3.2576e+00 l_g_gan: 2.2307e-01 l_d_real: 2.6382e-01 out_d_real: 4.7780e+00 l_d_fake: 3.1690e-01
out_d_fake: -1.9138e+00
2023-05-23 23:09:53,087 INFO: [Video..][epoch:1438, iter:  67,900, lr:(1.000e-04,)] [eta: 2 days, 17:15:02, time (data): 0.594 (0.001)]
l_g_pix: 1.5633e-02 l_g_percep: 4.1766e+00 l_g_gan: 2.5707e-01 l_d_real: 1.7011e-01 out_d_real: 4.9866e+00 l_d_fake: 3.2557e-01
out_d_fake: -2.2450e+00
2023-05-23 23:11:03,658 INFO: [Video..][epoch:1439, iter:  68,000, lr:(1.000e-04,)] [eta: 2 days, 17:13:48, time (data): 0.690 (0.089)]
l_g_pix: 1.9377e-02 l_g_percep: 4.4191e+00 l_g_gan: 3.3273e-01 l_d_real: 4.2711e-01 out_d_real: 4.2969e+00 l_d_fake: 1.4766e-01
out_d_fake: -3.1795e+00
2023-05-23 23:11:03,658 INFO: Saving models and training states.
2023-05-23 23:12:19,993 INFO: [Video..][epoch:1440, iter:  68,100, lr:(1.000e-04,)] [eta: 2 days, 17:14:26, time (data): 0.618 (0.001)]
l_g_pix: 2.0703e-02 l_g_percep: 4.1816e+00 l_g_gan: 2.5973e-01 l_d_real: 9.1856e-02 out_d_real: 7.6810e+00 l_d_fake: 2.7527e-01
out_d_fake: -2.3217e+00
```

Fig. 10.12 Real-ESRGAN Videox4 upscaling model training log

This figure illustrates the training log of Real-ESRGAN. We stopped training at iteration 68,000.

For us to use our newly trained AI upscaling model in Real-ESRGAN we have to first modify their source code to recognize our model which we named Videox4, the following adjustments were made to their inference_realesrgan_video.py source code (Fig. 10.13).

The modification mentions that if a user selects the Videox4 model and Real-ESRGAN finds that Videox4 is downloaded then it will use it to upscale a video by 4 times its native resolution otherwise it will download the Videox4 model from our GitHub repository and then use that model for the same purpose as we mentioned earlier (Fig. 10.14).

Real-ESRGAN as of now has a few issues, it takes a folder directory that consists of video frames as its input however Real-ESRGAN does not turn a video into frames so in the GUI of MonitorAI, we made sure to flush the tmp_frames directory first then we collected to video directory from the user file selection screen and passed it as a parameter to ffmpeg to turn that video into frames and then we pass the tmp_frames directory onto Real-ESRGAN to upscale each frame and then merge them

```python
# ---------------------- determine models according to model names ---------------------- #
args.model_name = args.model_name.split('.pth')[0]
if args.model_name == 'RealESRGAN_x4plus':  # x4 RRDBNet model
    model = RRDBNet(num_in_ch=3, num_out_ch=3, num_feat=64, num_block=23, num_grow_ch=32, scale=4)
    netscale = 4
    file_url = ['https://github.com/xinntao/Real-ESRGAN/releases/download/v0.1.0/RealESRGAN_x4plus.pth']
elif args.model_name == 'Videox4':  # x4 RRDBNet model
    model = RRDBNet(num_in_ch=3, num_out_ch=3, num_feat=64, num_block=23, num_grow_ch=32, scale=4)
    netscale = 4
    file_url = ['https://github.com/delltalal/MonitorAI/releases/tag/v1.1/Videox4.pth']
```

Fig. 10.13 Updated inference_realesrgan_video.py from Real-ESRGAN's source code

```
import os
import glob
import subprocess

import kivy
from tkinter import Tk
from tkinter.filedialog import askopenfilename
from kivy.app import App
from kivy.core.text import Label as CoreLabel
from kivy.properties import partial
from kivy.uix.boxlayout import BoxLayout
from kivy.uix.button import Button
from kivy.lang import Builder
from kivy.uix.screenmanager import ScreenManager, Screen

# kivy version Requirement
kivy.require('1.9.0')

def yolov5MainCamera(instance, *args):
    subprocess.call('python detect.py --source 0', cwd='yolov5')


def EsrganUpscale(instance, *args):
    Tk().withdraw()  # we don't want a full GUI, so keep the root window from appearing
    filename = askopenfilename()  # show an "Open" dialog box and return the path to the selected file

    #FlushTempFrames
    files = glob.glob('Real-ESRGAN-0.3.0/tmp_frames/*')
    for f in files:
        os.remove(f)

    # ExtractVideoFrames
    subprocess.call('ffmpeg -i '+filename+' -qscale:v 1 -qmin 1 -qmax 1 -vsync 0 tmp_frames/frame%08d.png',
                    cwd='Real-ESRGAN-0.3.0')
    #UpscaleVideoFrames
    subprocess.call('python inference_realesrgan_video.py -n Videox4 -i tmp_frames -o out_frames -s 2',
                    cwd='Real-ESRGAN-0.3.0')
    #FlushTempFrames
    files = glob.glob('Real-ESRGAN-0.3.0/tmp_frames/*')
    for f in files:
        os.remove(f)

# Load Design File

# Main App layout class
class MonitorAIApp(App):

    def build(self):
        button1 = Button(text="Detect Objects", size_hint=(0.25, 0.18), pos=(350, 100))
        button1.bind(on_press=partial(yolov5MainCamera, button1))
        button2 = Button(text="Enhance Video", size_hint=(0.25, 0.18), pos=(350, 200))
        button2.bind(on_press=partial(EsrganUpscale, button2))
        boxlayout = BoxLayout()
        boxlayout.add_widget(button1)
        boxlayout.add_widget(button2)
        return boxlayout


# run the App
if __name__ == '__main__':
    MonitorAIApp().run()
```

Fig. 10.14 GUI.py which is the main user interface that bridges the two AI architectures

together and save them in the out_frames directory and then finally we flush the tmp_frames directory.

10.4 Conclusion

In conclusion, this chapter presents a method that aims to enhance security capabilities of CCTV systems through the integration of custom object detection and video upscaling models. The chapter demonstrates a focus to provide extra layers of security while maintaining affordability and usability for a wider range of users.

Through the utilization of methodologies such as thorough threat analysis, alert systems, and custom AI models, the chapter aimed to accurately detect and identify potential threats while minimizing false positives. The system incorporated features like object detection, alerts through popups, sound, and email notifications, as well as an "Enhance" option for further verification of potential threats.

The implementation of the system involved the use of architectural frameworks and libraries such as Yolov5 for object detection and Real-ESRGAN for video upscaling. Custom datasets were curated for training these models, with a focus on including classes such as guns, knives, and people. Data collection and processing were facilitated by tools like ffmpeg and Roboflow, ensuring a balanced distribution of data among training, validation, and testing sets.

The technical analysis of the Yolov5 object detection model highlighted its utilization of convolutional neural networks and pooling layers for accurate detection and feature reduction. The evaluation of accuracy using the Intersection over Union metric demonstrated the effectiveness of the model in comparison to ground truth annotations. Similarly, the technical analysis of Real-ESRGAN showcased its ability to generate high-resolution data through a generative adversarial network approach.

Related literature reviewed in this chapter explored the theoretical background of object detection and image upscaling, emphasizing the use of deep learning techniques and various algorithms like YOLO. It also highlighted prior systems and their limitations, such as the costliness of hardware-based detectors and the subscription-based pricing of commercial upscaling software.

In summary, this chapter we successfully developed an enhanced security solution for CCTV systems by integrating custom object detection and video upscaling models. The approach presented in this chapter demonstrated cost-effectiveness, accessibility, and usability, making it a valuable contribution to the field of video surveillance systems. Future research can build upon this work by further refining the models and exploring additional security features.

References

Diederik P, Adam JL (2015) A method for stochastic optimization, 1–15, https://arxiv.org/abs/1412.6980

Gentle Introduction to the Adam Optimization Algorithm for Deep Learning, https://machinelearningmastery.com/adam-optimization-algorithm-for-deep-learning/

Guns Object Detection, https://www.kaggle.com/datasets/issaisasank/guns-object-detection

Intersection over Union (IoU) for object detection, https://pyimagesearch.com/2016/11/07/intersection-over-union-iou-for-object-detection/

Introduction to YOLO Algorithm for Object Detection, https://www.section.io/engineering-education/introduction-to-yolo-algorithm-for-object-detection/

Knife Dataset, https://www.kaggle.com/datasets/shank885/knife-dataset

Machine Learning Glossary | Google for Developers, https://developers.google.com/machine-learning/glossary

Muskan C, Sadanand S, Abhishek K, Vinay K, Nidhi S (2022) Object detection using YOLO models, 3785–3789, https://www.irjet.net/archives/V9/i5/IRJET-V9I5774.pdf

Monitor AI–v6 2023–05–11 3:53pm, https://app.roboflow.com/auk-lsntl/monitorai/6
People Detection Object Detection Dataset and Pre-Trained Model by People Detection, https://universe.roboflow.com/people-detection-qnzez/people-detection-kpqn3/
Rakshitha S, Revathi S (2020) Literature review on object detection with backbone network, 751–754 https://doi.org/10.22214/ijraset.2020.5118
The Generator | Machine Learning | Google for Developers, https://developers.google.com/machine-learning/gan/generator
Topaz Video AI, https://www.topazlabs.com/topaz-video-ai
xinntao/Real-ESRGAN: Real-ESRGAN aims at developing Practical Algorithms for General Image/Video Restoration., https://github.com/xinntao/Real-ESRGAN
YOLO Object Detection Explained: A Beginner's Guide, https://www.datacamp.com/blog/yolo-object-detection-explained
YOLO v5 model architecture [Explained], https://iq.opengenus.org/yolov5/

Chapter 11
Forecasting the COVID-19 End in India Using Machine Learning and Population Density Clustering

Karam Ratan Singh, Barenya Bikash Hazarika, and Deepak Gupta

Abstract In the literature, the traditional machine learning models are vastly explored in prognostic modelling for COVID-19. The goal of predictions is to make the citizens aware of potential consequences that can be caused by COVID-19. Nonetheless, conventional prediction models are unable to exploit several factors such as demographics and population density. This work addresses how these continuous and unprecedented factors lead us to develop complex models rather than just relying on them. In addition to that based on the obtained data till 3rd September 2020, the state-wise COVID-19 end prediction is also portrayed for India. Further, we have studied the relationship between the population densities and the number of infected cases. We have observed a strong correlation in Central zone, Eastern Zone (EZ) and North Eastern zone (NE-Z) and low correlation in Western Zone (WZ), Southern Zone (SZ), Northern Zone (NZ) and all the states of India. The popular yet powerful extreme learning machine (ELM) model is applied as a forecasting model. The estimation capacity of ELM is compared with the large margin distribution-machine based regression and the twin support vector regression based on R^2 and root mean square error. Experimental outcomes prove the proficiency of the ELM model.

Keywords COVID-19 · population density · Forecasting · Extreme learning machine · India · Zones

K. R. Singh
Department of Basic and Applied Science, National Institute of Technology Arunachal Pradesh, Jote, India

B. B. Hazarika
Faculty of Computer Technology, Assam down town University, Sankar Madhab Path, Gandhi Nagar, Panikhaiti, Guwahati, Assam, India

D. Gupta (✉)
Department of Computer Science and Engineering, Motilal Nehru National Institute of Technology Allahabad, Prayagraj, Uttar Pradesh, India
e-mail: deepakjnu85@gmail.com

D. K. Sharma et al. (eds.), *Machine Learning for Real World Applications*, Transactions on Computer Systems and Networks, https://doi.org/10.1007/978-981-97-1900-6_11

11.1 Introduction

Several researchers have tried to portend the progress of the COVID-19 pandemic through the usage of prediction-based models such as artificial intelligence (AI), machine learning (ML), statistical models and epidemiological analysis models. ML models are extensively used in a wide array such as market analysis (Chi-Hsien and Nagasawa 2019), oil-pipeline risk estimation (Priyanka et al. 2021), prediction of AFP based thermoplastics (Wanigasekera et al. 2021), sediment load prediction (Hazarika et al. 2021) and many more. COVID-2019 is an epidemiologic disease that has become a global pandemic and is rapidly spreading (Murugan and Goel 2020). The disease had spread to almost all countries, territories and areas by 8 October 2020. India first reported the case 2019-nCoV on 30 January 2020 in the state of Kerala and it spread to all states of India after 116 days in the state of Nagaland on 25 May 2020. Epidemiological and ML models are widely used to make valuable estimations on the number of infected, diseased and recovered are the Susceptible, Infectious or Recovered (SIR) (Ndiaye et al. 2020; Das et al. 2020) model and the Susceptible Exposed Infected Recovered (SEIR) model (Gupta et al. 2020).

There are several pieces of literature available for COVID-19 spread estimation. Herlawati (2020) predicted the COVID-19 spread pattern utilizing the support vector regression (SVR) using the benefit of various kernels. Simulation results show that the SVR model with radial basis function (RBF) kernel predicted the pattern of COVID-19 more precisely as compared to other kernels. Ardabil et al. (2020) estimated the COVID-19 spread by the multi-layer perceptron (MLP) and artificial neuro-fuzzy inference system (ANFIS) model. Satu et al. (2020) also used several machine learning algorithms viz., LR, polynomial regression, MLP, polynomial MLP, ridge regression, holt linear, holt-winter, ARIMA and prophet models for COVID-19 cases forecasting in Bangladesh. A wavelet coupled RVFL (WCRVFL) model was proposed by Hazarika and Gupta (2020) to predict the COVID-19 escalation in five worst-hit countries. Recently, Tandon et al. (2020) predict the disease spread and deaths using ARIMA models. Parbat and Chakraborty (2020) forecasted daily confirmed deaths and recovered cases using support vector regression. Few other prominent pieces of literature can be found in Singh et al. (2021), Javaid et al. (2020), Lu and Williams (2021) and Chowdhury et al. (2021).

Sun et al. (2020) used geographic location as well as population density as key attributes for predicting the disease spread in China. Coskun et al. (2020) used two different factors population density (PD) and wind speed for modelling the spread of COVID-19 in Turkey. They found that the humidity, air temperature, air pollution and PD, negatively impact the expansion of COVID-19. The relationship between COVID-19 spreads and the population density has been studied by many researchers namely Afshordi et al. (2020), Coskun et al. (2020), Bhadra et al. (2020) and others. However, countries like Algeria, Turkey have found a good relationship and some of the countries like China do not have any relationship between COVID-19 spreads and the PD. The virus is new, and we only know a few details regarding its properties. Authorities want to know when the pandemic will end and whether it will get worse.

Forecasting is thus crucial, even if the least indication can help to save the lives of people or aware them of the future situation. Hence, ML-based models must be deployed for the forecast. ELM (Huang et al. 2004) is a puissant, quick and effective ML model for prediction. Huynh et al. (2008) projected an improved ELM model for condensed single-layer feed-forward networks. Javid et al. (2020) adopted the ELM using a sliding window approach for COVID-19 TS foretelling and also to handle the non-stationary property in the time-series dataset. Yiu and Özkale (2020) suggested an enhanced ELM model based on Liu regression. Kadi and Khelfaoui (2020) in their work highlighted a few cities in Algeria with the highest COVID-19 disease-ridden cases and the maximum population densities using cluster analysis. They found a substantial link between population density (PD) and the total COVID-19 infected individuals. Thus, it is concluded that PD has a beneficial impact on COVID-19 spread. Inspired from the works of Kadi and Khelfaoui (2020) and Huang et al. (2004), the primary contributions of this work can be represented as follows:

1. An empirical study has been done to obtain the relationship between the zone-based population densities and the number of infected cases in India.
2. A state-wise end prediction is also forecasted using the popular ELM model.
3. Further, the COVID-19 end prediction for India is also shown using the ELM model.

The remaining paper is arranged as follows. Section 11.2 labels the data collection. Section 11.3 briefly explains the prediction models. Experimental analysis and state-wise prediction models are described in Sect. 11.4 and Sect. 11.5. Section 6 illustrates the relationship between PD and the number of infected cases. Section 7 describes the conclusion of the work.

11.1.1 The TSVR

Peng (2010) developed a novel TSVR model which seeks for the insensitive up-bound as well as the low-bound features to train two small quadratic programming problems (QPPs).

Consider, $\mathbf{T} = \{(\mathbf{x}_1, \mathbf{y}_1), (\mathbf{x}_2, \mathbf{y}_2), \ldots, (\mathbf{x}_n, \mathbf{y}_n)\}$ is the matrix of training examples where, $\mathbf{x}_i$ and $\mathbf{y}_i$ is the row and column matrix respectively. Further, consider $\mathbf{U} = (\mathbf{x}_1, \mathbf{x}_2, \ldots .\mathbf{x}_n)$ as well as $\mathbf{Y} = (\mathbf{y}_1, \mathbf{y}_2, \ldots \mathbf{y}_n)$. TSVR seeks for a pair of hyperplanes as –

$$f_1(\mathbf{x}) = k(\mathbf{x}^t, \mathbf{U}^t)\mathbf{u}_1 + \boldsymbol{\beta}_1 \quad \text{and} \quad f_2(\mathbf{x}) = k(\mathbf{x}^t, \mathbf{U}^t)\mathbf{u}_2 + \boldsymbol{\beta}_2.$$

Here, $\boldsymbol{\beta}_1, \boldsymbol{\beta}_2 \in R$ and $\mathbf{u}_1, \mathbf{u}_2 \in R^n$ are the vector of unknowns. $k(\mathbf{x}^t, \mathbf{U}^t)$ indicates the RBF kernel function. The primal expressions of TSVR can be stated as (Hazarika et al. 2020):

$$\min_{w_1,b_1,\xi_1} \frac{1}{2}\left\|\mathbf{Y} - \varepsilon_1\mathbf{e} - (k(\mathbf{x}^t, \mathbf{U}^t)\mathbf{u}_1 + \mathbf{e}\boldsymbol{\beta}_1)\right\|^2 + C_1\mathbf{e}^t\kappa_1,$$
$$s.t., \ \mathbf{Y} - (k(\mathbf{x}^t, \mathbf{U}^t)\mathbf{u}_1 + \boldsymbol{\beta}_1\mathbf{e}) \geq \varepsilon_1\mathbf{e} - \kappa_1, \ \kappa_1 \geq 0 \tag{11.1}$$

and

$$\min_{w_2,b_2,\xi_2} \frac{1}{2}\left\|\mathbf{Y} - \varepsilon_2\mathbf{e} - (k(\mathbf{x}^t, \mathbf{U}^t)\mathbf{u}_2 + \mathbf{e}\boldsymbol{\beta}_2)\right\|^2 + C_2\mathbf{e}^t\kappa_2,$$
$$s.t., \ (k(\mathbf{x}^t, \mathbf{U}^t)\mathbf{u}_2 + \boldsymbol{\beta}_2\mathbf{e}) - \mathbf{Y} \geq \varepsilon_2\mathbf{e} - \kappa_2, \ \kappa_2 \geq 0, \tag{11.2}$$

where, $C_1, C_2 > 0$ and $\varepsilon_1, \varepsilon_2 \geq 0$ represents user-defined parameters. κ_1 and κ_2 are slack vectors. **e** is the vector of ones. The final regression function of TSVR for an unknown input datapoint $\mathbf{x} \in R^N$ can be expressed as:

$$f(x) = \frac{1}{2}(\mathbf{u}_1 + \mathbf{u}_2)^t k(\mathbf{U}, \mathbf{x}) + \frac{1}{2}(\boldsymbol{\beta}_1 + \boldsymbol{\beta}_2). \tag{11.3}$$

11.1.2 The LDMR

Let the training sample $\{(\mathbf{x}_l, \mathbf{y}_l)\}_{l=1,2,\ldots,m}$ where input values and executed values are $\mathbf{x}_l = (\mathbf{x}_{l_2}, \mathbf{x}_{l_2}, \ldots, \mathbf{x}_{l_n}) \in R^n$ and $\mathbf{y}_l \in R$ respectively. Suppose, $\mathbf{U} = (\mathbf{U}_1, \mathbf{U}_2, \ldots, \mathbf{U}_n) \in R^{m\times n}$ is an input matrix with l^{th} value as $\mathbf{x}_l^t$ and $\mathbf{Y} = (\mathbf{y}_1, \mathbf{y}_2, \ldots, \mathbf{y}_n)$ is the resultant matrix. LDMR searches for a function $f(x) = k(\mathbf{x}^t, \mathbf{U}^t)\mathbf{w} + \boldsymbol{\beta}$.

The primal expression of LDMR (Rastogi et al. 2020; Hazarika et al. 2021) can be stated as:

$$\min_{(w,b,\xi_1,\xi_2)} \frac{c}{2}\left\|\mathbf{Y} - (K(\mathbf{U}, \mathbf{U}^t)\mathbf{u} + \mathbf{e}\boldsymbol{\beta})\right\|^2 + \frac{1}{2}C\mathbf{e}^t(\kappa_1 + \kappa_2) + \frac{\theta}{2}\|\mathbf{u}\|^2,$$
$$s.t., \ \mathbf{Y} - (K(\mathbf{U}, \mathbf{U}^t)\mathbf{u} + \mathbf{e}\boldsymbol{\beta}) \leq \mathbf{e}\varepsilon + \kappa_1, \ \kappa_1 \geq 0 \tag{11.4}$$

where $\varepsilon, c, \theta > 0$ are input parameters and $\mathbf{w} = \begin{bmatrix}\mathbf{u}\\ \boldsymbol{\beta}\end{bmatrix}$; $\|\mathbf{u}\|^2 = \mathbf{w}^t I_0 \mathbf{w}$ where

$\mathbf{I}_0 = \begin{bmatrix}\mathbf{I} & 0 & \\ \cdot & \cdot & \\ \cdot & \cdot & \\ 0 \ldots 0 & & \end{bmatrix}$; $\boldsymbol{I}^{m\times m}$ is the matrix of identity.

The decision function for LDMR for an unknown datapoint $\mathbf{x} \in R^N$ can be expressed as:

$$f(\mathbf{x}) = K(\mathbf{x}^t, \mathbf{U}^t)\mathbf{u} + \boldsymbol{\beta}. \tag{11.5}$$

11.1.3 The ELM

The ELM classifier has the basic advantage that the hidden layer (HL) weights need not be adjusted for the SLFNs (Huang et al. 2004; Gupta et al. 2020). ELM is a fast model based on least squares and it shows high generalization ability.

Let, the training data be $\{(\mathbf{x}_1, \mathbf{y}_1), \ldots, (\mathbf{x}_m, \mathbf{y}_m)\}$ (Huang et al. 2006), where the input samples and the corresponding output data-points are signified by $\mathbf{x}_i \in R^n$ and $\mathbf{y}_i \in R$ respectively for $i = 1, 2, \ldots, m$, where m specifies the total training data samples. The weights amid the HL, **H** and output vector i.e., **u**, are generated by resolving a system of linear equations:

$$\mathbf{u} = \mathrm{H}^{\dagger}\mathbf{y}, \tag{11.6}$$

where $\mathrm{H}^{\dagger}$ indicates the Moore–Penrose generalized pseudoinverse of **H**. Finally, the regressor of ELM for a new example $\mathbf{x} \in R^N$ can be expressed as:

$$f(\mathbf{x}) = (\mathbf{U}(\mathbf{a}_1, \boldsymbol{\beta}_1, \mathbf{x}), \ldots, \mathbf{U}(\mathbf{a}_L, \boldsymbol{\beta}_L, \mathbf{x}))\mathbf{u}. \tag{11.7}$$

11.2 Experimental Setup and Analysis

In this research work, the required data is collected from: https://api.COVID19india.org/, https://www.telangana.gov.in/About/State-Profile, https://www.censusindia.gov.in/2011Census, https://www.censusindia.gov.in/2011census/PCA/PCA_Highlights/pca_highlights_file/India/Chapter-1.pdf, https://www.who.int/, https://api.COVID19india.org, https://www.mygov.in, https://en.wikipedia.org/wiki/Administrative_divisions_of_India.

However, Lakshadweep is not included in our study due to the non-availability of COVID 19 data. The optimum values of the model parameters are obtained using the tenfold CV method. The key advantage of this method is that every portion of the dataset is trained at least once. The efficiency of the models is computed using root mean square error (RMSE) and R^2 which can be given by the following equations:

$$RMSE = \sqrt{\frac{1}{N}\sum_{i=1}^{N}(f_i - o_i)^2} \tag{11.8}$$

$$R^2 = \frac{\sum_{i=1}^{N}(o_i f_i) - (\sum o)(\sum f)}{\sqrt{\left[\sum_{i=1}^{N} o_i^2 - (\sum o)^2\right]\left[\sum_{i=1}^{N} f_i^2 - (\sum f)^2\right]}} \tag{11.9}$$

Table 11.1 Prediction errors based on RMSE and R^2

Country	RMSE			R^2		
India	TSVR	LDMR	ELM	TSVR	LDMR	ELM
	0.24783	0.0659	0.06140	0.61679	0.9941	0.9974

where, N is nothing but the total amount of samples. O is an original and f represents the predicted samples. In this work, we use the multiquadric activation function for ELM. The multiquadric activation function can be defined as:

$$\text{Multiquadric}: f(\boldsymbol{\alpha}, \boldsymbol{\beta}, \mathbf{x}) = \sqrt{\|\mathbf{x} - \boldsymbol{\alpha}\|_2^2 + \boldsymbol{\beta}^2} \tag{11.10}$$

where $f(\boldsymbol{\alpha}, \boldsymbol{\beta}, \mathbf{x})$ denotes the resultant function for $\boldsymbol{\alpha}, x$ and $\boldsymbol{\beta}$. The Gaussian kernel is selected for TSVR and LDMR which can be denoted by $k(\mathbf{x}_i, \mathbf{x}_j) = \exp\left(-\mu\|\mathbf{x}_i - \mathbf{x}_j\|^2\right)$ where $i, j = 1, \ldots, m$ and $\mu > 0$ is the kernel parameter. $\|\cdot\|$ is the norm. In TSVR, the parameters (ps) C and μ are carefully chosen from $\{10\text{^} - 5, 10\text{^} - 4, \ldots, 10\text{^}5\}$ and $\{2\text{^} - 5, 2\text{^} - 4, \ldots, 2\text{^}5\}$ correspondingly. For ELM, the hidden layer ps are selected from $L = \{10, 20, 40, 50, 80, 100, 200, 500\}$. For LDMR the C, θ ps are chosen from $\{10^{-5}, \ldots, 10^5\}$ and μ is taken from $\{2^{-5}, \ldots, 2^5\}$. The ε parameter is taken as 0.1 for TSVR and LDMR. The parameter c of LDMR is considered as 1.

Table 11.1 illustrates the experimental outcomes on the INDIA dataset. From the above Table 11.1, we observe that ELM finds the best fitting distribution model than LDMR and TSVR for the data corresponding to COVID-19. The prediction plot is shown in Fig. 11.1 and forecasting in Fig. 11.2 where X-axis presents the original and Y-axis represent the predicted value.

11.3 Zone-Based Prediction for Indian States

In 2021, the population of India is more than 1.4 billion with the highest populated country. PD of India is 382 per sq. km and a decadal growth rate of 17.64. As per UN estimates, India registered a higher growth rate during 2000–2010 with an annual grew rate of 1.23% world population. In this research, we have studied the different zones in India for our experiments. The various zone-wise distribution is given below:

I. Southern Zone: Andhra Pradesh, Goa, Kerala, Karnataka, Lakshadweep, Puducherry, Tamil Nadu and Telangana,
II. Eastern Zone: A & N Islands, Bihar, Jharkhand, Odisha and West Bengal,
III. North Eastern Zone: Assam, Arunachal Pradesh, Nagaland, Mizoram, Manipur, Sikkim, Tripura and Meghalaya.
IV. Northern Zone: Chandigarh, Delhi, Haryana, Himachal Pradesh, Jammu &Kashmir, Ladakh, Punjab, and Rajasthan

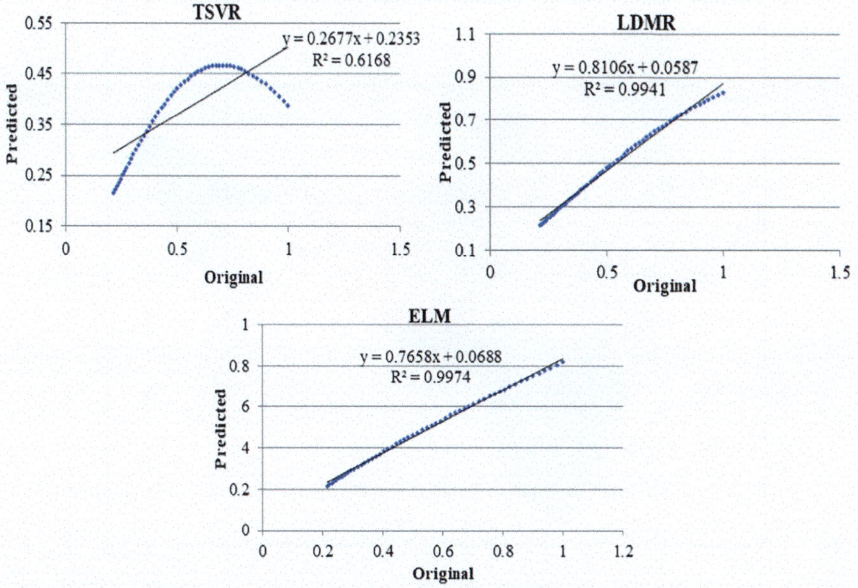

Fig. 11.1 Prediction plot over INDIA dataset using TSVR, LDMR and ELM

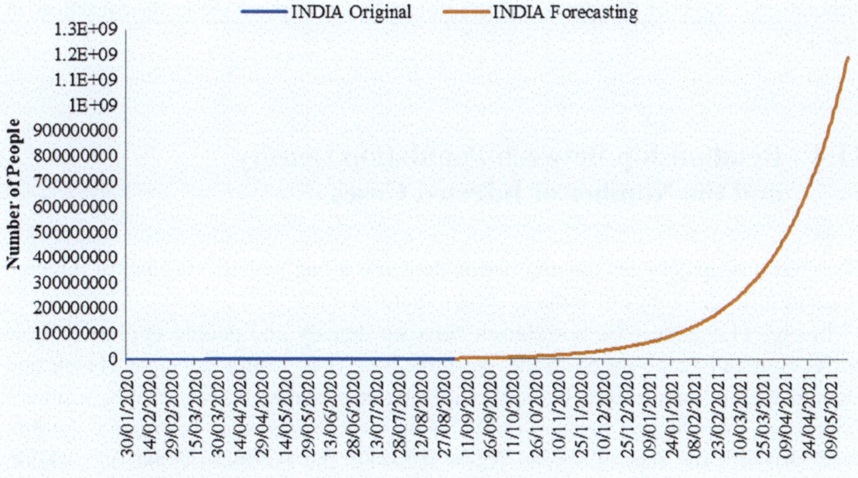

Fig. 11.2 End forecasting for India

V. Western Zone: Daman & Diu and D & N Haveli, Maharashtra and Gujarat
VI. Central Zone: Chhattisgarh, Madhya Pradesh, Uttarakhand and Uttar Pradesh

The geographical map of India showing different zones is shown in Fig. 11.3.

From the above Table 11.2, we have predicted the number of days along with the date of the last case to reach 98.616% (99% approximate) of the census 2011

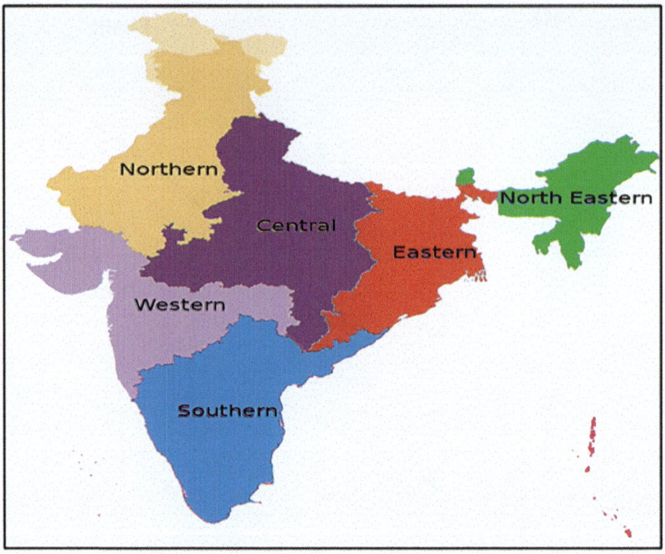

Fig. 11.3 Geographical map showing different zones in India (https://commons.wikimedia.org/wiki/File:Zonal_Councils.svg)

population. Graph of the states and union territories as per zone is shown below in Fig. 11.4.

11.4 Relationship Between Population Density and the Number of Infected Cases

The relationship between the population densities along with the number of infected cases in different zones of India is shown in Table 11.3.

In Fig, 11.5 zone-wise correlation between density and people is plotted. Low correlation can be observed in the Western Zone (i.e., Dadra & Nagar Haveli and Daman & Diu, Maharashtra, Gujarat) with the rate of spread R2 = 0.27538, Southern Zone (i.e., Puducherry, Kerala, Tamil Nadu, Goa, Karnataka, Telengana, Andhra Pradesh) with the rate of spread R2 = 0.24882 and Northern Zone (i.e., Delhi, Chandigarh, Haryana, Punjab, Rajasthan, Himachal Pradesh, Jammu and Kashmir) with the rate of spread R2 = 0.23766.

Table 11.2 Predicted infection population along with no. of days, RMSE and R^2

Zones	State/Union Territory	Estimated date of infection	Estimated infected population	Number of days From 4th Sept 2020	RMSE	R^2
Southern	Puducherry	26–12-2020	1,247,700	114	0.23022	0.8967
	Kerala	4/3/2021	32,546,900	182	0.28599	0.9836
	Tamil Nadu	18–07-2021	71,194,400	318	0.02216	0.9998
	Goa	30–01-2021	1,450,270	149	0.13759	0.9991
	Karnataka	22–02-2021	60,019,500	172	0.52775	0.7211
	Telengana	15–05-2021	34,442,700	254	0.32751	0.8705
	Andhra Pradesh	6/2/2021	49,561,000	156	0.35405	0.9944
Eastern	Bihar	18–06-2021	103,595,000	288	0.2612	0.9979
	West Bengal	6/6/2021	89,811,800	276	0.02462	0.9995
	Jharkhand	18–02-2021	32,587,400	168	0.84885	0.7468
	Odisha	17–02-2021	40,824,400	167	0.70577	0.8896
	Andaman and Nicobar Islands	8/4/2021	376,443	217	0.46119	0.8301
North Eastern	Assam	1/5/2021	30,669,900	240	0.13679	0.9892
	Tripura	17–02-2021	3,590,650	167	0.72964	0.9569
	Meghalaya	24–04-2021	2,960,760	233	0.20538	0.7488
	Manipur	12/6/2021	2,570,310	282	0.3894	0.9805
	Nagaland	31–12-2021	1,975,650	484	0.18753	0.9896
	Sikkim	2/6/2021	610,567	272	0.088	0.9829
	Mizoram	16–10-2021	1,087,660	408	0.03096	0.9837
	Arunachal Pradesh	10/4/2021	1,353,810	219	0.17566	0.9573
Northern	Delhi	28–07-2022	16,765,600	692	0.01922	0.9991
	Chandigarh	17–12-2020	1,010,840	105	0.07161	0.9987
	Haryana	14–06-2021	24,923,800	284	0.01858	0.9999
	Punjab	17–07-2021	27,708,300	317	0.32425	0.9812
	Rajasthan	17–08-2021	68,381,800	348	0.21526	0.9009
	Himachal Pradesh	25–05-2021	6,863,830	264	0.11996	0.9973
	Jammu and Kashmir	4/7/2021	12,095,500	304	0.13871	0.9952
	Ladakh	22–04-2021	273,614	231	0.23067	0.9746
Western	DNH and Daman and Diu	21–12-2021	585,579	474	0.02718	0.9991
	Maharashtra	6/6/2021	111,896,000	276	0.31668	0.8485
	Gujarat	20–03-2022	60,237,400	563	0.07047	0.9978

(continued)

Table 11.2 (continued)

Zones	State/Union Territory	Estimated date of infection	Estimated infected population	Number of days From 4th Sept 2020	RMSE	R^2
Central	Uttar Pradesh	29–04-2021	194,932,000	238	0.26917	0.7295
	Madhya Pradesh	6/7/2021	71,053,800	306	0.05332	0.9964
	Uttarakhand	12/3/2021	9,969,490	190	0.16796	0.9883
	Chhattisgarh	19–01-2021	25,061,900	138	0.38698	0.9986

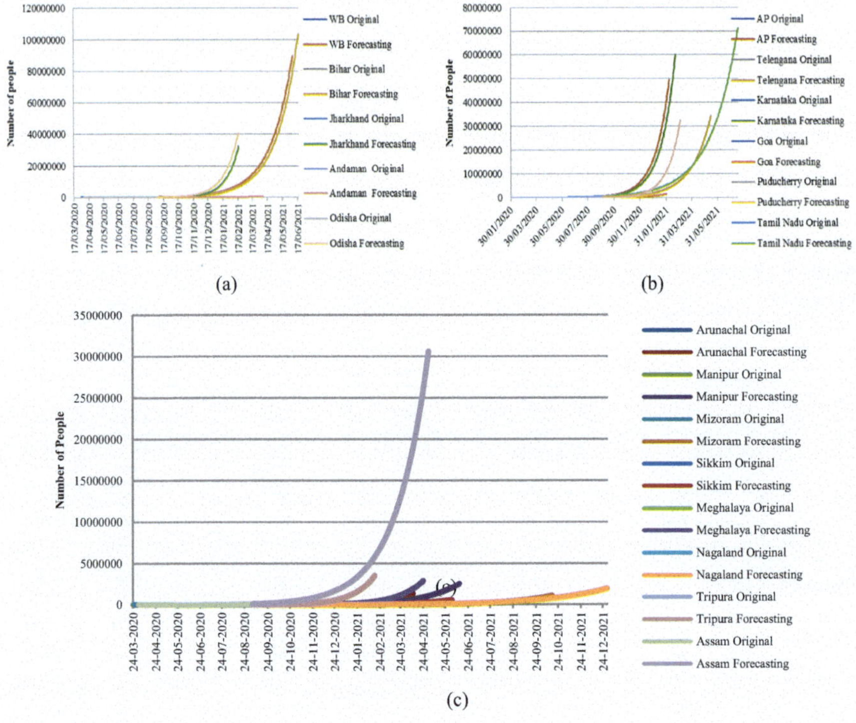

Fig. 11.4 End forecasting for **a** Southern, **b** Eastern **c** North-east **d** Northern **e** Western and **f** Central zone in India

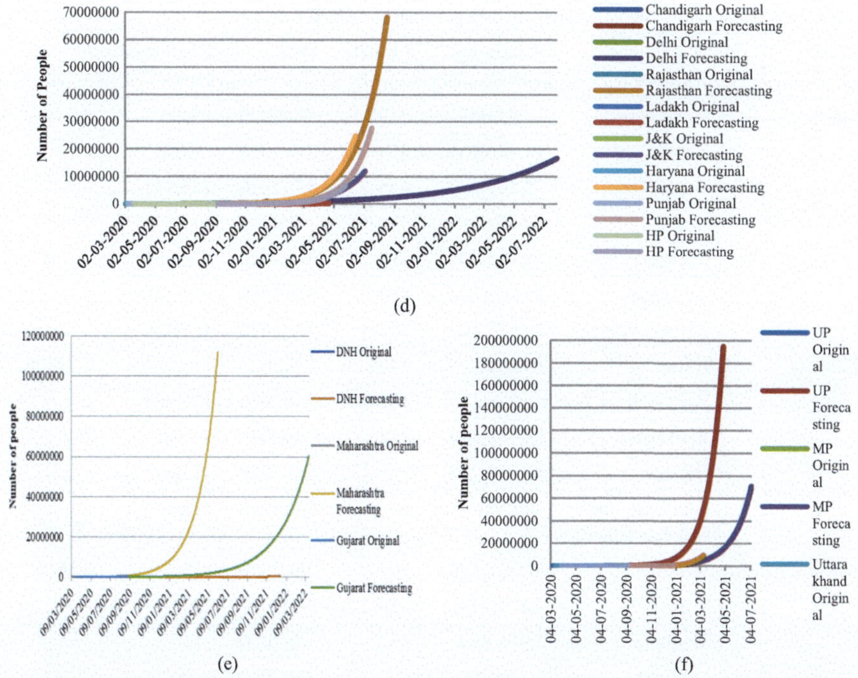

Fig. 11.4 (continued)

11.5 Conclusion

The COVID-19 pandemic has spread all over the globe. As of April 2020, it had expanded to over 180 nations. Scientists have worked tirelessly to discover a cure. ML is being used widely for COVID-19 analysis. In this research study, the forecast of COVID-19 using ELM is analyzed. It can be noticed that ELM is the best fitting distribution model among ELM, LDMR and TSVR for the data corresponding to COVID-19 of Indian states and union territories. This prediction might help manage the COVID-19. Further, we have studied the relationship between the population densities along with the total number of infected cases in India. A strong correlation can be noticed in the CZ, EZ and NE-Z of India. Further state-wise COVID-19 end forecasting is also portrayed. This work can be useful for machine learning researchers to estimate not only the end but also other cases and in different countries. If more datasets are provided, the model's performance could be boosted or that the model can deliver more accurate classification outcome. The model computes classified results based on data that has been obtained by health agencies. Forecasting may not be perfect, but it may certainly be utilized as a preventative step. However, there are several limits to our research, such as the optimum hidden layer nodes, tuning the uniform randomization range scaling, and reliable data availability. For the time being, the ELM network predicts using just samples of TS data. A potential

Table 11.3 Correlation between density and number of COVID-19

Zones	State/Union Territory	Population	Density	Date	Confirmed Cases	R^2
Southern	Puducherry	1,247,953	2598	3/9/2020	15,581	0.24882
	Kerala	33,406,061	859	3/9/2020	79,626	
	Tamil Nadu	72,147,030	555	3/9/2020	445,851	
	Goa	1,458,545	394	3/9/2020	19,355	
	Karnataka	61,095,297	319	3/9/2020	370,206	
	Telangana	35,003,674	312	3/9/2020	133,406	
	Andhra Pradesh	49,577,103	303	3/9/2020	465,730	
Eastern	Bihar	104,099,452	1102	3/9/2020	142,156	0.71538
	West Bengal	91,276,115	1029	3/9/2020	171,681	
	Jharkhand	32,988,134	414	3/9/2020	44,862	
	Odisha	41,974,218	269	3/9/2020	113,411	
	Andaman and Nicobar Islands	380,581	46	3/9/2020	3186	
North Eastern	Assam	31,205,576	397	3/9/2020	115,280	0.56669
	Tripura	3,673,917	350	3/9/2020	13,309	
	Meghalaya	2,966,889	132	3/9/2020	2517	
	Manipur	2,570,390	122	3/9/2020	6609	
	Nagaland	1,978,502	119	3/9/2020	4066	
	Sikkim	610,577	86	3/9/2020	1704	
	Mizoram	1,097,206	52	3/9/2020	1040	
	Arunachal Pradesh	1,383,727	17	3/9/2020	4360	
Northern	Delhi	16,787,941	11,297	3/9/2020	182,306	0.23766
	Chandigarh	1,055,450	9252	3/9/2020	5065	
	Haryana	25,351,462	573	3/9/2020	70,099	
	Punjab	27,743,338	550	3/9/2020	58,515	
	Rajasthan	68,548,437	201	3/9/2020	85,379	
	Himachal Pradesh	6,864,602	123	3/9/2020	6416	
	Jammu and Kashmir	12,267,032	297	3/9/2020	39,943	
	Ladakh	274,289	2.8	3/9/2020	2785	
Western	DNH and Daman and Diu	585,764	970	3/9/2020	2459	0.27538
	Maharashtra	112,374,333	365	3/9/2020	843,844	
	Gujarat	60,439,692	308	3/9/2020	100,375	
Central	Uttar Pradesh	199,812,341	828	3/9/2020	247,101	0.98448

(continued)

Table 11.3 (continued)

Zones	State/Union Territory	Population	Density	Date	Confirmed Cases	R^2
	Madhya Pradesh	72,626,809	236	3/9/2020	68,586	
	Uttarakhand	10,086,292	189	3/9/2020	21,234	
	Chhattisgarh	25,545,198	189	3/9/2020	36,520	

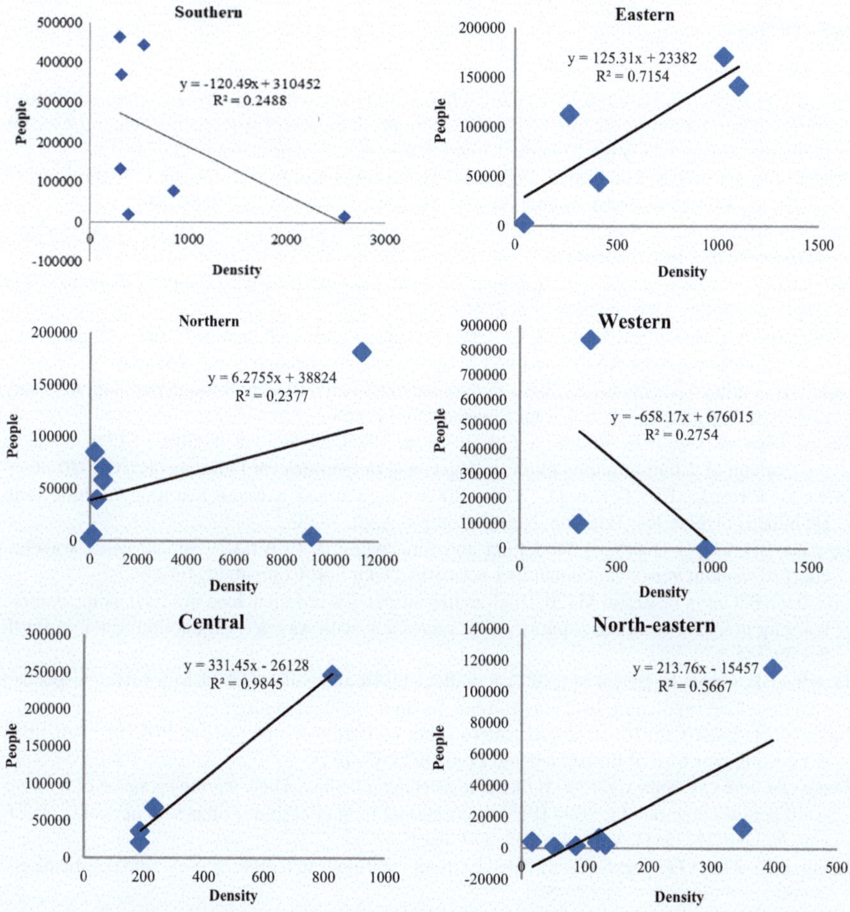

Fig. 11.5 Correlation between density and infected population for Southern, Eastern, Central zone, North-east, Northern and Western

future approach is to include more qualities such as the number of death due to COVID-19, vaccinated or not vaccinated information in time series data so that the data may be mined for the most information possible and the prediction performance can be enhanced. Moreover, an enhanced ELM approach can be adopted in the future. In addition to that, few managerial implications can be undertaken.

References

Afshordi N, Holder B, Bahrami M, Lichtblau D (2020) Diverse local epidemics reveal the distinct effects of population density, demographics, climate, depletion of susceptibles, and intervention in the first wave of COVID-19 in the United States. arXiv preprint arXiv:2007.00159

Ardabili SF, Mosavi A, Ghamisi P, Ferdinand F, Varkonyi-Koczy AR, Reuter U, Atkinson PM (2020) COVID-19 outbreak prediction with ML. Available at SSRN: 3580188

Bhadra A, Mukherjee A, Sarkar K (2020) Impact of population density on COVID 19 infected and mortality rate in India. medRxiv

Chi-Hsien K, Nagasawa S (2019) Applying machine learning to market analysis: knowing your luxury consumer. J Manag Analyt 6(4):404–419

Chowdhury AA, Hasan KT, Hoque KKS (2021) Analysis and prediction of COVID-19 pandemic in bangladesh by using ANFIS and LSTM network. Cogn Comput 13(3):761–770

Coşkun H, Yıldırım N, Gündüz S (2020) The spread of COVID-19 virus through population density and wind in Turkey cities. Sci Total Environ 751:141663

Das A, Dhar A, Goyal S, Kundu A (2020) Covid-19: analysis of a modified SEIR model, a comparison of different intervention strategies and projections for India. medRxiv, 2020-06

Gupta D, Hazarika BB, Berlin M (2020) Robust regularized extreme learning machine with asymmetric Huber loss function. Neural Comput Appl, 1–28

Hazarika BB, Gupta D (2020) Modelling and forecasting of COVID-19 spread using wavelet-coupled random vector functional link networks. Appl Soft Computing, 106626

Hazarika BB, Gupta D, Berlin M (2020) Modeling suspended sediment load in a river using extreme learning machine and twin support vector regression with wavelet conjunction. Environ Earth Sci 79:234

Hazarika BB, Gupta D, Berlin M (2021) A coiflet LDMR and coiflet OB-ELM for river suspended sediment load prediction. Int J Environ Sci Technol 18(9):2675–2692

Herlawati H (2020) COVID-19 spread pattern using support vector regression. PIKSEL: Penelitian Ilmu Komputer Sistem Embedded and Logic 8(1):67–74

Huang GB, Zhu QY, Siew CK (2004) Extreme learning machine: a new learning scheme of feedforward neural networks. In: 2004 IEEE international joint conference on neural networks (IEEE Cat. No. 04CH37541), Vol 2, pp 985–990. IEEE

Huang GB, Zhu QY, Siew CK (2006) Extreme learning machine: theory and applications. Neurocomputing 70(1–3):489–501

Huynh HT, Won Y, Kim JJ (2008) An improvement of extreme learning machine for compact single-hidden-layer feedforward neural networks. Int J Neural Syst 18(05):433–441

Javaid M, Haleem A, Singh RP, Haq MIU, Raina A, Suman R (2020) Industry 5.0: potential applications in COVID-19. J Indus Integrat Manag 5(4):507–530

Javid AM, Liang X, Venkitaraman A, Chatterjee S (2020) Predictive analysis of COVID-19 time-series data from Johns Hopkins University. arXiv preprint arXiv:2005.05060

Kadi N, Khelfaoui M (2020) Population density, a factor in the spread of COVID-19 in Algeria: statistic study. Bull Nat Res Centre 44:1–7

Lu Y, Williams TL (2021) Modeling analytics in COVID-19: prediction, prevention, control, and evaluation. J Manag Analyt 8(3):424–442

Murugan R, Goel T, E-DiCoNet: Extreme learning machine based classifier for diagnosis of COVID-19 using deep convolutional network. J Ambient Intell Humanized Comput, 1–12
Ndiaye BM, Tendeng L, Seck D (2020) Analysis of the COVID-19 pandemic by SIR model and machine learning technics for forecasting. arXiv preprint arXiv:2004.01574
Parbat D, Chakraborty M (2020) A python based support vector regression model for prediction of COVID19 cases in India. Chaos, Solitons Fractals 138:109942
Peng X (2010) TSVR: an efficient twin support vector machine for regression. Neural Netw 23(3):365–372
Priyanka EB, Thangavel S, Gao XZ, Sivakumar NS (2021) Digital twin for oil pipeline risk estimation using prognostic and machine learning techniques. J Indus Info Integrat, 100272
Rastogi R, Anand P, Chandra S (2020) Large-margin distribution machine-based regression. Neural Comput Appl 32(8):3633–3648
Satu M, Howlader KC, Islam SMS (2020) Machine learning-based approaches for forecasting COVID-19 cases in Bangladesh. Machine learning-based approaches for forecasting COVID-19 cases in Bangladesh
Singh RP, Haleem A, Javaid M, Kataria R, Singhal S (2021) Cloud computing in solving problems of COVID-19 pandemic. J Indus Integrat Manag
Sun Z, Zhang H, Yang Y, Wan H, Wang Y (2020) Impacts of geographic factors and population density on the COVID-19 spreading under the lockdown policies of China. Sci Total Environ 746:141347
Tandon H, Ranjan P, Chakraborty T, Suhag V (2020) Coronavirus (COVID-19): ARIMA based time-series analysis to forecast near future. arXiv preprint arXiv:2004.07859
Wanigasekara C, Oromiehie E, Swain A, Prusty BG, Nguang SK (2021) Machine learning-based inverse predictive model for AFP based thermoplastic composites. J Ind Inf Integr 22:100197
Yıldırım H, Özkale MR (2020) An enhanced extreme learning machine based on liu regression. Neural Process Lett 52:421–442

Chapter 12
A Machine Learning-Based Analysis of Internet Addiction Among Children and Adolescents During Covid-19 Lockdown

Satyananda Swain, Manas Ranjan Patra, and Bonomali Khuntia

Abstract The increase in Internet usage during the lockdown period seems to have a serious impact on people in different age groups. This has led to unstable mental conditions, depression, and stress, anxiety, and many other similar mental illnesses among individuals. In this study, we explore the use of machine learning classification algorithms to detect and classify Chinese children and adolescents with internet addiction during the Covid-19 period based on Young's Internet Addiction Test (IAT-20 test). Using a dataset of 2050 participants, an attempt has been made to classify their addiction behavior into different addiction levels (Normal, Problematic, and Addictive). The classification algorithms considered are KNN, Naive Bayes, Decision Tree, and Support Vector Machines. It is observed that SVM is the most reliable method for the classification of Internet addiction with an accuracy of 0.987. The goal of the study is to build a classification model for the accurate categorization of unknown samples into appropriate internet addiction levels.

Keywords Decision Tree · IAT-20 · KNN · Naive Bayes · Support Vector Machines

S. Swain · M. R. Patra · B. Khuntia (✉)
Department of Computer Science and Engineering, Centurion University, Bhubaneswar, India
e-mail: bonomalikhuntia@gmail.com

S. Swain
e-mail: satyananda.swain@cutm.ac.in

M. R. Patra
e-mail: mrpatra@nist.edu

M. R. Patra
Department of Computer Science & Engg, NIST University, Berhampur, India

B. Khuntia
Department of Computer Science, Berhampur University, Berhampur, India

D. K. Sharma et al. (eds.), *Machine Learning for Real World Applications*, Transactions on Computer Systems and Networks, https://doi.org/10.1007/978-981-97-1900-6_12

12.1 Introduction

Nobody can deny the importance of the Internet in this modern era. Life without the internet is just unimaginable. But, continuous use of the internet can adversely affect the mental as well as physical health of individuals. In recent years, constant use of the internet has led to mental illnesses like stress, depression, and internet addiction, the most affected groups being children and adolescents (Shaikh et al. 2018; Paulus et al. 2018).

According to Young, internet addiction is a new psychological disorder (Young 2009). A test called, Internet Addiction Test (IAT) has been devised to measure the presence and severity of internet dependency among adults and adolescents on a 20-item scale. In Young's Internet Addiction Test (YIAT20), one is asked to answer the following questions on the Likert scale: 1 = rarely, 2 = occasionally, 3 = frequently, 4 = often, 5 = always. The questions used for the YIAT20 test are described in Table 12.1 (Young 2009).

The basic characteristics of YIAT-20 include the extent to which internet usage of students has an influence on their productivity, daily routine, sleeping pattern, social life, and feelings. The score ranges from 20 to 100. According to Young, a score of 20–39 points describes an average online user, whereas a score of 40–69 indicates frequent problems in a user due to internet usage, and a score of 70–100 reflects major problems due to heavy internet use.

12.2 Literature Review

Young revealed that addiction to the internet has become as severe as drug addiction. In many cases, internet addiction has led to deterioration in academic performance among young students, social distancing, and sometimes occupational impairments. Young's study investigates the extent to which one can be impacted by internet addiction (Young 2009).

Moreno, et al. studied the emergent area of problematic internet use among US youth. They systematically reviewed and studied the problems associated with Internet use in a population of US adolescents and college students in order to develop diagnostic criteria for internet addiction or problematic Internet use (Moreno et al. 2011).

Frangos et al. demonstrates the percentage of university students in Greek having internet addiction (Frangos et al. 2010).

S. Chen, et al. report their study on the development of the Chinese Internet addiction scale (CIAS). Using factor analysis, correlation analysis, and t-test, they have analyzed symptoms and related issues due to internet addiction of students. Through their work, they have shown that CIAS is an appropriate test to find internet addiction. A subsequent study provides the revised version of CIAS, called the CIAS-R framework (Chen et al. 2003).

Table 12.1 YIAT20 question set with question type

Question No	Questions
IAT 1	"How often do you find that you stay online longer than you intended?"
IAT 2	"How often do you neglect household chores to spend more time online?"
IAT 3	"How often do you prefer the excitement of the Internet to intimacy with your partner?"
IAT 4	"How often do you form new relationships with fellow online users?"
IAT 5	"How often do others in your life complain to you about the amount of time you spend online?"
IAT 6	"How often do your grades or school work suffer because of the amount of time you spend on-line?"
IAT 7	"How often do you check your email before something else that you need to do?"
IAT 8	"How often does your job performance or productivity suffer because of the Internet?"
IAT 9	"How often do you become defensive or secretive when anyone asks you what you do online?"
IAT 10	"How often do you block out disturbing thoughts about your life with soothing thoughts of the Internet?"
IAT 11	"How often do you find yourself anticipating when you will go online again?"
IAT 12	"How often do you fear that life without the Internet would be boring, empty, and joyless?"
IAT 13	"How often do you snap, yell, or act annoyed if someone bothers you while you are on-line?"
IAT 14	"How often do you lose sleep due to late-night log-ins?"
IAT 15	"How often do you feel preoccupied with the Internet when offline, or fantasize about being online?"
IAT 16	"How often do you find yourself saying "Just a few more minutes" when online?"
IAT 17	"How often do you try to cut down the amount of time you spend online and fail?"
IAT 18	"How often do you try to hide how long you've been online?"
IAT 19	"How often do you choose to spend more time online over going out with others?"
IAT 20	"How often do you feel depressed, moody, or nervous when you are offline, which goes away once you are back online?"

Source Adopted from Reference (Young 2009)

In (Young 1998), Young discusses the problem of the addictive potential of the internet in recent times. His work teaches how to guard against internet overuse.

Widyanto and Mcmurran observes that Egyptian adolescents are at high risk due to internet use, and are highly prone to Psychiatric disorders (Widyanto and McMurran 2004).

Adiele and Olatokun studied the presence of internet addiction among adolescents and tried to find whether it is different from offline behavioral addiction. They found that internet addiction was mainly due to extrinsic reasons (Adiele 2014).

Chakraborty et al. have done an interesting review of the literature on internet addiction during the period 1970–2010. Their study mainly focused on finding whether behavioral addictions are true addictions or not, what actually leads to internet addiction, and how one can diagnose internet addiction in a person (Chakraborty et al. 2010).

Zhao et al. observed too much internet usage among Chinese children and adolescents during the Covid-19 period. Based on their study on internet usage characteristics of Chinese participants, they tried to classify participants into different levels of internet addiction as normal, problematic, and addictive internet users (Zhao et al. 2017).

12.3 Methodology

12.3.1 Data Set Description

A survey was conducted among Chinese children and adolescents in the age group of 6–18 years. A total of 2050 participants responded to questionnaires containing Young's internet demographic information and other internet usage characteristics (Dong et al. 2020). The demographic measures include gender, age, DOB, frequency of internet use, and many others. Young's IAT was used to analyze symptoms of internet addiction. Accordingly, an IAT total score of 70 or above was classified as an addictive internet user, a total score of 40–69 was classified as a problematic user, and scores of 39 or below were classified as a normal internet user (Fig. 12.1).

```
In [38]: print(df1.head())

   IAT1  IAT2  IAT3  IAT4  IAT5  IAT6  IAT7  IAT8  IAT9  IAT10  ...  IAT12  \
0     2     2     2     2     2     2     2     2     2      1  ...      2
1     3     3     3     3     3     3     3     3     3      3  ...      3
2     2     2     1     1     1     1     1     1     2      1  ...      1
3     2     1     2     2     3     1     2     3     2      3  ...      2
4     2     1     1     1     2     1     1     1     3      2  ...      2

   IAT13  IAT14  IAT15  IAT16  IAT17  IAT18  IAT19  IAT20  \
0      1      2      2      2      2      2      2      2
1      3      3      3      3      3      3      3      3
2      1      1      1      1      1      1      1      1
3      2      2      2      3      1      2      1      2
4      1      2      1      2      2      1      1      2

   IAT_classification_3
0                     4
1                     1
2                     4
3                     4
4                     4

[5 rows x 21 columns]
```

Fig. 12.1 Data set description for IAT-20

12.3.2 Implementation Procedure

Step-1: The problem is described as a classification problem that attempts to classify persons with internet addiction into normal, problematic, or addictive users on the basis of the demographic data gathered through the IAT-20 test.

Step-2: The original data sample pertains to school-aged children and adolescents from three different areas (central, east, and north) of China. In addition to details like age, gender, education, parent details, income, etc., Young's Internet addiction test (YIAT-test) was used to observe different symptoms of Internet addiction. IAT is a self-rated test with 20 basic questions and each response is recorded on a 1–5-point scale. According to Young's criteria, participants were classified into Addictive (1), Problematic (2), and normal (3) user group labels.

Step-3: From the raw data available, a comprehensive analysis has been done to evaluate each feature, its respective values, and an attempt has been made to find parameters that can be used as classification criteria. The features that appeared prominent, promising, and selected for learning were the 20 items of the IAT test. Based on these features, classification labels were defined to categorize individuals into different class labels.

Step-4: After feature selection, the data set is split in the ratio 70:30 for training and testing, respectively.

Step-5: Various machine learning classification algorithms, viz., KNN, NB, SVM, and DT are employed for training and testing.

Step-6: The performance of the selected classification algorithms were measured in terms of accuracy, precision, recall, and F1 score.

Step-7: A comparative study of the classifiers is done to judge their relative performance on the dataset.

Step-8: The classification model is then applied to unlabeled data samples to categorize the samples into Normal, Problematic, or addictive user categories (Fig. 12.2).

12.3.3 Feature Reduction

In addition to evaluating different classification techniques, we also demonstrate the strategy of feature reduction to show the importance of the individual questions as well as their association with other questions for classification in the YIAT test. To check the significance of individual questions we strategically reduce one question at a time from the set of 20 questions and check the classification result to determine the impact of removing a question on classification. To check the association among the questions and to observe the impact that an association of questions can have on classification, we reduced one question at a time from the YIAT-20 question set and test it for classification in groups using the same classification algorithm. Results were evaluated and the effect of feature reduction on classification procedures was studied.

12.4 Result and Discussion

12.4.1 Classification Result

Experiments were conducted using Python programming and Jupyter notebook to implement different supervised machine learning algorithms, in order to study Internet addiction behavior in children and adolescents. The algorithms considered were SVM, KNN, NB, and DT. Classification matrices are developed in terms of true and false positives and true and false negatives. Table 12.2 depicts the performance of ML algorithms in terms of the following measurement parameters:

$$\text{Accuracy} = (\text{TP} + \text{TN})/(\text{TP} + \text{FP} + \text{FN} + \text{TN}). \tag{12.4.1}$$

$$\text{Precision} = \text{TP}/(\text{TP} + \text{FP}). \tag{12.4.2}$$

$$\text{Recall} = \text{TP}/(\text{TP} + \text{FN}). \tag{12.4.3}$$

$$\text{F1 Score} = 2 * (\text{Recall} * \text{Precision})/(\text{Recall} + \text{Precision}). \tag{12.4.4}$$

Support defines the number of samples with the true responses in each class of target values.

Overall, the SVM algorithm demonstrates the highest accuracy of 0.987 as compared to other classification algorithms. KNN and NB yield an accuracy of 0.924 and 0.917 respectively. The lowest performance was observed in the case of DT with a minimal accuracy score of 0.785.

Figure 12.3 provides a graphical representation of the comparison indicating the supremacy of SVM over other classifiers with an accuracy score of 0.987.

12.4.2 Impact of Feature Reduction

Next, experiments were conducted to study the impact of feature reduction on the classifiers by dropping one feature at a time (a question represents one feature). Since the classification accuracy of the DT algorithm is comparatively poor it is not considered in our subsequent experiments.

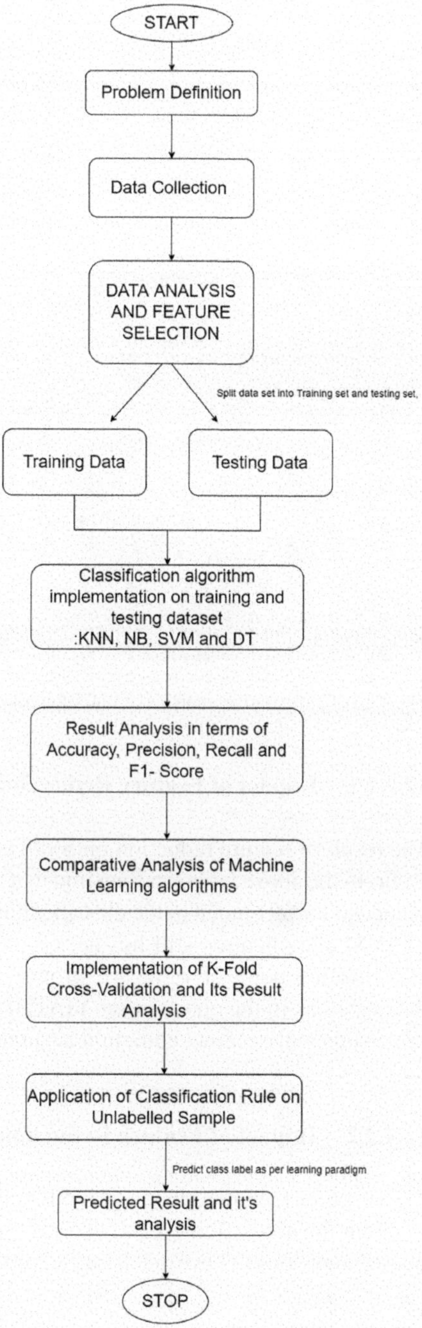

Fig. 12.2 Schematic diagram for internet addiction behavior classification

Table 12.2 Internet addiction behavior classification results using SVM, KNN, NB, and DT

Serial no	Algorithm/method name	Accuracy	Precision	Recall	F1 score	Support
1	SVM	0.987	0.987	0.987	0.987	615
2	KNN	0.924	0.924	0.924	0.922	615
3	NB	0.917	0.931	0.917	0.919	615
4	DT	0.785	0.783	0.785	0.783	615

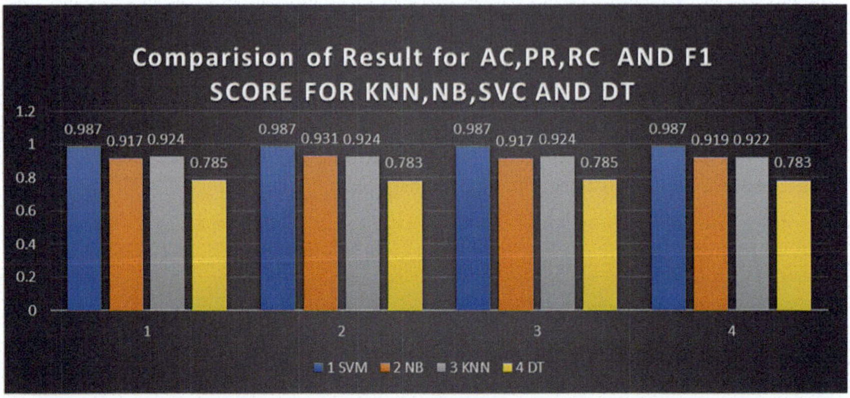

Fig. 12.3 Comparative analysis of classification results of SVM, KNN, NB, and DT

12.4.2.1 Impact of Feature Reduction on SVM

The result of feature reduction on SVM is depicted in Table 12.3 (Fig. 12.4).

From the above table, one can find that the removal of a question from the question set doesn't affect much to the classification results in the case of SVM. The accuracy of SVM classification still lies between 0.95 and 0.97 which is quite good from a classification point of view. We strongly recommend that the association between the questions in the question set YIAT20 test strongly determine the addiction level of persons with internet addiction accurately.

12.4.2.2 Impact of Feature Reduction on NB

The result of feature reduction on NB is depicted in Table 12.4.

From the above Table 12.4, one can find that the removal of a question from the question set doesn't have much impact on the classification done using NB (Fig. 12.5).

Table 12.3 Feature reduction results for SVM (reduction of one feature at a time)

Serial no	Removed feature Name	Accuracy	Precision	Recall	F1 score
1	IAT1	0.977	0.977	0.977	0.977
2	IAT2	0.972	0.973	0.972	0.972
3	IAT3	0.966	0.966	0.966	0.966
4	IAT4	0.971	0.971	0.971	0.971
5	IAT5	0.969	0.97	0.969	0.969
6	IAT6	0.977	0.978	0.977	0.977
7	IAT7	0.966	0.966	0.966	0.966
8	IAT8	0.976	0.976	0.976	0.976
9	IAT9	0.966	0.966	0.966	0.966
10	IAT10	0.958	0.958	0.958	0.958
11	IAT11	0.966	0.966	0.966	0.966
12	IAT12	0.964	0.964	0.964	0.964
13	IAT13	0.984	0.984	0.984	0.984
14	IAT14	0.971	0.971	0.971	0.971
15	IAT15	0.977	0.979	0.977	0.977
16	IAT16	0.977	0.977	0.977	0.977
17	IAT17	0.958	0.959	0.958	0.958
18	IAT18	0.964	0.965	0.964	0.964
19	IAT19	0.967	0.968	0.967	0.968
20	IAT20	0.974	0.974	0.974	0.974
21	No removal	0.987	0.987	0.987	0.987

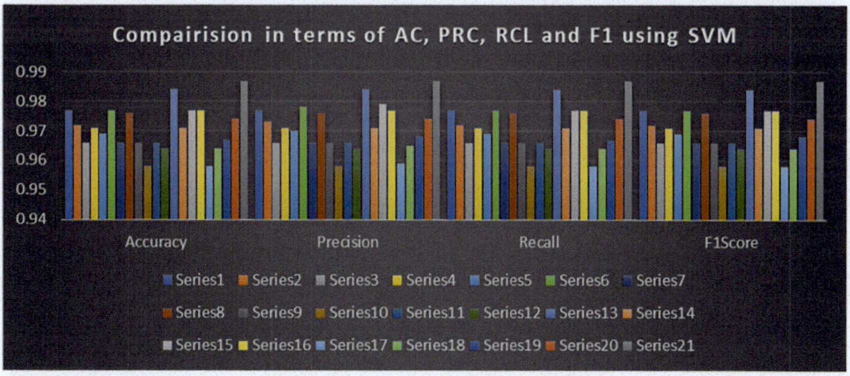

Fig. 12.4 Comparative analysis of feature reduction result for SVM

Table 12.4 Feature reduction results for NB (reduction of one feature at a time)

Serial no	Removed feature name	Accuracy	Precision	Recall	F1 score
1	IAT1	0.912	0.925	0.912	0.915
2	IAT2	0.915	0.929	0.915	0.918
3	IAT3	0.919	0.929	0.919	0.921
4	IAT4	0.919	0.93	0.919	0.921
5	IAT5	0.92	0.933	0.92	0.922
6	IAT6	0.919	0.931	0.919	0.921
7	IAT7	0.914	0.926	0.914	0.916
8	IAT8	0.924	0.936	0.924	0.926
9	IAT9	0.911	0.925	0.911	0.913
10	IAT10	0.912	0.926	0.912	0.915
11	IAT11	0.914	0.925	0.914	0.916
12	IAT12	0.915	0.929	0.915	0.918
13	IAT13	0.925	0.937	0.925	0.927
14	IAT14	0.924	0.932	0.924	0.925
15	IAT15	0.917	0.931	0.917	0.919
16	IAT16	0.912	0.926	0.912	0.915
17	IAT17	0.912	0.927	0.912	0.915
18	IAT18	0.917	0.931	0.917	0.919
19	IAT19	0.917	0.931	0.917	0.919
20	IAT20	0.92	0.934	0.92	0.923
21	No removal	0.917	0.931	0.917	0.919

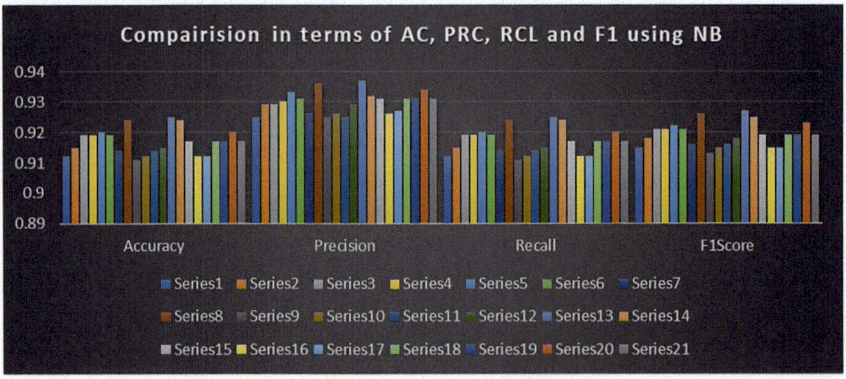

Fig. 12.5 Comparative analysis of feature reduction result for NB

Table 12.5 Feature reduction results for KNN (reduction of one feature at a time)

Serial no	Removed feature name	Accuracy	Precision	Recall	F1 score
1	IAT1	0.918	0.919	0.919	0.917
2	IAT2	0.909	0.909	0.909	0.907
3	IAT3	0.915	0.915	0.915	0.914
4	IAT4	0.909	0.909	0.909	0.908
5	IAT5	0.912	0.911	0.912	0.91
6	IAT6	0.92	0.92	0.92	0.919
7	IAT7	0.914	0.913	0.914	0.913
8	IAT8	0.912	0.912	0.912	0.911
9	IAT9	0.927	0.928	0.927	0.925
10	IAT10	0.909	0.908	0.909	0.907
11	IAT11	0.917	0.916	0.917	0.916
12	IAT12	0.917	0.917	0.917	0.916
13	IAT13	0.919	0.918	0.919	0.917
14	IAT14	0.911	0.911	0.911	0.909
15	IAT15	0.917	0.917	0.917	0.915
16	IAT16	0.907	0.908	0.907	0.905
17	IAT17	0.907	0.907	0.907	0.905
18	IAT18	0.92	0.92	0.92	0.919
19	IAT19	0.925	0.924	0.925	0.924
20	IAT20	0.919	0.918	0.919	0.917
21	No removal	0.924	0.924	0.924	0.922

12.4.2.3 Impact of Feature Reduction on KNN

The result of feature reduction on KNN is depicted in Table 12.5 (Fig. 12.6).

Once again removal of a question from the question set doesn't affect much the classification results in the case of KNN.

12.4.2.4 Impact of Feature Association on SVM

Next, experiments were conducted to study the impact of groups of features on the classifiers by considering a set of questions. The result of feature association on SVM is depicted in Table 12.6. The findings are depicted below.

From the above Table 12.6, one can find that removal of a set of questions (1 or more) from the question set radically affects the classification results in the case of SVM. The accuracy of SVM classification declines steadily from 0.97 to 0.73, which is noteworthy. We strongly recommend that the association between the questions in the question set YIAT20 test strongly determine the addiction level of persons

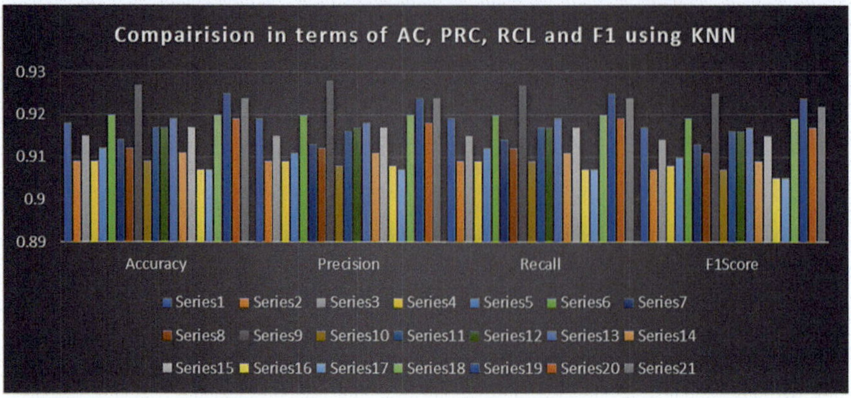

Fig. 12.6 Comparative analysis of feature Reduction result for KNN

Table 12.6 Feature association results for SVM (group reduction of feature)

Serial No	Considered feature name	Accuracy	Precision	Recall	F1 score
1	IAT1-IAT20	0.977	0.977	0.977	0.977
2	IAT2-IAT20	0.977	0.977	0.977	0.977
3	IAT3-IAT20	0.961	0.961	0.961	0.961
4	IAT4-IAT20	0.953	0.953	0.953	0.953
5	IAT5-IAT20	0.946	0.946	0.946	0.946
6	IAT6-IAT20	0.933	0.934	0.933	0.933
7	IAT7-IAT20	0.927	0.927	0.927	0.927
8	IAT8-IAT20	0.924	0.923	0.924	0.924
9	IAT9-IAT20	0.909	0.91	0.909	0.909
10	IAT10-IAT20	0.901	0.9	0.901	0.901
11	IAT11-IAT20	0.886	0.886	0.886	0.885
12	IAT12-IAT20	0.87	0.871	0.87	0.869
13	IAT13-IAT20	0.868	0.87	0.868	0.867
14	IAT14-IAT20	0.841	0.842	0.841	0.84
15	IAT15-IAT20	0.842	0.849	0.842	0.841
16	IAT16-IAT20	0.833	0.837	0.833	0.832
17	IAT17-IAT20	0.81	0.811	0.81	0.806
18	IAT18-IAT20	0.778	0.792	0.779	0.778
19	IAT19-IAT20	0.75	0.744	0.75	0.743
20	IAT20	0.739	0.742	0.74	0.729

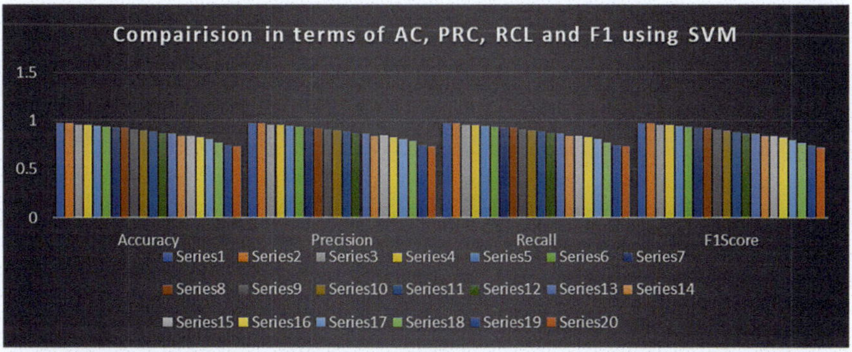

Fig. 12.7 Comparative analysis of feature reduction group results for SVC

with internet addiction accurately. However, if features are reduced gradually then the classification accuracy also falls drastically (Fig. 12.7).

12.4.2.5 Impact of Feature Association on NB

The result of feature association on NB is depicted in Table 12.7 (Fig. 12.8).

As evident from the above table, the classification accuracy of the NB classifier declines steadily from 0.91 to 0.74.

12.4.2.6 Impact of Feature Association on KNN

The result of feature association on KNN is depicted in Table 12.8 (Fig. 12.9).

The same is in the case of KNN where the classification accuracy falls steadily from 0.92 to 0.7, indicating the importance of the association of features.

Table 12.7 Feature association results for NB (group reduction of feature)

Serial no	Considered feature name	Accuracy	Precision	Recall	F1 score
1	IAT1-IAT20	0.917	0.931	0.917	0.919
2	IAT2-IAT20	0.912	0.925	0.912	0.915
3	IAT3-IAT20	0.911	0.922	0.911	0.913
4	IAT4-IAT20	0.911	0.921	0.911	0.913
5	IAT5-IAT20	0.909	0.919	0.909	0.911
6	IAT6-IAT20	0.904	0.915	0.904	0.907
7	IAT7-IAT20	0.896	0.906	0.896	0.899
8	IAT8-IAT20	0.894	0.903	0.894	0.897
9	IAT9-IAT20	0.885	0.894	0.885	0.887
10	IAT10-IAT20	0.88	0.89	0.88	0.882
11	IAT11-IAT20	0.863	0.872	0.863	0.866
12	IAT12-IAT20	0.863	0.87	0.863	0.865
13	IAT13-IAT20	0.857	0.863	0.857	0.859
14	IAT14-IAT20	0.859	0.863	0.859	0.86
15	IAT15-IAT20	0.854	0.854	0.854	0.854
16	IAT16-IAT20	0.837	0.835	0.837	0.836
17	IAT17-IAT20	0.808	0.802	0.808	0.803
18	IAT18-IAT20	0.793	0.785	0.793	0.787
19	IAT19-IAT20	0.766	0.754	0.766	0.755
20	IAT20	0.745	0.738	0.745	0.74

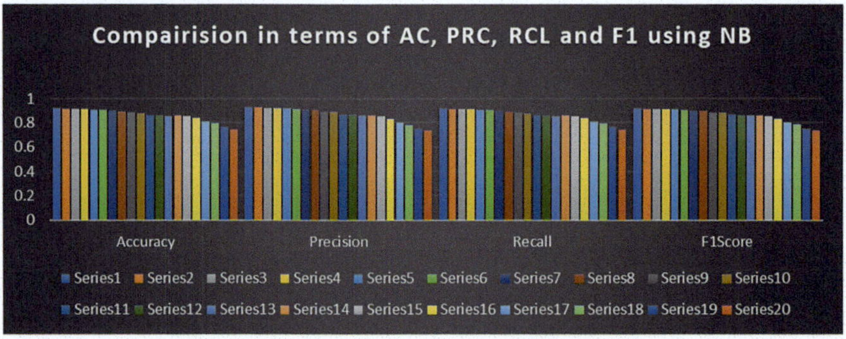

Fig. 12.8 Comparative analysis of feature reduction group results for NB

Table 12.8 Feature reduction results for KNN (group reduction of feature)

Serial no	Considered feature name	Accuracy	Precision	Recall	F1 score
1	IAT1-IAT20	0.924	0.924	0.924	0.922
2	IAT2-IAT20	0.919	0.919	0.919	0.917
3	IAT3-IAT20	0.917	0.916	0.917	0.915
4	IAT4-IAT20	0.906	0.905	0.906	0.904
5	IAT5-IAT20	0.898	0.896	0.898	0.895
7	IAT7-IAT20	0.881	0.88	0.881	0.879
8	IAT8-IAT20	0.893	0.893	0.893	0.891
9	IAT9-IAT20	0.885	0.883	0.885	0.881
10	IAT10-IAT20	0.873	0.873	0.873	0.87
11	IAT11-IAT20	0.846	0.841	0.846	0.841
12	IAT12-IAT20	0.846	0.844	0.846	0.842
13	IAT13-IAT20	0.844	0.843	0.844	0.84
14	IAT14-IAT20	0.826	0.835	0.826	0.823
15	IAT15-IAT20	0.839	0.839	0.839	0.834
16	IAT16-IAT20	0.813	0.814	0.813	0.81
17	IAT17-IAT20	0.798	0.805	0.798	0.797
18	IAT18-IAT20	0.784	0.785	0.784	0.774
19	IAT19-IAT20	0.754	0.75	0.754	0.751
20	IAT20	0.74	0.742	0.74	0.729

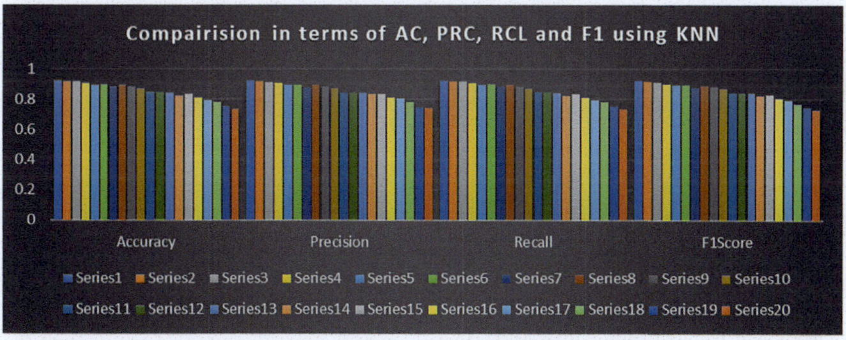

Fig. 12.9 Comparative analysis of feature Reduction group results for KNN

12.5 Conclusion

The COVID-19 Pandemic situation has completely changed the behavior of internet users because of the shift from offline to online modes of work. Our study investigates excessive internet use by Chinese children and adolescents during the lockdown period. We have classified Internet users into normal, problematic, and addictive user categories based on Young's Internet addiction test which comprises 20 questions related to the Internet activity of users. The classification algorithms considered were SVM, KNN, NB, and DT which provide interesting results with a maximum accuracy of 0.987 in the case of SVM. Thus, one can use these algorithms to classify new samples accurately for predicting internet addiction. Our next study will focus on the impact of other factors such as depression, anxiety, and stress in predicting internet addiction issues among children and adults.

References

Adiele WO (2014) Prevalence and determinants of Internet addiction among adolescents. Comput Human Behav 31:100–110

Chakraborty K, Basu D, Kumar KV (2010) Internet addiction: consensus, controversies, and the way ahead. 20(3):123–132

Chen SH, Weng LJ, Su YJ, Wu HM, Yang PF (2003) Development of a Chinese Internet addiction scale and its psychometric study. Chin J Psychol

Dong H, Yang F, Lu X, Hao W (2020) Internet addiction and related psychological factors among children and adolescents in China during the coronavirus disease 2019 (COVID-19) epidemic. Front Psych 11:751

Frangos CC, Fragkos KC, Kiohos A (2010) Internet addiction among Greek university students: demographic associations with the phenomenon, using the Greek version of Young's internet addiction test. Int J Econ Sci Appl Res 3(1):49–74

Moreno MA, Jelenchick L, Cox E, Young H, Christakis DA (2011) Problematic internet use among youth: a systematic review. Arch Pediatr Adolesc Med 165(9):797–805

Paulus FW, Ohmann S, Von Gontard A, Popow C (2018) Internet gaming disorder in children and adolescents: a systematic review. Dev Med Child Neurol 60(7):645–659

Shaikh BM, Doke PP, Gothankar JS (2018) Depression, anxiety, stress, and stressors among rural adolescents studying in Pune and a rural block of Nanded district of Maharashtra, India. Indian J Public Health 62(4):311

Widyanto L, McMurran M (2004) The psychometric properties of the internet addiction test. Cyberpsychol Behav 7(4):443–450

Young KS (1998) Caught in the net: how to recognize the signs of internet addiction—and a winning strategy for recovery. John Wiley & Sons

Young KS (2009) Internet addiction: the emergence of a new clinical disorder. Cyberpsychol Behav 1(3)

Zhao F, Zhang ZH, Bi L, Wu XS, Wang WJ, Li YF, Sun YH (2017) The association between life events and internet addiction among Chinese vocational school students: the mediating role of depression. Comput Hum Behav 70:30–38

Chapter 13
Building Architectural Styles Classification Using Convolutional Neural Networks Models

Aaron Rasheed Rababaah

Abstract An intelligent model for classifying architectural styles is presented in this chapter. Traditional machine vision methods face difficulties in image classification of architectural styles, particularly in the feature extraction phase where many visual features need to be extracted, refined, and optimized. This step is typically conducted manually in traditional Machine Learning (ML) models. However, with the emergence of Deep Learning (DL), these challenging operations can now be automated. In this study, a DL-based machine vision model was developed to assess the effectiveness of DL in solving the classification problem. The proposed model was trained and validated on a publicly available dataset containing over 5000 images of 8 different architectural styles. The results indicate that the proposed model achieved a classification accuracy of 95.44%, demonstrating its reliability.

Keywords Architectural styles classification · Machine intelligence · Machine vision · Deep learning · Feature extraction · Impact of number kernels/features

13.1 Introduction

Architectural style can be defined as a set of categories recognized and classified by architectural historians based on time, location, and context (Crisman 2022). Another interesting definition is given by (Archisoup 2021) as a set of factors that influence the characteristics of form, material, and construction method of a building. Also, Yoshimura et al. (2018), define it as the fundamental format to design a particular building in a given geographical location and in a historic era. For space brevity, we show four of the most prominent architectural styles explored in (Arora 2021): Classical architecture which spans 7th–4th BC century. These include Greek and Roman styles as shown in Fig. 13.1. The second style is NeoClassical Architecture which spans between 1750 and 1920 AD. Sample of this style is shown in Fig. 13.2.

A. R. Rababaah (✉)
American University of Kuwait, Salmiya, Kuwait
e-mail: haroun01@gmail.com

D. K. Sharma et al. (eds.), *Machine Learning for Real World Applications*, Transactions on Computer Systems and Networks, https://doi.org/10.1007/978-981-97-1900-6_13

The 3rd style is Modernism Architecture which spans between 1917 and 1965 AD and is depicted in Fig. 13.3. The fourth style is Parametricism architecture style which spans 1997-Present and is shown in Fig. 13.4.

Looking at the aforementioned styles, we can observe some significant differences with respect to visual appearance such as lines, curves, windows, roof shapes, and construction features like columns, domes, etc. For few styles, a nonexpert may be able to distinguish among them but, knowing that there are 10 s of architectural

Fig. 13.1 Example on classical architecture

Fig. 13.2 Example on neoclassic architecture

Fig. 13.3 Sample of modernism architecture

Fig. 13.4 Sample of parametricism architecture

styles in the history and evolution of building architecture, the problem is evidently challenging. Furthermore, these different styles have sophisticated features that make classification more challenging such as towers, buttresses, arches, domes, irregular shapes, symmetry, order, proportion, decorative, glass elements, texture, piers, bold contrasts, pediments, balconies, and exaggerated ornamentation (Tan et al. 2016; Arora 2021; Zhao et al. 2018). A number of previous studies have reported that building architectural style classification is more challenging than popular problems in the field of machine vision such as face and object recognition (Tan et al. 2016; Zhao et al. 2018).

Traditional machine vision algorithms depend on a multi-stage process model that includes image acquisition, image conditioning, feature extraction, feature optimization, traditional machine learning, training, and classification (Rababaah 2020; Sharma et al. 2021; Kalyan et al. 2021). This process is depicted in Fig. 13.5. The most challenging stage in the traditional machine learning is hand-crafting the features. This challenge has been in the area of pattern recognition for decades until the birth of Deep Learning (DL) models in 2012 (Alom et al. 2018).

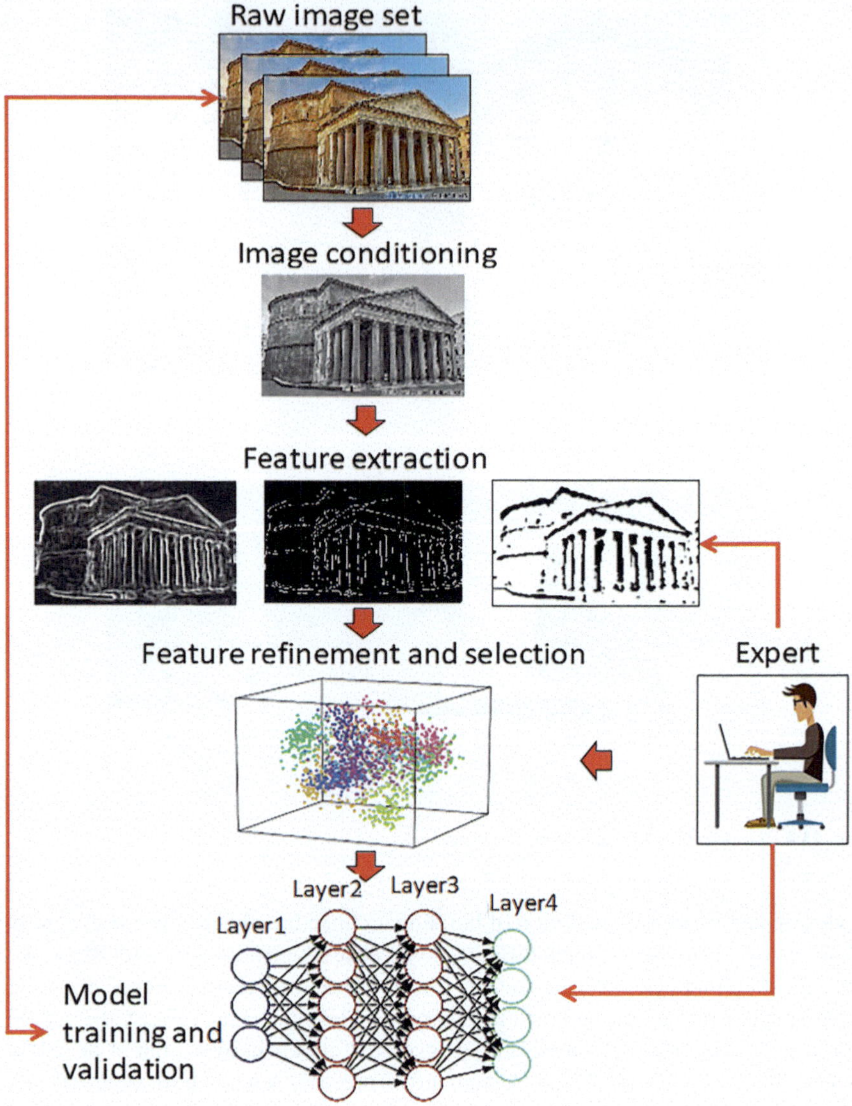

Fig. 13.5 Traditional machine vision model

Due to the complexity of the features of the architectural styles and the difficulty of hand-crafting these features in traditional ML models, we will investigate the effectiveness of a DL model based on Convolution Neural Networks (CNN) architecture to solve the automatic feature extraction and pattern classification problems.

The main contribution of Deep Learning (DL) models is to automate optimal feature learning. As it is shown in Fig. 13.6, DL model architecture includes, typically, the following layers: convolution kernels that are initially randomly generated and

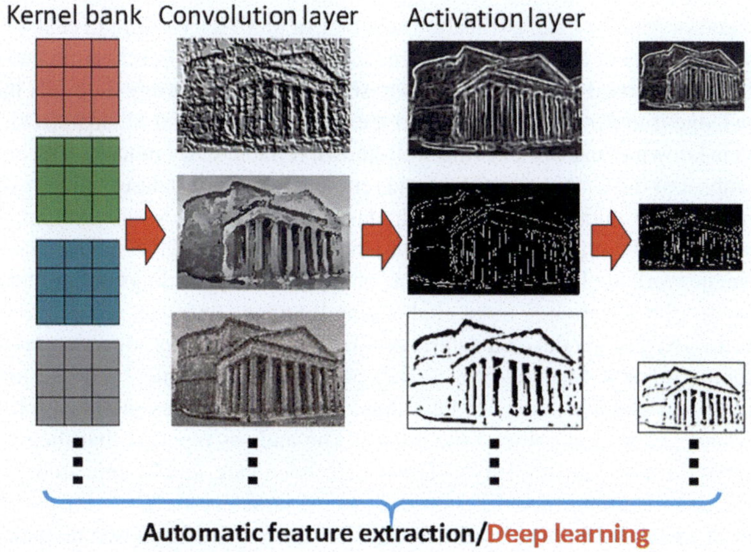

Fig. 13.6 Deep learning model

gradually learned to generate optimal features in the input images; convolution layer that processes the input images using the kernel bank; activation layer that suppresses negative responses after convolution; pooling layer that down-samples the activated image matrices dimensionality reduction and computational efficiency; flatten layer to linearize the resulting data structure as a vector space; and finally the traditional neural network model for typical pattern classification tasks (Rababaah 2022).

In Sect. 13.2, we explore many previous related works and highlight our windows of opportunity to contribute to problem of interest at hand. In Sect. 13.3, we present the theoretical background of the used methods and algorithms. In Sect. 13.4, we present our experimental work and discuss the results to validate the proposed model.

13.2 Literature Review

In this section, we present previous work closely related to the scope of our study in this paper.

A CNN model was used in the work of (Dautov and Astafeva 2021). The authors trained the model on 15 different building styles. The dataset was collected by the authors using the publicly available images on the web. In the work of (Montoya et al. 2016), a CNN model was proposed for Mexican building image classification using sparse features. The study considered three classes modern, colonial, and pre-Hispanic. The authors reported facing the problem of uneven distribution of training samples.

The authors reported a classification accuracy of 88.01%. Yoshimura et al. (2018), presented a survey of architectural design classifications for different architects. The authors also proposed their own method to solve the problem using the CNN model. A knowledge transfer method is used in the work of (Zoph and Shlens 2018). This method is known to use pre-existing well-known reliable CNN models. The method uses a subset of the CNN architecture and retrains it on a new dataset of interest. The Authors reported improvements in both classification accuracy and computational efficiency. A study of fine art painting classification using the CNN model was introduced in the work of (Tan et al. 2016). One of the interesting findings of the study was that fine art classification is more challenging than face or object recognition. The authors trained their model on a dataset of 80,000 samples and produced a classification accuracy of 68%. Shalunts et al. (2012), presented a hierarchical heuristic method to analyze and classify domes of buildings. The set of features included: physical dimensions, shape, and color of the domes. The authors reported that their model considered three architectural styles Russian, Renaissance, and Islamic architectural styles. A deep learning model is introduced in (Liamas et al. 2017) for the classification of images of heritage architecture. The study considered two methods, the first is to design a new CNN and the second that finetunes a pre-existing model. The authors created their own dataset and made it public. Furthermore, the used methods were reported to have a promising accuracy and are a good contribution to the digital documentation of images of heritage buildings. A data mining method was used in (Goel et al. 2012) to address the problem of content-based building images retrieval. The authors employed feature-based analysis and characterization of input images to train the proposed model. Existing public datasets were used to verify the proposed method. The authors reported that the methods are promising in the future of image retrieval applications. A clustering technique is proposed in (Haxhimusa et al. 2012) to categorize Gothic and Baroque elements in buildings. These elements are known as tracery, balustrade, and pediment. The main feature utilized in the analysis was shape gradients. The authors reported reliable performance and high classification rates. A machine learning and computer vision method for building architectural styles was presented in (Chen 2018). The proposed method is based on edge detection and wavelet transformation of image features. The authors used a simulated dataset for model training and validation and found that their proposed method is reliable and applicable. The work of (Meltser et al. 2017) presented a method to solve two problems of building architectural style classification and content-based image retrieval. The authors used a pre-trained CNN model to extract features of input images. Furthermore, the authors claimed a novel model by proposing several preprocessing and classification techniques to enhance the performance of the model. A data mining method was introduced in (Mercioni and Holban 2019) to address the problem of building architectural style classification. The proposed method is based on hierarchical clustering based on a number of distance-based characterizations. The main purpose of the study was to investigate the effectiveness of clustering techniques in architectural styles classification. The work of (Marzouk et al. 2018) established a classification system for Egyptian heritage building styles based on several features including geometric properties, condition, cultural value, construction method, color, and material. Based on these features, a systematic, heuristic, and expert-based, the building input space is clustered. The authors reported their observations and challenges in the study and they concluded that the established system can be used in similar environmental settings. An ensemble approach was implemented by (Zhao et al. 2018) to address the challenges in architectural styles classification. The authors proposed a feature extraction method to minimize intra-class association and maximize inner-class association. For the classifier, they employed a Support Vector Machine (SVM) and reported a reliable a competitive performance. The work of (Xia et al. 2020) established characteristics of architectural styles based on experts' input and surveys. They used salient morphological features such as richness of tones, curves, and roof shapes. The authors trained and validated a neural network model for classification which produced an average accuracy of 77.2%. An automatic building style classifier was introduced by (Mathias et al. 2012). The proposed model works for façade images of buildings. The Gabor filter was used for feature extraction and SVM was used for the classification phase. The classification average accuracy of the model was 77.6%.

Main contribution of our work

Firstly, based on our literature review, the research work conducted in the area of Architectural Styles Classification is far less than any other common areas such as signal pattern recognition, object recognition, face recognition, figure print recognition, human activity classification, etc. Therefore, we believe that our study contributes to the enrichment of the literature of this research area. Secondly, our study is unique in being the only one that investigates the impact of the number of features on the model reliability as it will be shown in the experimental work section. Finally, our model has demonstrated superior performance to comparable existing models as it will be demonstrated in the discussion section.

13.3 Technical Background

This section presents the main algorithms used in the proposed process model according to the presented architecture in Sect. 13.1, Fig. 13.6.

13.3.1 Input Image

The input to the CNN model is a colored, red, green, and blue—RGB, image with normalized size for all input images of all structural style classes. Each image is saved as a 3D byte matrix. The first dimension is the x-axis or the column index of a pixel; the second dimension is for the y-axis or the row index of a pixel; and the third dimension is for the RGB buffers. This structure is depicted in Fig. 13.7.

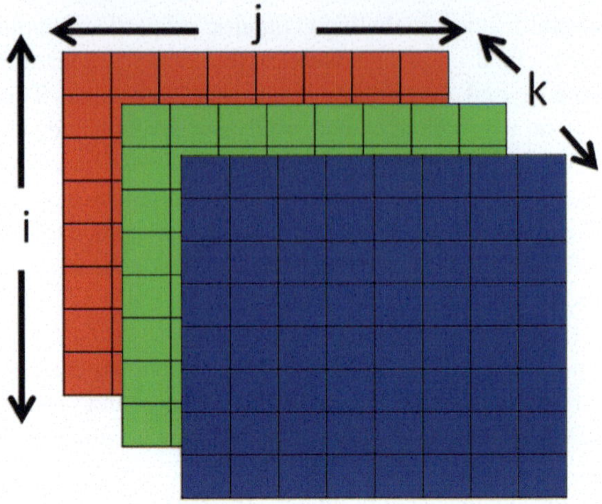

Fig. 13.7 Input image 3D representation

Although some studies consider only gray-scale images for computational efficiency, we believe that the classification reliability is enhanced by the color information. In this stage, several preprocessing may be considered such as image filtration, sharpening, and color normalization. We don't consider any extra image processing only if we observe that noise may be negatively influencing the model performance.

13.3.2 Convolution Layer

This layer is concerned with producing effective features used in training the CNN model. It does that by the convolution operation. Convolution is a mathematical operation that computes an output 2D matrix out of an input 2D matrix and a kernel with various coefficients. These kernels are designed to extract some features of interest including noise filtration, intensity normalization, edges, corners, color, texture, etc. The convolution operation is expressed by the mathematical model in Eq. (13.1).

$$P = \frac{\sum_{i=1}^{N} \sum_{j=1}^{N} K_{ij} X_{ij}}{S} \tag{13.1}$$

where

$P =$ the final value of the output pixel
$K =$ the kernel
$N =$ the size of the kernel
$X =$ the convolved image
$i,j =$ indexes of the pixel
$S =$ the summation of all coefficients of the kernel.

13.3.3 Activation Layer

The resulting matrices from the convolution layer may contain negative pixel values due to the convolution formula that allows this outcome. To eliminate these negative values, the activation layer applies different types of functions including hard threshold, sigmoid, and rectified linear function. In our model, we use the Rectified Linear Unit (ReLU) as the activation function in this layer. The ReLU is expressed in Eq. (13.2).

$$ReLU(X[i,j]) = \begin{cases} X[i,j], & X[i,j] \geq 0 \\ 0, & X[i,j] < 0 \end{cases} \tag{13.2}$$

where

$X =$ the input image to ReLU
$i,j =$ the indexes of current pixel.

13.3.4 Pooling Layer

The number of features is controlled by the number of used kernels in the convolution layer. Every kernel added overloads the computational time in all of the consequent layers. Therefore, the pooling layer is designed at this stage to sample-down the resulting matrices from the Activation layer to improve the computational time efficiency of the process. Pooling can be implemented using a variety of mathematical models including mean, median, and max. In our model, we implemented the Max-Pooling function as it aggregates all overlapping pixel values under the pooling kernel to a single max value that represents the response of that kernel. This operation is expressed mathematically in Eq. (13.3).

$$\mathrm{P}(X[i,j]) = \mathrm{Max}\left(X\left[i - \frac{k}{2} : i + \frac{k}{2}, j - \frac{k}{2} : j + \frac{k}{2}\right]\right) \tag{13.3}$$

where

$X =$ the input image to max-pooling
$i,j =$ the indexes of the central pixel under the kernel
$k =$ the kernel size.

We used the ":" notation to express a range of indexes in the kernel with respect to the indexes of the central pixel as explained in Fig. 13.8.

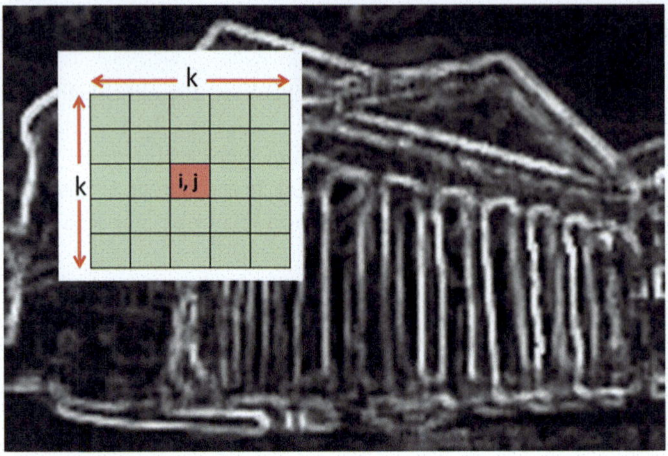

Fig. 13.8 Max-pooling kernel

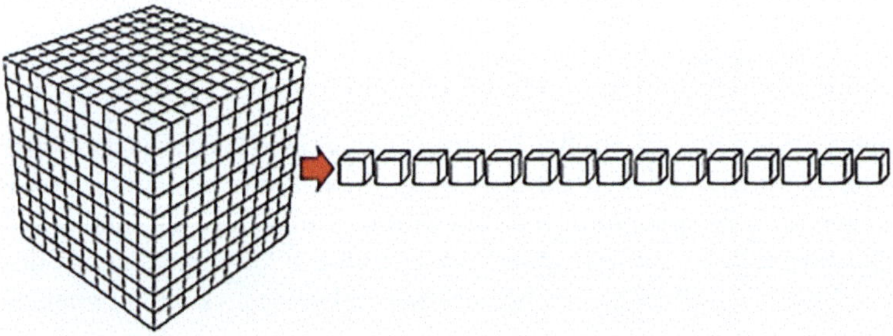

Fig. 13.9 Flattening the 3D data volume to a linear vector

13.3.5 Flatten Layer

So far, all layers produced stacks of 2D matrices. This form of data structure is not compatible as an input to a typical neural network designed for pattern recognition. The typical input to a neural network is a vector; therefore, a flatten layer is needed to flatten the 3D data structure to a 1D array (vector) that can be fed to the classifier neural network. This operation is depicted in Fig. 13.9.

13.3.6 Fully Connected Layer

The previous layers {convolution, activation, pooling, and flatten} accomplish automatic feature learning so researchers do not have to struggle with hand-crafting features. The outcome of this sequence of layers is vector space that is used to train and validate a traditional pattern recognition neural network. For this stage, we used a Multi-Layer Perceptron Neural Network (MLP). The training algorithm of MPL is illustrated in Fig. 13.10.

The training algorithm in Fig. 13.10 is called the Back-Propagation algorithm. This is the most important algorithm in the entire process where it is used to train the classifier and to learn the best features that can represent the presented patterns. The results of the MLP are propagated back across all layers reaching the kernel bank.

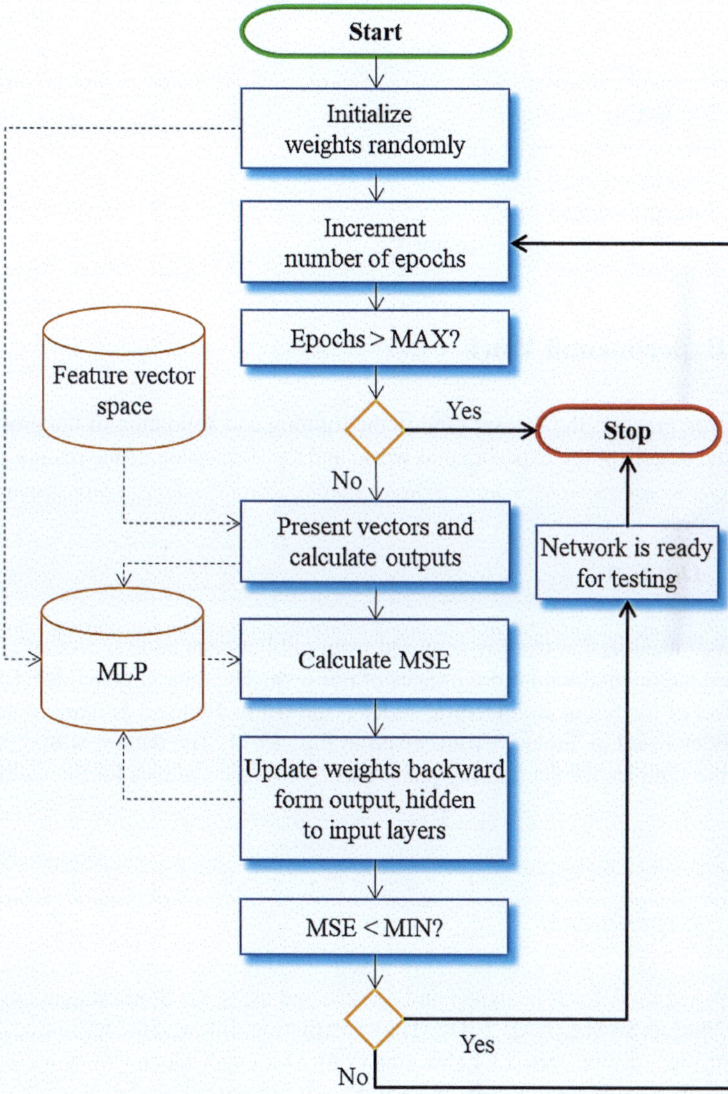

Fig. 13.10 Training algorithm of MLP

The final step in the classification process is to apply Soft-Max function to the resulting neuron scores at the output layer. The Soft-Max function normalizes the output scores vector to a probability distribution that maps the scores to the interval [0, 1]. This operation eliminates the negative values and ensures the scores add up to one and each score is within the interval [0, 1]. The mathematical model of Soft-Max is expressed in Eq. (13.4).

$$\text{SMax}(V[i]]) = \frac{e^{V[i]}}{\sum_{j=1}^{n} e^{V[j]}} \tag{13.4}$$

where

SMax =	Soft-Max function
$V =$	the scores vector at the output layer
$i =$	the ith element in V
$j =$	the jth element in V
$n =$	the size of the scores vector V.

13.4 Experimental Work

This section presents the dataset used in the training and validation of the proposed model, the details of the experimental work, and the discussion of the results.

13.4.1 Dataset

The dataset used in this work is acquired from publicly available source of (Danci 2021) and augmented with more images of our own. The dataset contained 10,802. The subset of the target architectural styles is shown in Table 13.1. Sample images of the styles listed in Table 13.1 are given in Fig. 13.11. The dataset was split into 80% of the samples for the training phase and 20% of the samples for the validation phase.

13.4.2 Experiments

The proposed model was designed and constructed using the development environment of MatLab® (MATLAB 2018). The construction followed the presented architecture in Fig. 13.6 in Sect. 13.1 (Introduction). Our experimental work included 8 independent experiments to train and validate the proposed model. In each experiment, we ran the model for 20 epochs each of which had 10 iterations. In each

Table 13.1 Sample sizes of all architectural styles

No	Style name	Samples
1	Achaemenid architecture	638
2	American craftsman style	501
3	Baroque architecture	531
4	Bauhaus architecture	784
5	Chicago school architecture	772
6	Colonial architecture	594
7	Gothic architecture	488
8	Palladian architecture	648
Total		4956

experiment, a different number of kernels was configured to investigate the impact of the number of features on the classification accuracy, hence the reliability of the model. The number of kernels/features was graduated from 1 to 8. The training and validation results of each experiment are shown in a pair of consecutive Figures. For example, Figs. 13.12 and 13.13 illustrate the training and validation results of the model with the number of kernels = 1. The remaining figures namely Figs. 13.14, 13.15, 13.16, 13.17, 13.18, 13.19, 13.20, 13.21, 13.22, 13.23, 13.24, 13.25, 13.26 and 13.27 follow the same scheme. Figure 13.28 summarizes the training results of all experiments. Finally, Fig. 13.29 aggregates all the validation results in one chart comparing the model classification accuracy of the model among all experiments.

13.4.3 Discussion of Results

This section highlights our observations on the experimental work and discusses the results. When it is mentioned, training figures refer to even-numbered figures between (Figs. 13.12, 13.13, 13.14, 13.15, 13.16, 13.17, 13.18, 13.19, 13.20, 13.21, 13.22, 13.23, 13.24, 13.25 and 13.26) and validation figures refer to odd-numbered figures between (Figs. 13.13, 13.14, 13.15, 13.16, 13.17, 13.18, 13.19, 13.20, 13.21, 13.22, 13.23, 13.24, 13.25, 13.26 and 13.27).

- In the 8 training experiments, the training was done using 80% of the dataset and 20 epochs with 10 iterations per epoch as it can be seen in all of the training figures. The training results are plotted using a line chart showing the training progress with iteration/epoch versus current model accuracy.
- In the training experiments, it is observed that the model mostly did well with training classification accuracy between 85.2 and 100%, as shown in Fig. 13.28. Although reaching 100% in training is a good sign, only validation can prove that the model is reliable.
- In the 8 validation experiments, testing of the model was done using 20% of the dataset and the results were plotted using a step chart as it is more appropriate

Fig. 13.11 Samples of the 8 architectural styles listed in Table 13.1. The listing order of the styles in this figure follows the same order in Table 13.1

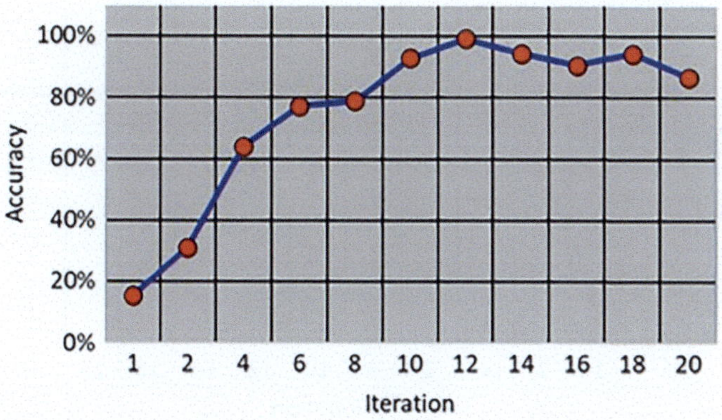

Fig. 13.12 Training results with number of kernels = 1

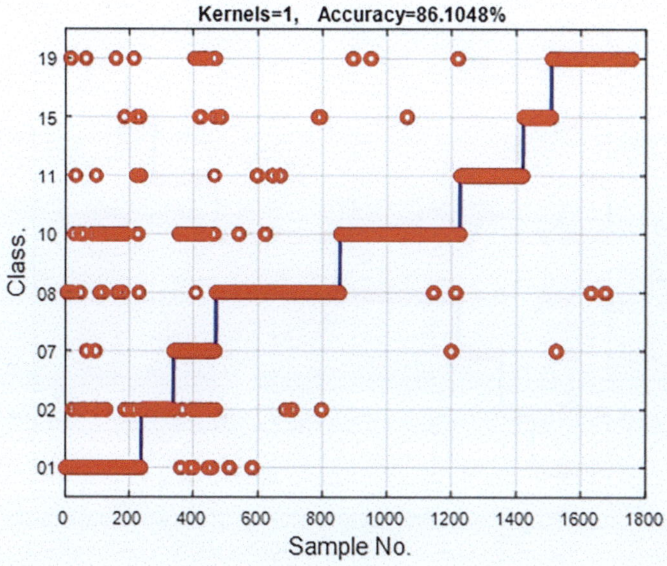

Fig. 13.13 Validation results with number of kernels = 1

to show the correctly classified samples appear on the line of the step whereas the misclassified samples are scattered around the edges. So, the more scattered samples in the validation figure, the lower the model accuracy.

- In Fig. 13.29, the summary of all 8 validation experiments is depicted and it can be observed that the accuracy ranged between 84.23 to 95.44%. Therefore, the best-performing model was with the number of features/kernels of 8.
- Fig. 13.29 shows an interesting result of the impact of the number of features on the accuracy of the model. With the number of features = 3, the accuracy

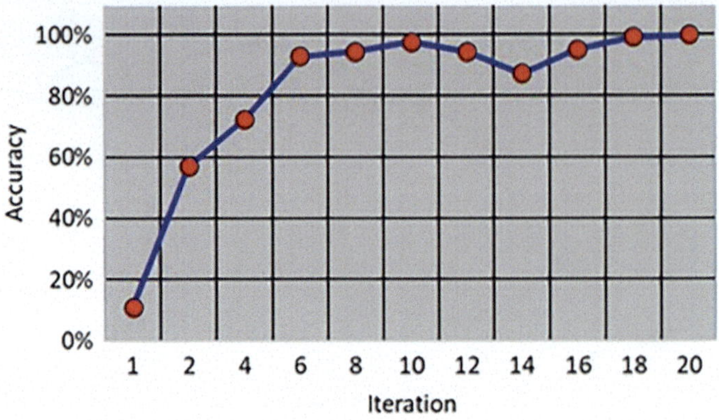

Fig. 13.14 Training results with number of kernels = 2

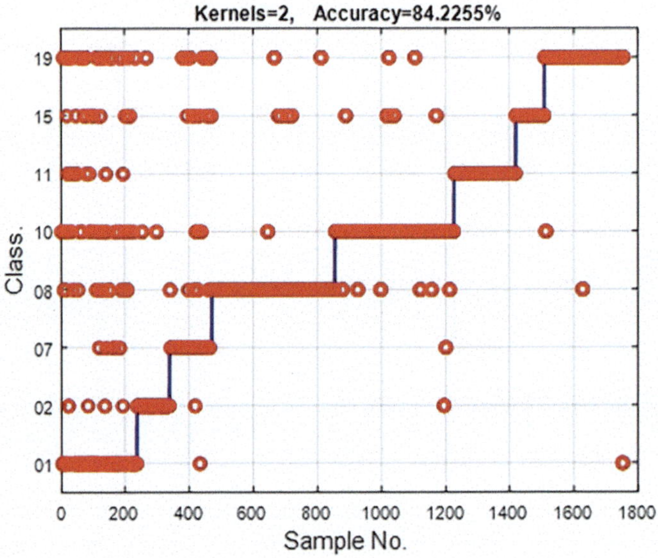

Fig. 13.15 Validation results with number of kernels = 2

was 94.87, increasing the number of kernels afterward seems to have a marginal impact on the accuracy. At the highest number of kernels ($k = 8$), the accuracy reached the highest at 95.44%. if we calculated the gain in accuracy, we found that it is (95.44–94.87)/94.87 = 0.06% which is negligible. This is a typical tradeoff between efficiency and accuracy in intelligent models.

- In the best-performing model (Fig. 13.27), it is clear that the three styles {Style-2: American craftsman style, style-4: Bauhaus architecture, and style-5: Chicago school architecture} were the most challenging as the scattering occurred the

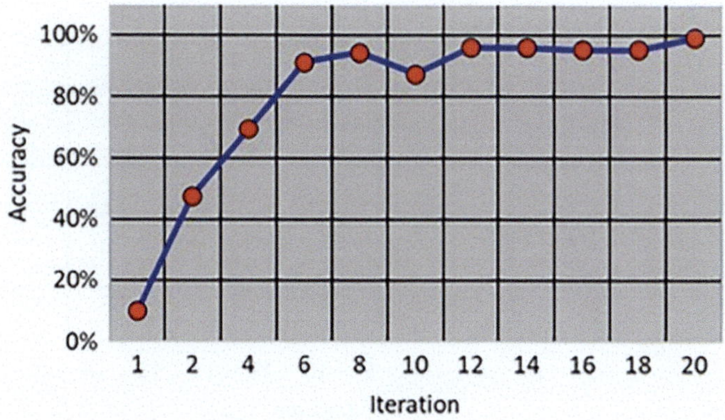

Fig. 13.16 Training results with number of kernels = 3

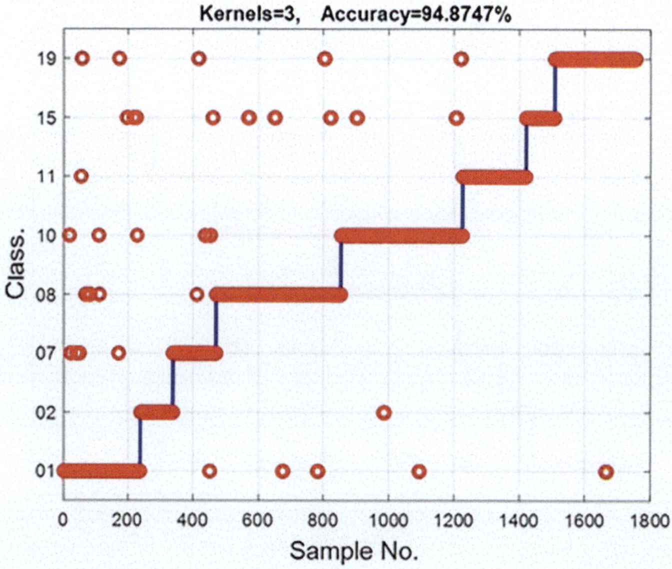

Fig. 13.17 Validation results with number of kernels = 3

most in their regions. Reviewing the original sample images, we observed that in all of these three styles, there are common factors that may contributed to the challenges including inconsistent background as some images include grass and trees whereas others do not and some buildings occupy the full image space whereas others partially occupy the image space. These factors confuse the model since training samples need to be as consistent as possible.

- Comparing our results to related work in Table 13.2, we find that related work has used a smaller number of styles/classes in some cases and more styles/classes in

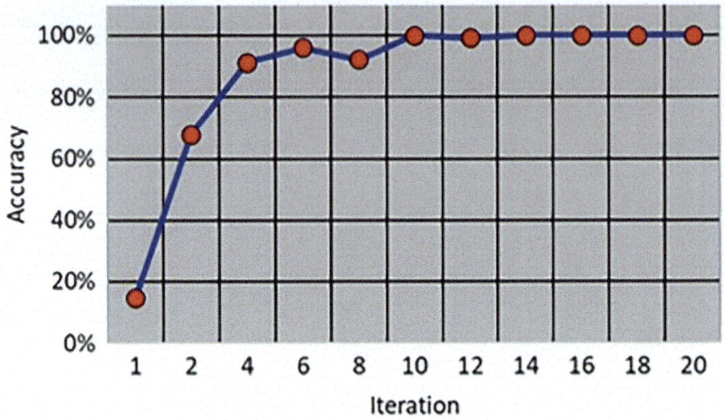

Fig. 13.18 Training results with number of kernels = 4

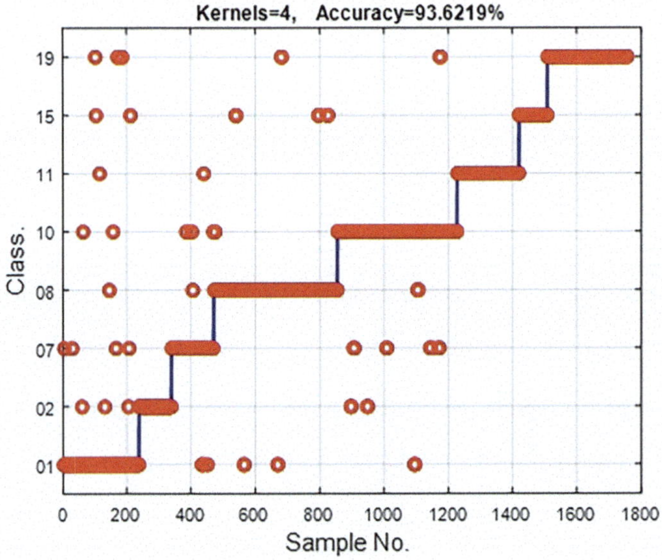

Fig. 13.19 Validation results with number of kernels = 4

others. To be fair, we look at those works who used equal or a smaller number of styles as in (Montoya Obeso et al. 2016; Xia et al. 2020). In all of these works, our results were superior as it is clearly shown in Table 13.2.

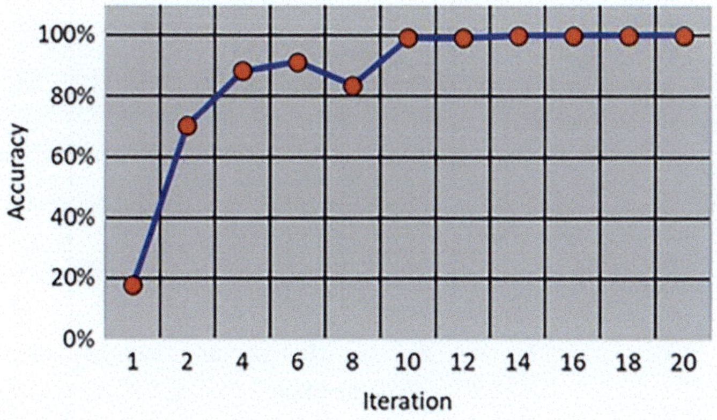

Fig. 13.20 Training results with number of kernels = 5

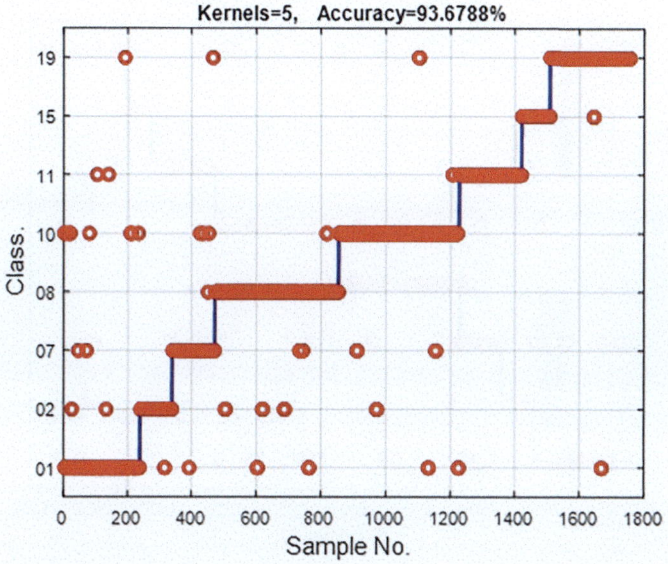

Fig. 13.21 Validation results with number of kernels = 5

13.5 Conclusions

This work presented a proposed intelligent machine vision model to classify building architectural styles. The main objective of the study was to investigate the effectiveness of the Deep Learning (DL) models in providing a reliable solution to the problem at hand. Furthermore, the study investigated the impact of the number of kernels/features on the classification accuracy. A DL model based on Convolution Neural

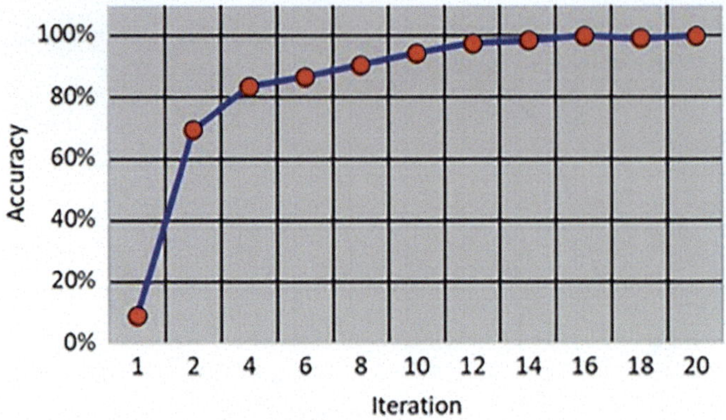

Fig. 13.22 Training results with number of kernels = 6

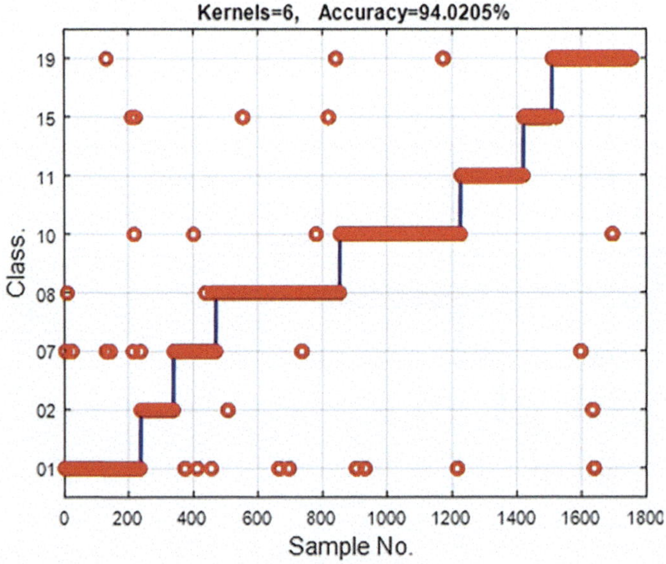

Fig. 13.23 Validation results with number of kernels = 6

Networks was constructed and a publicly accessible dataset was used to train and validate the proposed model.

Several observations were made throughout the experimental work including the lowest model performance was at 84.23% and the highest performance of the model was at 95.44%; the impact of the number of kernels is marginal and it was observed that with the number of kernels = 3 we get very close performance to the number of kernels = 8. Comparing our results to related work, it was shown that our model is superior to other existing models. We believe that our study adds more evidence

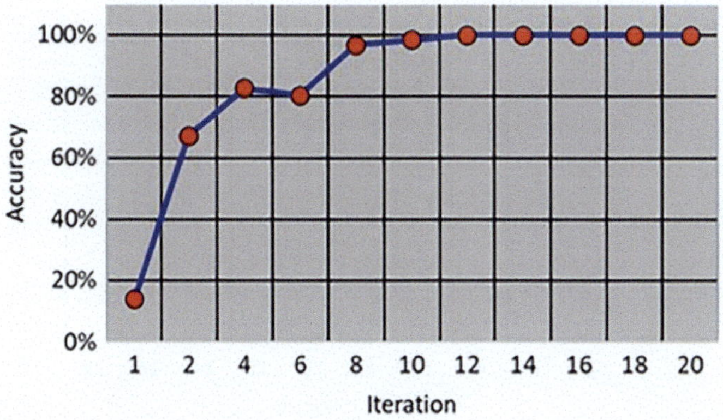

Fig. 13.24 Training results with number of kernels = 7

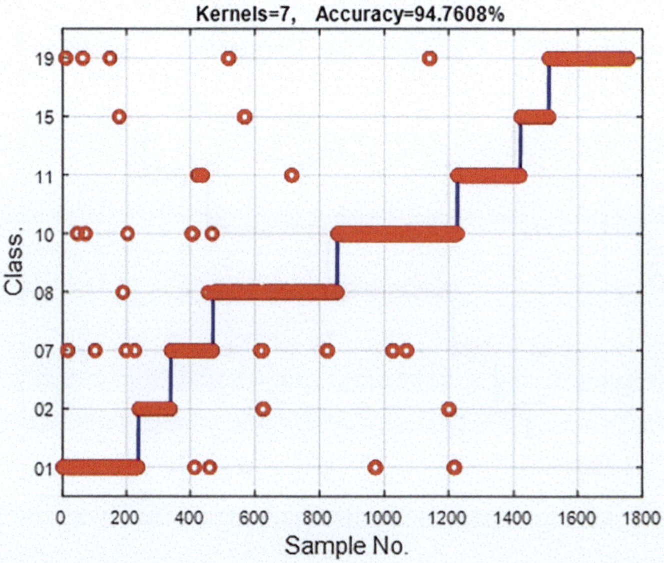

Fig. 13.25 Validation results with number of kernels = 7

to the effectiveness of DL models to the pattern recognition applications in general and to the scope of the study in particular. As a future work, it is recommended that the number of styles be expanded to at least double of investigated one in this study and to evaluate the model accordingly. Furthermore, the largest examined number of features was 8, and it would be interesting to extend this number as well and observe if it improves the classification accuracy any further.

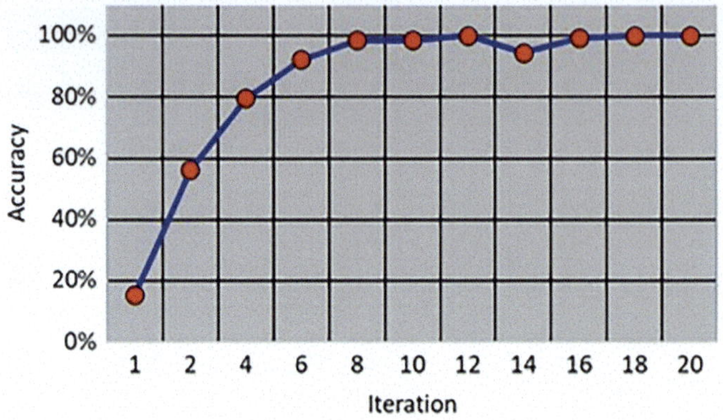

Fig. 13.26 Training results with number of kernels = 8

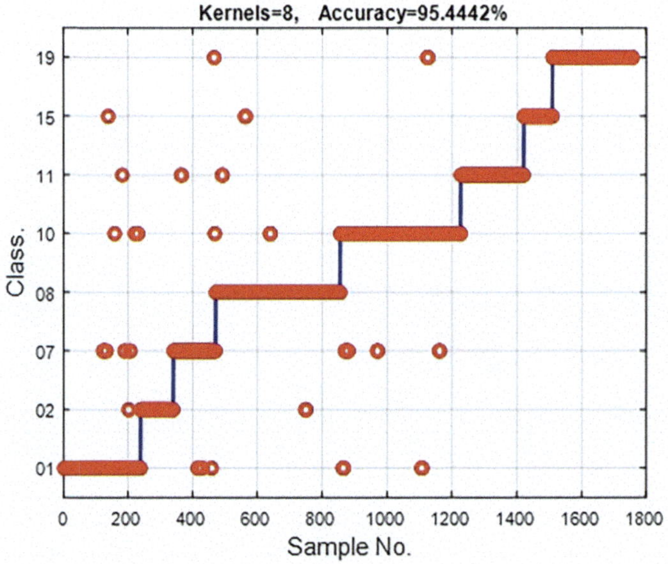

Fig. 13.27 Validation results with number of kernels = 8

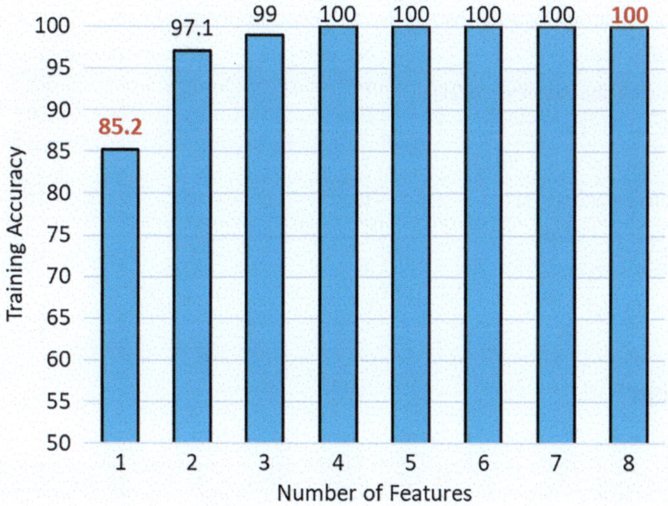

Fig. 13.28 Summary of training results

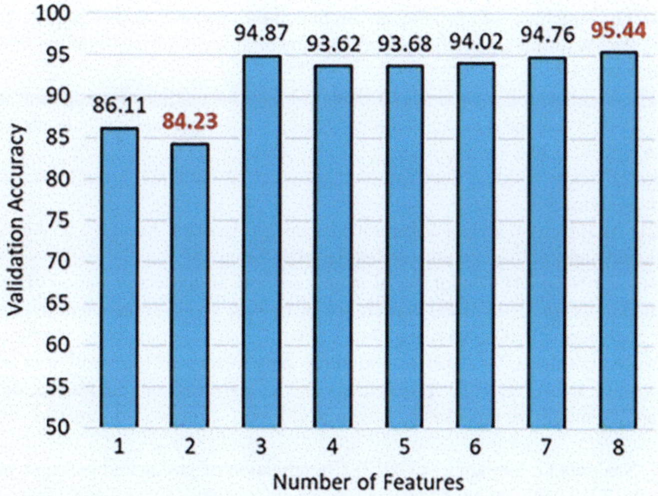

Fig. 13.29 Summary of validation results

Table 13.2 Comparison between our results and the results of related work

No	References	No. of styles	Classification accuracy (%)
1	Dautov and Astafeva (2021)	15	90.4
2	Montoya et al. (2016)	3	88.01
3	Yoshimura et al. (2018)	34	73
4	Tan et al. (2016)	10	68
5	Xia et al. (2020)	8	72.4
6	Mathias et al. (2012)	3	77.6
7	Our model	8	95.44

References

Alom M et al (2018) The history began from AlexNet: a comprehensive survey on deep learning approaches. In arxiv.org, Accessed 01 Dec 2021, from https://arxiv.org/ftp/arxiv/papers/1803/1803.01164.pdf

Archisoup (2021) Definition of architectural styles. In: Archisoup.com, an online learning resource and platform for architecture students. Accessed 28 Nov 2021, from https://www.archisoup.com/architectural-styles

Arora J (2021) Timeline of prominent architectural styles. Accessed 28 Nov 2021, from https://www.re-thinkingthefuture.com/architectural-styles/a5161-timeline-of-prominent-architectural-styles

Chen W (2018) Architectural style analysis method based on intelligent computing technology. In: 2018 10th international conference on measuring technology and mechatronics automation (ICMTMA), pp 321–324

Crisman P (2022) Whole building design guide. Accessed 22 Jan 2022. From https://www.wbdg.org/resources/style

Danci M (2021) Architectural style recognition is an image classifier. Accessed 05 Aug 2021, from https://github.com/dumitrux/architectural-style-recognition

Dautov E, Astafeva N (2021) Convolutional neural network in the classification of architectural styles of buildings. In: 2021 IEEE conference of russian young researchers in electrical and electronic engineering (ElConRus), pp 274–277

Goel A, Juneja M, Jawahar CV (2012) Are buildings only instances? Exploration in architectural style categories. In: ICVGIP'12: proceedings of the eighth indian conference on computer vision, graphics and image processing, Article No.: 1, pp 1–8. https://doi.org/10.1145/2425333.2425334

Haxhimusa Y, Shalunts G, Sablatnig R (2012) Classification of gothic and baroque architectural elements. In: 2012 19th international conference on systems, signals and image processing (IWSSIP), pp 316–319

Kalyan J, Sourav B, Soumya N, Chinmaya P (2021) Machine learning-based classification: an analysis based on COVID-19 transmission electron microscopy images. Int J Comput Appl Technol 66(3/4):350–361

Llamas J, Lerones PM, Medina R, Zalama E, Gómez-GarcíaBermejo J (2017) Classification of architectural heritage images using deep learning techniques. Appl Sci 7(10):992

Marzouk M, Saleeb N, El Sharkawy M, Eid A, Ali M, Metawie M (2018) Classification system for egyptian heritage buildings. In: 2018 IEEE international conference on industrial engineering and engineering management (IEEM), pp 854–858

Mathias M, Martinovic A, Weissenberg J, Haegler S, Van Gool L (2012) Automatic architectural style recognition. In: ISPRS—international archives of the photogrammetry, remote sensing and spatial information sciences, XXXVIII-5/W16, pp 171–176

MATLAB (2018) 9.4.0.813654 (R2018a). The MathWorks Inc, Natick, Massachusetts, USA. Accessed from https://www.mathworks.com/. Access date 5 Oct 2021

Meltser R, Banerji S, Sinha A (2017) What's that style? A CNN-based approach for classification and retrieval of building images. In: 2017 ninth international conference on advances in pattern recognition (ICAPR), pp 1–6

Mercioni M, Holban S (2019) A study on hierarchical clustering and the distance metrics for identifying architectural styles. In: 2019 international conference on energy and environment (CIEM), pp 49–53

Montoya Obeso A, Benois-Pineau J, Ramirez A, Vázquez M (2016) Architectural style classification of Mexican historical buildings using deep convolutional neural networks and sparse features. J Electron Imaging 26(1):011016

Rababaah A (2020) Angle histogram of hough transform as shape signature for visual object classification—(AHOC). Int J Comput Vis Robot 10(4):312–336

Rababaah A (2022) Deep learning solution for machine vision problem of vehicle body damage classification. Int J Comput Vis Robot 12(4):85–100. https://doi.org/10.1504/IJCVR.2021.10041717

Shalunts G, Haxhimusa Y, Sablatnig R (2012) Architectural style classification of domes. In: Bebis G, Boyle R, Parvin B, Koracin D, Fowlkes C, Wang S, Choi MH, Mantler S, Schulze J, Acevedo D, Mueller K, Michael P (eds) Advances in visual computing. ISVC 2012. Lecture notes in computer science, vol 7432. Springer, Berlin, Heidelberg, New York, pp. 420–429

Sharma D, Hota H, Rababaah A (2021) Forecasting US stock price using hybrid of wavelet transforms and adaptive neuro fuzzy inference system. Int J Syst Assur Eng Manage

Tan WR, Chan CS, Aguirre HE, Tanaka K (2016) Ceci n'est pas une pipe: a deep convolutional network for fine-art paintings classification. In: 2016 IEEE international conference on image processing (ICIP), Phoenix, AZ, pp 3703–3707

Wikipedia contributors (2021) List of architectural styles. In: Wikipedia, the free Encyclopedia. Accessed 28 Nov 2021, from https://en.wikipedia.org/w/index.php?title=List_of_architectural_styles&oldid=1052596474

Xia B, Li X, Shi H, Chen S, Chen J (2020) Style classification and prediction of residential buildings based on machine learning. J Asian Arch Build Eng 19(6):714–730. https://doi.org/10.1080/13467581.2020.1779728

Yoshimura Y, Cai B, Wang B, Ratti C (2018) Deep learning architect: classification for architectural design through the eye of artificial intelligence. (n.d.). Accessed 27 Nov 2021, from https://arxiv.org/ftp/arxiv/papers/1812/1812.01714.pdf

Zhao P, Miao Q, Song J, Qi Y, Liu R, Ge D (2018) Architectural style classification based on feature extraction module. IEEE Access 6:52598–52606

Zoph B, Shlens J (2018) Learning transferable architectures for scalable image recognition. arXiv preprint arXiv:1707.07012. Accessed 27 Nov 2021, from https://arxiv.org/abs/1707.07012

Chapter 14
Development of Clinical Decision Support System Using Genetically Optimized Artificial Neural Network

Anamika Shukla Sharma and H. S. Hota

Abstract Clinical Decision Support System (CDSS) is a computerized framework designed to aid healthcare professionals in clinical decision-making. One of the most promising applications of computational intelligence is using Artificial Intelligence (AI) and Machine Learning (ML) to make medical data predictions. When solving a complicated optimization issue with a vast and unattainable number of parameters, a Genetic Algorithm (GA) can be a helpful tool. This research work incorporates the use of GA to construct a novel model GA_BPNN which is a hybrid of GA and Back Propagation Neural Network (BPNN) for predicting various diseases like Diabetes, Breast Cancer, and Parkinson's Disease (PD), where GA is used for optimizing learning rate, momentum, and hidden layers of Neural Network, and neural networks is employed for the purpose of data classification. The main objective of this research work is to create and implement a robust CDSS that can accurately forecast the occurrence of diabetes, breast cancer, and PD. It was found that the BPNN with GA optimization performed better than the BPNN approach with no GA optimization. Four different models were developed for the diagnosis of diseases through CDSS. The first model is the All-Features Back Propagation Neural Network (AF_BPNN); the Second model is the All Features with Optimized Back Propagation Neural Network (AF_OBPNN); the third one used for this study is the Selected Features Back Propagation Neural Network (SF_BPNN). Moreover, the fourth model used is Selected Features with Optimized Back Propagation Neural Network (SF_OBPNN). The hybrid optimized GA-BPNN with selected features improved classification accuracy over GA-BPNN alone with redundant inputs. In the case of diabetes, SF_OBPNN yielded the highest accuracy (92.3%) and precision (90.9%) when compared with the other three models. For Breast cancer, SF_OBPNN yielded the highest accuracy (98.5%) and precision (97.0%). Similarly, the SF_OBPNN model yielded the highest accuracy (98.1%) and precision (95.8%) for PD. It is evident from the results that when GA is used for optimization, the accuracy achieved by the model is the highest. Finally a CDSS named

A. S. Sharma
Govt. E. R. Rao PG Science College, Bilaspur, Chhattisgarh, India

H. S. Hota (✉)
Atal Bihari Vajpayee Vishwavidyalaya, Bilaspur, Chhattisgarh, India
e-mail: proffhota@gmail.com

D. K. Sharma et al. (eds.), *Machine Learning for Real World Applications*, Transactions on Computer Systems and Networks, https://doi.org/10.1007/978-981-97-1900-6_14

as HealthScan+ was developed which provides interactive way to upload patient data and diagnose diseases.

Keywords Breast Cancer · Diabetes · Parkinson's Disease (PD) · Genetic Algorithm (GA) · Back Propagation Neural Network (BPNN)

14.1 Introduction

It has been reported by the World Health Organization (WHO) that 77 million Indians over the age of 18 have type 2 diabetes, with another 25 million at high risk of developing the disease (*Diabetes—India*). Damage to the cardiovascular system, blood vessels, eyes, kidneys, and nerves are only some long-term effects of diabetes, a metabolic condition characterized by high blood glucose levels. Around 422 million people have diabetes, most residing in low- and middle-income countries; annually, diabetes is directly responsible for 1.5 million deaths. The incidence and prevalence of diabetes have both skyrocketed in recent decades. An early diagnosis is essential if someone will have a good quality of life despite having diabetes. The longer a person goes misdiagnosed and untreated for diabetes, the more likely their health will deteriorate.

WHO also states that breast cancer is responsible for more disability-adjusted life years (DALYs) lost by women worldwide than any other type of cancer. WHO reports 627,000 deaths and 2.09 million cases worldwide due to breast cancer. It is the most frequent cancer in women in India, accounting for 14% of all cases. Breast cancer begins in the epithelial lining of the ducts (85% of cases) or lobules (15%) in the breast's glandular tissue. In its first stages, the malignant growth is localized ("in situ") within the duct or lobule, where it does not often produce any symptoms and has a low risk of metastasizing (spreading to other parts of the body). These in situ malignancies (stage 0) may get more advanced over time and infiltrate the surrounding breast tissue, known as invasive breast cancer. The cancer may then migrate to the lymph nodes in the area (known as regional metastasis) or travel to other organs in the body (known as distant metastasis).

Widespread metastases are the cause of breast cancer deaths in women. Around 685,000 people died worldwide, and 2.3 million women were diagnosed with breast cancer in 2020 (WHO 2023). The most common cancer globally as of the end of 2020 was breast cancer, diagnosed in 7.8 million women in the previous five years (*Breast Cancer*). Treatment for breast cancer can be pretty successful, especially if the disease is discovered early.

Disability and death rates resulting from Parkinson's disease (PD) are rising at a rate higher than any other neurological ailment. PD is a degenerative brain condition that is associated with motor symptoms (slow movement, tremor, rigidity, and walking imbalance) and a wide variety of non-motor sequelae (cognitive impairment, mental health disorders, pain, and other sensory disturbances). Motor symptoms of PD include slow movement, tremors, rigidity, and walking imbalance (Parkinson's

Disease). Motor impairments, such as painful involuntary muscle contractions (dystonia) and involuntary movements (dyskinesias), contribute to limitations in speech and mobility, which lead to constraints in many facets of daily living. The progression of these symptoms and complications leads to a significant decline in functionality and quality of life, which leads to high rates of disability and the need for care, in addition to stress and load for caregivers. According to the most recent estimations, about 8.5 million people will live with PD in 2019. 5.8 million disability-adjusted life years were lost due to PD, an increase of 81% since 2000. Additionally, 329,000 people died due to PD, an increase of over 100% since 2000.

These life-threatening diseases demand accurate and faster automation of diagnosis systems. The utilization of ML and AI for disease prediction based on diagnostic reports and pathological images is now one of the rapidly developing applications (Malakar et al. 2022). Artificial Neural Network (ANN) is a widespread technique that maps input patterns with corresponding output patterns and is used for making medical data predictions. Many researchers are using ANN in designing CDSS. In this research work, GA is integrated to develop a hybrid GA-BPNN model for predicting Diabetes, Breast Cancer, and PD, where GA is used for optimizing parameters learning rate, momentum, and hidden layers, and Neural Network is used to classify the data.

The primary objective of this study is to design and develop a CDSS for predicting diabetes, breast cancer, and PD. The research paper is systematized in 6 sections. Related work is presented in Sect. 14.2. Details of datasets, methods, and materials are discussed in Sect. 14.3. Section 14.4 presents an overview of the proposed methodology and model architecture. The performance of the model is discussed in Sect. 14.5. Section 14.6 offers the clinical decision support system (CDSS) developed, and the paper concludes with the last, i.e., Sect. 14.7, with a discussion on the future scope.

Four models were constructed in the course of this study. After comparing the outcomes of four proposed models for the diabetes dataset, it was discovered that SF_OBPNN excels the others with an accuracy of 92.3%, surpassing SF_BPNN (75.8%), AF_BPNN (71.4%), and AF_BPNN (67.0%). For the purpose of diagnosing breast cancer, the SF_OBPNN model achieved the maximum accuracy of 98.5%, while the SF_BPNN model achieved 95.5%, the AF_OPBPNN model achieved 93.9%, and the AF_BPNN model achieved 81.8% accuracy. When SF_OBPNN was used on the PD dataset, it achieved a maximum accuracy of 98.1%. In comparison, SF_BPNN achieved an accuracy of 85.2%, AF_OBPNN achieved 81.5% accuracy, and AF_BPNN achieved 68.5% accuracy. A CDSS named as HealthScan+ provides facility to diagnose diseases either through single patient data or multiple patient data through interactive GUI.

14.2 Related Work

The research study analyzed articles published in peer-reviewed journals on the designing of CDSS and disease diagnosis using AI. The search technique was designed to retrieve material from several databases according to the key fields of computer science and healthcare because our objective is cross-disciplinary. It was discovered that much work had been done in this area. Authors (Vijayan and Anjali 2015; Yahyaoui et al 2019; Farran et al. 2013) investigated the implementation of Support Vector Machine (SVM) in the diagnosis of diabetes, whereas authors (Übeyli 2007; Hota 2014) utilized SVM for diagnosis of breast cancer and discovered that SVM outperforms other models. The performance of SVM was evaluated and compared to that of Random Forest (RF), Logistic Regression (LR), Decision Tree (DT), K Nearest Neighbor (KNN), Multi-layer Perceptron (MLP), Stacking classifier (SC), and an ensemble model by Mall et al. (2022). Authors Vijayan and Anjali (2015) utilized Naïve Bayes (NB), SVM, DT, and AdaBoost techniques on the diabetes dataset. They concluded that with the decision stump serving as the base classifier, the AdaBoost method achieved an accuracy of 80.72%. Mansour (2018) applied AlexNet Deep Neural Network (DNN), Principal Component Analysis (PCA), and Linear Discriminant Analysis (LDA) on the diabetic retinopathy fundus dataset and discovered that AlexNet DNN yielded higher accuracy. Researchers Temurtas et al. (2009) employed Multi-layer NN trained by the Levenberg–Marquardt (LM) algorithm and probabilistic neural network on the diabetes dataset and concluded that the accuracy of MLNN and LM was better than others. Yahyaoui et al. (2019) used SVM, RF, and Convolutional Neural Network (CNN) for experimentation on the diabetes dataset. They discovered that Deep Learning (DL), SVM, and RF yielded an accuracy of 76.81%, 65.38%, and 83.67%, respectively.

On the other hand, optimization of various parameters and architecture of ANN is very important and needs to be taken care while developing CDSS. Arabasadi et al. (2017) proposed a highly accurate hybrid method for diagnosing coronary artery disease. The researchers claim that the proposed method can improve the performance of neural networks by around 10% by boosting the weights of the networks' initial nodes with GA, which implies improved weights for neural networks. By utilizing such a system, it is possible to reach accuracy, sensitivity, and specificity rates of 93.85%, 97%, and 92%, correspondingly. The researchers Karegowda et al. (2011) used a hybrid model that incorporates both the Back Propagation Neural Network (BPNN) and GA, and their findings demonstrate that the GA-optimized BPNN strategy is superior to the BPNN technique that does not include GA optimization for diabetes dataset. Hosseinpoor (2022) utilized GA and Feed-Forward Neural Networks. The accuracy of the proposed method was 93% in predicting gestational diabetes. Alharbi and Alghahtani (2019) developed a fully automated approach for the detection of type 2 diabetes mellitus at an early stage using an Extreme Learning Machine neural network for classification and evolutionary genetic algorithms for feature extraction. With an accuracy of 97.5% and using only six of the eight features provided in the dataset, the hybrid GA-ELM algorithm has delivered an optimum diagnosis of type 2 diabetes patients. Amma (2012) introduced a GA and neural network-based medical diagnosis system for estimating the likelihood of cardiovascular illness. For complex classification tasks, a multi-layered feed-forward neural network is optimal. Training is done using a genetically based neural network. The neural network's final weights are saved in the weighted base and utilized to predict cardiovascular disease's likelihood. The accuracy of classification that may be achieved by utilizing this method is 94.17%. Ahmad et al. (2013) found that GA-ANN and Resilient Back Propagation (RP) achieved 99.43% and 98.29% best and average accuracy.

Monirujjaman Khan et al. (2022) examined the results of many ML algorithms for detecting breast cancer. LR, KNN, DT, and RF were used for prediction. RF yielded 95.61%, DT 94.73%, whereas KNN yielded 90.35% accuracy. The accuracy of LR was found to be 98%, which is higher than the accuracy achieved by the previously published approach.

Hota (2013) used supervised ANN, unsupervised ANN, statistical methods, and a DT-based approach to classifying breast cancer data. An ensemble model was produced after each model was constructed, evaluated, and integrated. The results reveal that the accuracy gained by the ensemble model is more significant than that obtained by any of the individual models; nevertheless, the counter propagation network (CPN) model is competitive among the other individual models, and its accuracy is very close to that obtained by the ensemble model. A ranking-based feature selection strategy is employed with the best ensemble model to lower the dimensionality of the breast cancer dataset. However, the experimental results demonstrate that the model is less accurate with fewer features. With the Training dataset, Ensemble of Bayesian and C5.0 yielded an accuracy of 99.24%, whereas it yielded an accuracy of 95.45% with testing data. In his research, Hota (2014) investigated an ensemble model constructed using SVM and DT methods, paying particular attention to feature selection. Based on the rank of the breast cancer data, a rank-based feature selection technique reduced the features one by one. Using a smaller selection of only five features, a support vector machine, and a C5.0 decision tree approach combination generated high accuracy of 92.59%, 100% sensitivity, and 60% specificity. The primary aim of the research conducted by Übeyli (2007) was to provide a comprehensive resource for anyone interested in constructing an automated DSS for breast cancer diagnosis. The accuracies achieved by the classifiers MLPNN, CNN, SVM, Recurrent Neural Network (RNN), and probabilistic neural network were compared by the authors. The classifiers utilized in this study were trained on the features of each individual record from the Wisconsin breast cancer database. The results of this investigation suggest that the SVM exhibited superior diagnostic accuracies compared to other automated diagnostic systems.

In another work, Nilashi et al. (2022) discovered that in comparison to DTs, neuro-fuzzy, and SVR used alone or in combination with other clustering techniques, expectation–maximization with SVR ensembles can offer superior prediction accuracy for the prediction of PD. Authors Templeton et al. (2022) found that the classification of sensor-based features allowed for the discrimination of Parkinson's disease from healthy controls with an accuracy of 92.6% and early and advanced stages of

Parkinson's disease with an accuracy of 73.7%, in comparison to the current gold standard tools which have an accuracy of 78.3% and functional movement assessments which have an accuracy of 70%. Mall et al. (2022) revealed that the ensemble model outperforms other models with 94.87% accuracy and 94.52% f1 score. Authors Grover et al. (2018) used DNN in their study and found that DNN has shown better accuracy than others, with classification accuracy of 94.4% and 62.7% for train and test datasets. Choi et al. (2017) utilized CNN. Their research work showed high accuracy in differentiating PD from non-PD.

To identify the presence of PD and Alzheimer's disease, Noella (2020) used PET images and ANN (with a combination of feed-forward and backpropagation). With a higher accuracy of 93.14% compared to other current systems like SVM, ID3, and NB classifiers, they came to the conclusion that the categorization groups include AD, PD, and Healthy brain. The researchers applied the ANN and KNN algorithms (Asmae et al. 2020) to differentiate between people with Parkinson's disease and healthy individuals. According to the findings of the experiments, the ANN classifier had an overall average performance that was superior to that of the KNN classifier in terms of accuracy. With an accuracy of 96.7%, the created system can differentiate between healthy persons and an acceptable range of people with PD by applying ANN. The objective of the study carried out by Lin et al. (2022) was to develop neural network models that can recognize PD at its early stage. The data from the inertial measurement unit (IMU) sensor were utilized to create neural network models, which, when validated, had an average accuracy of 92.72% in identifying individuals at an advanced stage of Parkinson's disease. With an accuracy of 99.67%, the models could also discriminate patients with early-stage PD from normal older subjects. Success in differentiating those with PD from those without the disease and those at varying stages of the disease was validated by testing the models on a second, independent sample of participants.

The literature review reflected that researchers are mainly interested in developing more efficient classifiers for CDSS. Most authors have used ANN, ML, and Feature Selection Techniques (FST) to develop CDSS.

14.3 Material and Methods

An ANN is a computational model that mimics the structure and behavior of real neural networks in the living world. It is programmed to identify patterns and formulate predictions depending on the data fed into it. A hybrid of the GA-BPNN model is developed in this experimentation for predicting Diabetes, Breast Cancer, and PD.

Table 14.1 Summary of datasets used

S. no	Disease	Data source	Features	Features selected	Class 1
1	Diabetes	Kaggle	8	5	2 (0 or 1) Diabetic or non-diabetic
2	Breast cancer	Kaggle	30	12	2 (0 or 1) Benign or malignant
3	Parkinson's disease	UCI machine learning repository	23	6	2 (0 to 1) Parkinson's or healthy

14.3.1 Dataset

Developing Clinical Decision Support Systems (CDSS) using intelligent techniques involves a combination of healthcare datasets, computational resources, models, and a systematic set of methods. High-quality healthcare data is the fundamental material for CDSS development. In this research work, numerical data related to Parkinson's was collected from the UCI Machinery Repository. Data for Breast Cancer and diabetes were collected from Kaggle. The dataset of diabetes has 768 instances with eight attributes. The breast cancer dataset has Breast mass features computed from a digital image of a fine needle aspirate (FNA) specimen. They describe the attributes of the cell nuclei that may be seen in the image. A total of 569 instances are there, along with 30 attributes. The dataset collected from the UCI repository related to PD has 197 instances and 23 attributes. Table 14.1 provides a concise summary of the dataset utilized for this research work.

14.3.2 Artificial Neural Network

ANN technique is inspired by the human brain architecture of brain cells or neurons. Neurons (nodes) are processing elements which are interconnected, constituting a structure in which the weight of these connections is adjusted to train the network. It is usually a layered structure (Rajasekaran and Pai 2003).

The input layer, the hidden layer, and the output layer are the three primary categories of layers that are present in a standard ANN. The preliminary data is delivered to the input layer. Each neuron in the network takes information from the layer beneath it as input, applies some mathematical function to that information, and then generates output. After that, these outputs are sent to the subsequent layer, and the process is repeated until the final output is obtained.

During training, the network modifies its parameters to minimize errors and improve its predictions. This allows it to learn the strength of the connections between neurons called weights. The number of hidden layers is going to be established based on the structure of the neural network, which could be anywhere from one to multiple. The inputs of a hidden layer and their connections to those of other hidden layers ultimately define its performance. The activity of the hidden layers and the relative weights of the hidden and output units both affect how well the output layer performs. Through trial and error, the designer decides on the optimal number of layers and neurons in each layer.

The error backpropagation technique is the most frequent approach used to train networks and alter weights to attain the lowest possible error among all other available ways (Mitchell). Multi-layer perceptron networks (MLP networks) and radial basis function networks (RBF networks) are the two primary classifications of ANN. A backpropagation algorithm in an ANN with an MLP structure is used in the current work.

The main focus of the authors (Huang et al. 2009) was on the collection and re-evaluation of the weight matrices of a BPNN during the processing of GA operations in each generation, with the goal of optimizing the weight matrices. The suggested model's value was assessed by a Monte Carlo analysis, which compares it to earlier studies. This assessment was conducted using in-sample, interpolation, and extrapolation data for six test functions.

The suggested GA-BPNN model demonstrates that the GA may function as both a proficient search method and as a means to optimize the parameters and topology of the neural network by treating them as decision variables, resulting in improved outcomes. The GA employs a two-phase evaluation procedure to assess the quality of the solution, while the BPNN is subjected to testing using both interpolation and extrapolation data.

The outcome further demonstrated that the GA-BPNN model is a viable forecasting technique for intricate nonlinear issues. The proposed GA-BPNN utilizes an extended topology and parameters of a BPNN to effectively explore a larger search space. This allows for the identification of more possible solutions and increases the likelihood of finding superior solutions for complicated nonlinear problems.

Numerous researchers have endeavored to tackle the issues related to the training algorithm by applying constraints on the search for area or by restructuring the architecture of the neural network. (Sexton et al. 1998) showed in their research study that limitations and restructuring can be avoided by using a sophisticated initial architecture and a suitable global search algorithm. They also showed that the GA may work as a global search algorithm. Additionally, by properly setting the objective function, it can achieve a parsimonious architecture at the same time. The superiority of employing the GA over the BPNN for optimizing neural networks was demonstrated in a Monte Carlo study. This study compared the performance of each algorithm on in-sample, interpolation, and extrapolation data for seven different test functions.

14.3.3 Genetic Algorithm

GA refers to computerized search and optimization techniques that have their foundation in the principles of natural genetics and natural selection. In the middle of the sixties, Professor Holland from the University of Michigan conceived of the idea behind these algorithms and published his key work on the subject (Holland,1975). GAs exhibit significant divergence from conventional optimization techniques. The transformation of design space into genetic space is crucial. GA functions by modifying variables using coding techniques. Utilizing a coding scheme that incorporates variable space provides the advantage of discretizing the search space, even in cases where the underlying function remains continuous. This computational method simulates the evolutionary procedure by iteratively generating and evaluating populations of potential solutions to a problem. For optimization, it is applied to a population, where each individual stands for a different solution to the problem. Each individual is encoded as a chromosome, a string of genes representing the solution's parameters. The algorithm manipulates these genes through mutation, crossover, and selection to generate new, potentially better solutions.

14.4 Methodology

BPNN is widely used and has been applied to classification problems. This technique is the most frequent approach used to train networks and alter weights to attain the lowest possible error among all other available ways (Mitchell 1997). The learning algorithm utilized in BPNN (Rumelhart et al.; Arabgol and Ko 2013) is a form of gradient descent technique that involves propagating the backward error (gradient) to explore the error surface and minimize the error criterion.:

$$E = \frac{1}{2}\sum (Tj - Oj)^2 \tag{14.1}$$

Tj represents the target output, while Oj represents the output computed by the network. A BPNN calculates its weights using a gradient search technique, which might lead to the issue of getting stuck in a local minimum. On the contrary, GA has been proven to excel at efficiently discovering satisfactory solutions to problems within a very small number of iterations. An important factor affecting an ANN's performance is the selection of an appropriate feature subset. Only some features should be given equal weight; others should be ignored because they are noisy, redundant, or unimportant to the task at hand. This provides motivation for combining GA and BPNN.

This research work uses GA for the simultaneous selection of BPNN features and the optimization of BPNN parameters. Selecting an appropriate number of hidden layers and hidden nodes (HN), as well as connection weights and learning rate, is difficult when designing an ANN. A smaller number of hidden neurons typically leads to the issue of under-fitting. Conversely, if the HN are very large, the network will encounter the issue of over-fitting. This means that it will be able to generate accurate predictions only during the training phase, but will exhibit subpar performance during the testing phase. Over-fitting can lead to exaggerated forecasts even when the data is free of noise. Automated simultaneous optimization of input characteristics and NH size is essential.

The GA-BPNN model is trained using feed-forward neural networks and the back-propagation method. The neurons within the MLP are trained using the backpropagation learning technique, while the parameters of the MLP classifier are optimized through the use of a GA specifically designed for classification. Here, the chromosomes labeled as "1" and "0" were employed to indicate whether a specific feature was selected or not. The chromosome has additional bits that reflect the optimization parameter. After initializing the population size, fitness values are calculated. In a GA, the fitness value is a measure of how well a candidate solution (individual) performs the task at hand during the selection process of the GA.

Fitness values are used to create new generations using selection, crossover, and mutation. The selection operator is also called the reproduction operator. There exists a number of selection operators in GA. In this study, the Tournament selection method is used for selecting chromosomes for parents to cross over. The advantage of tournament selection is that it allows weaker individuals to participate in the selection process, which can help to maintain diversity in the population. Mutation is an opt-out process whereby a tiny alteration in a chromosome allows an individual to experience something novel with an extremely low probability of mutation.

The GA has been applied in conjunction with BPNN for many purposes, including calculating the appropriate number of hidden nodes and hidden layers, selecting relevant feature subsets, optimizing the learning rate and momentum, as well as initializing and optimizing the network connection weights of BPNN. In this research work, the optimized parameters are learning rate, momentum, number of hidden layers, and number of neurons at the hidden layer.

14.5 Result and Discussions

Four different models using BPNN were developed as follows:

AF_BPNN: All-Features Back Propagation Neural Network.
AF_OBPNN: All Features with Optimized Back Propagation Neural Network.
SF_BPNN: Selected Features Back Propagation Neural Network.
SF_OBPNN: Selected Features with Optimized Back Propagation Neural Network.

The outcomes of the models were evaluated using confusion matrices, measures of accuracy, precision, recall, and F1 Score.

The confusion matrix helps understand a classification model's strengths and weaknesses and provides insights into its performance. A confusion matrix is a table that is often used to describe the performance of a classification model. It allows us to visualize the performance of a classification algorithm by displaying the number of correct and incorrect predictions made by the model compared to the actual outcomes. It has two dimensions: the predicted classes, which represent the label predicted by the classification model for the corresponding data instances, and the actual classes, which represent the true labels or classes of the data instances. The confusion matrix helps understand a classification model's strengths and weaknesses and provides insights into its performance.

Figures 14.1, 14.2, and 14.3 illustrate the confusion matrices at the testing stage of SF_OBPNN for Diabetes, Breast Cancer, and PD, respectively.

The hybrid GA_BPNN shows substantial improvement in the classification accuracy of BPNN. Significant features selected using GA further enhanced the classification accuracy of GA-BPNN. A comparative study of evaluation matrices for proposed models is shown in the form of Tables 14.2, 14.3, and 14.4:

Fig. 14.1 Confusion Matrix for SF_OBPNN (Diabetes)

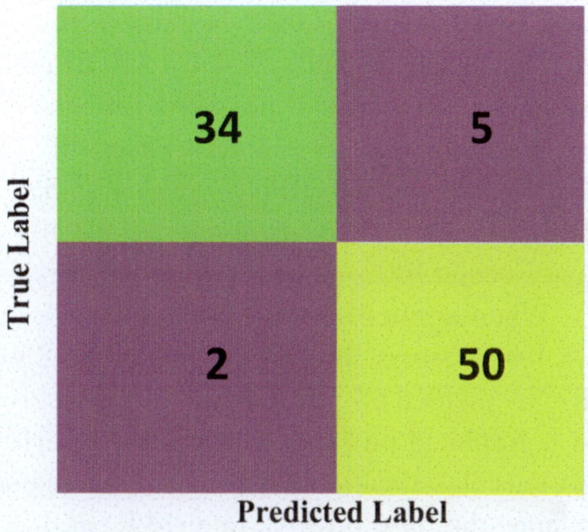

Fig. 14.2 Confusion Matrix for SF_OBPNN (Breast Cancer)

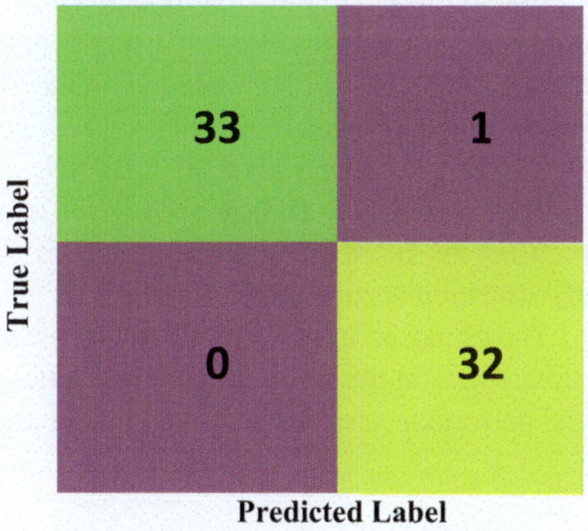

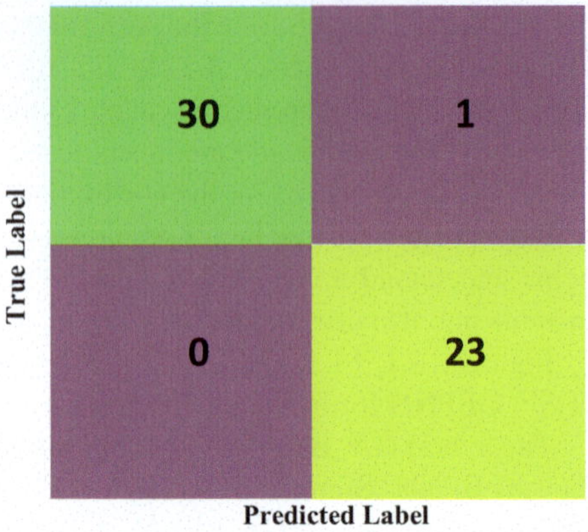

Fig. 14.3 Confusion Matrix for SF_OBPNN (PD)

Comparing the results of four proposed models for the diabetes dataset, it was found that SF_OBPNN outperforms with an accuracy of 92.3% as compared to 75.8%(SF_BPNN), 71.4% (AF_BPNN), and 67.0% (AF_BPNN). SF_BPNN yielded good values of precision (90.9%) and Recall (96.2%).

When applied for Breast Cancer, SF_OBPNN yielded the highest accuracy of 98.5%, whereas SF_BPNN yielded 95.5%, AF_OPBPNN yielded 93.9%, and AF_BPNN yielded 81.8% accuracy. SF_BPNN yielded good values of precision (97.0%) and Recall (100%).

When applied to the PD dataset, SF_OBPNN yielded the highest accuracy of 98.1%, whereas SF_BPNN yielded 85.2%, AF_OBPNN yielded 81.5%, and AF_BPNN yielded 68.5% accuracy. The Receiver Operating Characteristic (ROC) curve is a graphical plot that illustrates the performance of a classification model across various classification thresholds. It is created by plotting the true positive rate (TPR) against the false positive rate (FPR) at different threshold settings. The ROC curve provides a visual representation of the trade-off between the sensitivity (recall) and specificity of a classification model. The sensitivity (TPR) is plotted on the y-axis, representing the proportion of true positive cases correctly classified. In contrast, the specificity (1—FPR) is plotted on the x-axis, representing the proportion of true negative cases correctly classified. Figures 14.4, 14.5, and 14.6 show the Receiver Operating Characteristic of SF_OBPNN.

The hybrid GA_BPNN shows substantial improvement in the classification accuracy of BPNN. Significant features selected using GA further enhanced the classification accuracy of GA-BPNN.

The proposed model's performance was compared with those proposed by other researchers on the same or different datasets. As seen in Table 14.5, the efficiency and performance of the proposed model are better than other research works.

Table 14.2 Comparative chart of models for diabetes

Method	Accuracy	Precision	Recall	F1 Score	Layers	NH1	NH2	NH3	NH4	NH5	LR	MO
AF_BPNN	67.0	70.5	64.6	67.4	5	10	10	10	10	10	0.001	0.90
AF_OBPNN	71.4	72.9	72.9	72.9	5	8	4	5	6	4	0.072	0.23
SF_BPNN	75.8	73.9	77.3	75.6	5	10	10	10	10	10	0.001	0.90
SF_OBPNN	92.3	90.9	96.2	93.5	5	4	3	6	8	4	0.046	0.68

Table 14.3 Comparative chart of models for Breast Cancer

Method	Accuracy	Precision	Recall	F1 Score	Layers	NH1	NH2	NH3	NH4	NH5	LR	MO
AF_BPNN	81.8	79.1	91.9	85.0	5	10	10	10	10	10	0.001	0.90
AF_OBPNN	93.9	91.4	97.0	94.1	5	7	9	7	5	5	0.099	0.19
SF_BPNN	95.5	93.9	96.9	95.4	5	10	10	10	10	10	0.001	0.90
SF_OBPNN	98.5	97.0	100	98.5	5	8	3	5	4	4	0.056	0.35

Table 14.4 Comparative chart of models for Parkinson's Disease

Method	Accuracy	Precision	Recall	F1 Score	Layers	NH1	NH2	NH3	NH4	NH5	LR	MO
AF_BPNN	68.5	77.8	65.6	71.2	5	10	10	10	10	10	0.001	0.90
AF_OBPNN	81.5	90.0	69.2	78.3	5	8	9	4	5	7	0.035	0.75
SF_BPNN	85.2	94.1	69.6	80.0	5	10	10	10	10	10	0.001	0.90
SF_OBPNN	98.1	95.8	100	97.9	5	5	8	4	4	6	0.073	0.32

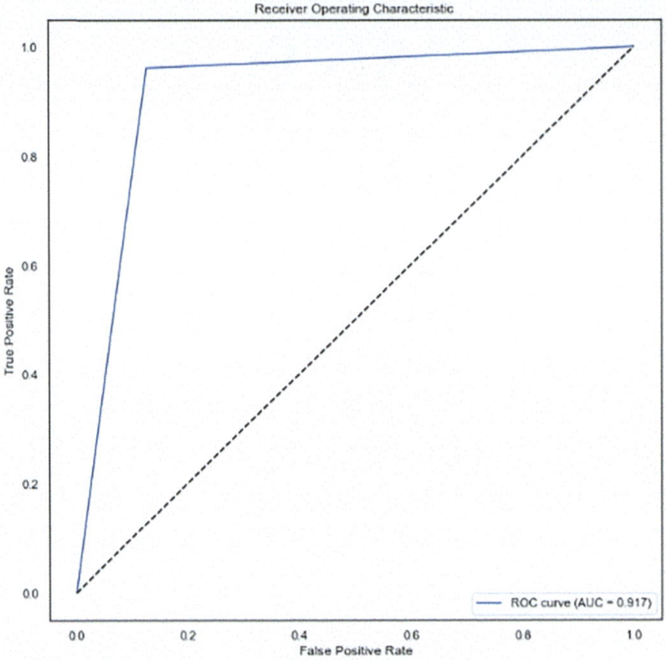

Fig. 14.4 ROC Curve for SF_OBPNN **(Diabetes)**

14.6 Clinical Decision Support System

CDSSs are sophisticated systems that effectively integrate and present clinical information specific to individual patients in a standardized fashion. Although the healthcare industry has long included a DSS, only over the course of the previous twenty years, have CDSSs aided in prognosis and diagnosis.

One of the applications that are developing at the most rapid pace in recent times is designing CDSS utilizing artificial intelligence (AI) and ML.

CDSS is a computerized framework designed to aid healthcare professionals in clinical decision-making. By supplying clinicians with information that is timely and pertinent, this subfield of clinical informatics works toward the goal of enhancing patient safety, in addition to improving the quality of care and clinical outcomes. Management planning, screening patients at various stages, informing policy, and speeding up clinical trials are examples of how ML may be applied in the healthcare industry better to use limited resources and a wealth of data. Consequently, it might change the way doctors practice medicine. The application of ML techniques has facilitated the improvement of CDSS.

This research employs GA and BPNN to develop robust, reliable, accurate, and scalable CDSS (HealthScan+) for the prediction of three diseases. The graphical user interface (GUI) plays a significant role in CDSS by facilitating the interaction

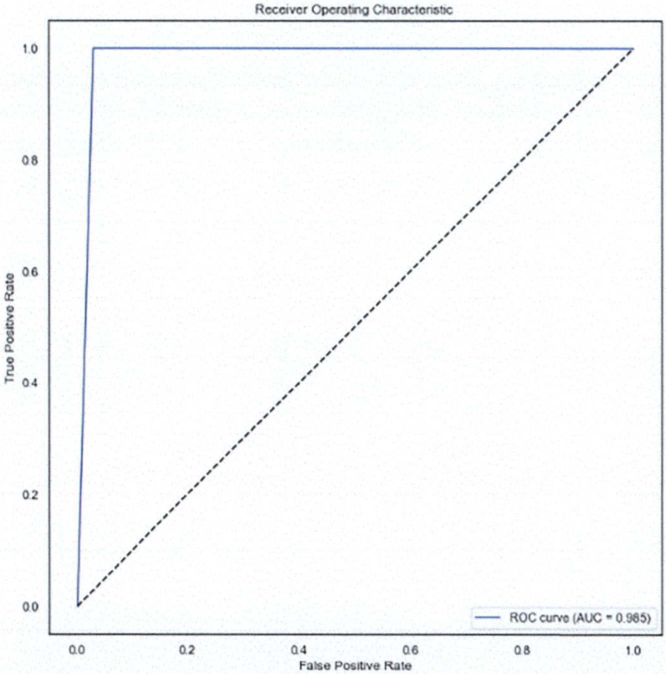

Fig. 14.5 ROC Curve for SF_OBPNN **(Breast Cancer)**

between healthcare professionals and the system. The critical roles of a GUI in CDSS are Information Presentation, Efficient Information Access, User-Friendly Interaction, and providing Decision Support tools.

A user-friendly Graphical User Interface (GUI), HealthScan+ is designed and developed for users to interact with the system. To use the proposed CDSS GUI, the user must log in through a valid id, as shown in Fig. 14.7.

After successful login, the user must select one of the three diseases. After the processing system displays the result, the user can upload bulk records or data about a single record as required. HealthScan + provides the facility to user to register new user and change the password. Figure 14.8 illustrates GUI for Diabetes Prediction for a single record and Prediction in bulk (Fig. 14.9).

Figures 14.10 and 14.11 display GUI to predict breast cancer for single and multiple records.

Figures 14.12 and 14.13 display GUI to predict PD for single and multiple records.

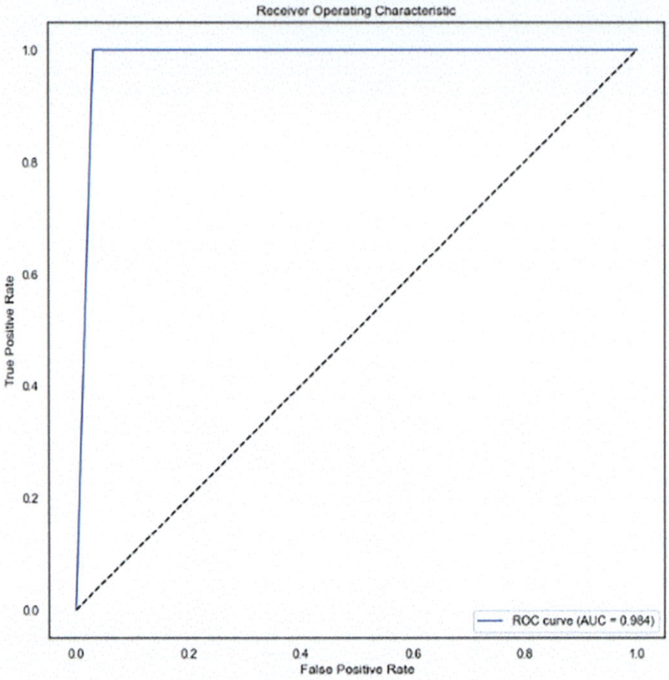

Fig. 14.6 ROC Curve for SF_OBPNN **(PD)**

Table 14.5 Performance of the proposed model and other models

Author	Technique	Dataset Used	Outcome
Karegowda et al. (2011)	BPNN, GA	Diabetes database	GA-BPN yielded 77.71% accuracy with all attributes, 84.08% accuracy with inputs identified by DT, and 84.71% accuracy for GA-Correlation-Based Feature Selection (CFS), whereas BPN yielded 72.88% accuracy with all attributes, 78.21% accuracy with inputs identified by DT, and 79.5% accuracy with inputs identified by GA-CFS
Amma (2012)	Multilayered Feed-Forward NN and GA	Cardiovascular dataset	The proposed model achieved an accuracy of 94.17%
Ahmad et al. (2013)	ANN, GA	Breast Cancer dataset	GA-ANN and Resilient Back Propagation (RP) achieved 99.43% and 98.29% best and average accuracy
Arabasadi et al. (2017)	GA and EBPN with MLP structure	Heart dataset	The proposed model (GA + NN) yielded higher 93.85% accuracy, 97% sensitivity, and 92%specificity compared to 84.62% accuracy, 86% sensitivity, and 83% specificity achieved by classic NN
Hosseinpoor (2022)	GA and Feed-Forward NN	Diabetes data	The proposed model yielded 93.2% accuracy
Proposed Model (SF_OBPNN)	GA, BPNN	Diabetes, Breast Cancer, PD's dataset	The proposed model achieved 92.3% accuracy for Diabetes, 98.5% accuracy for Breast Cancer, and 98.1% accuracy for Parkinson's Disease

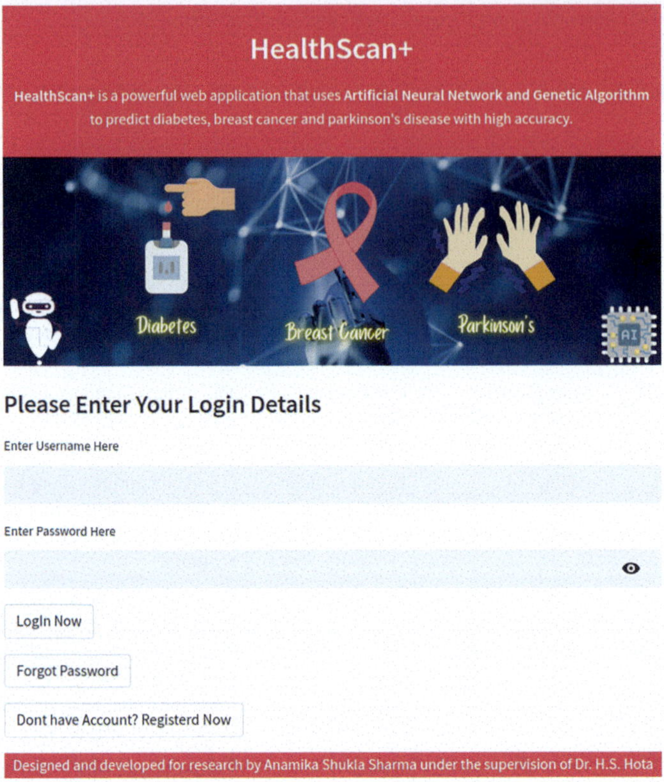

Fig. 14.7 Login page

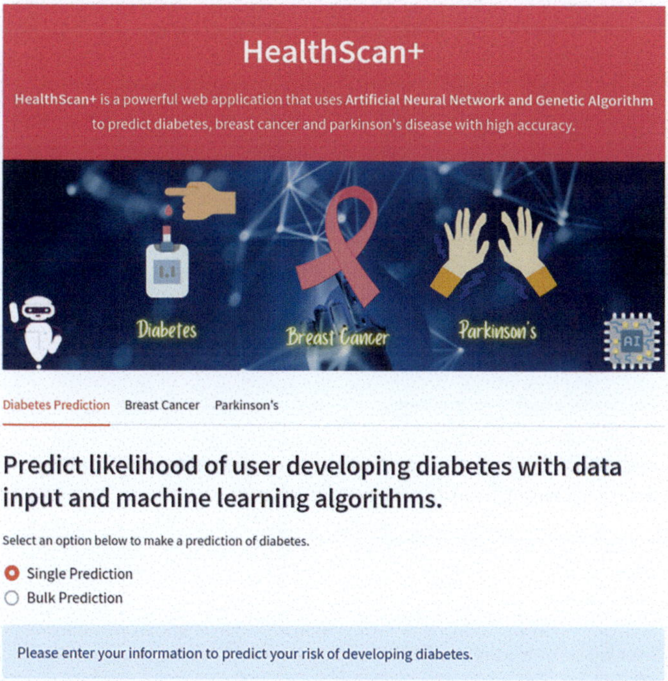

Fig. 14.8 GUI to upload a single record for prediction of diabetes

Fig. 14.9 GUI to upload multiple records for prediction of diabetes

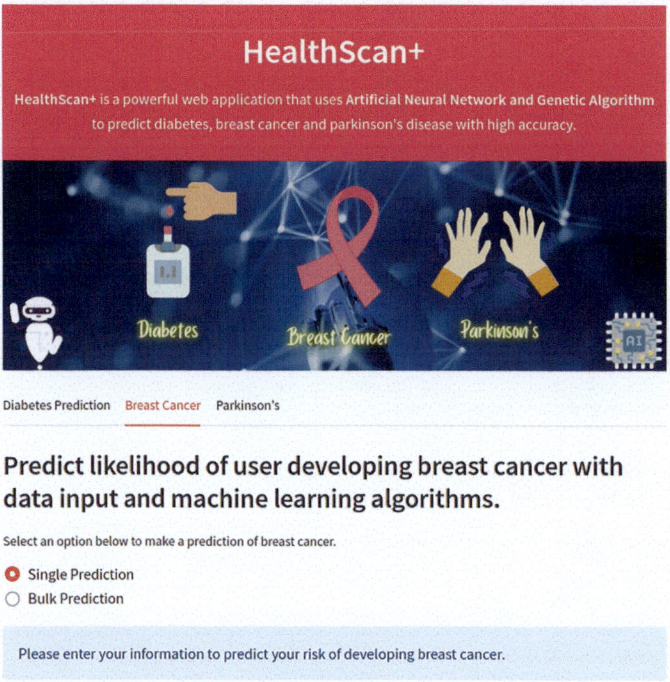

Fig. 14.10 GUI to upload a single record for prediction of breast cancer

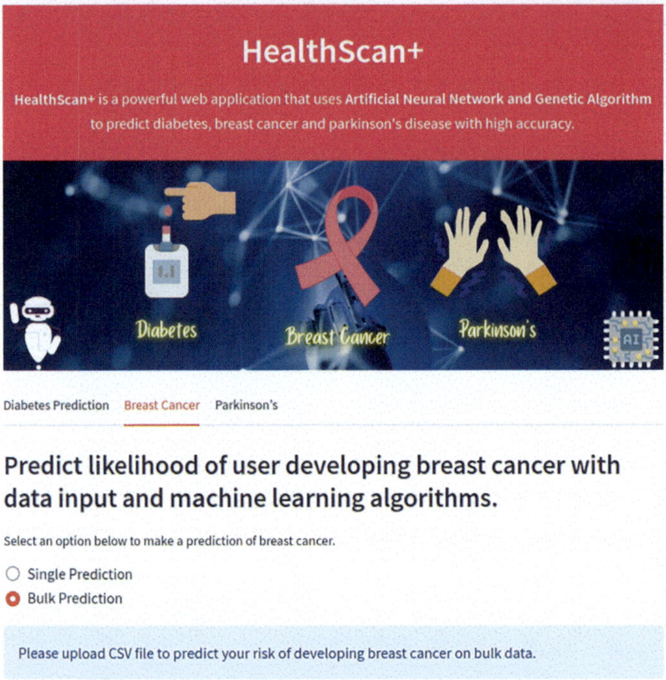

Fig. 14.11 GUI to upload multiple records for prediction of breast cancer

Fig. 14.12 GUI to upload a single record for prediction of PD

Fig. 14.13 GUI to upload multiple records for prediction of PD

14.7 Conclusion

This research employs GA and BPNN to construct reliable, accurate, and scalable CDSS for the prediction of three diseases: diabetes, Breast Cancer, and Parkinson's Disease (PD). The proposed models achieved promising results in predicting the diseases with excellent accuracy. Compared to the results provided by GA-BPNN alone with some redundant inputs, those produced by the hybrid GA-BPNN with relevant inputs led to further improved categorization accuracy. The model proposed during this research work outperformed the baseline models. For the PIMA diabetes dataset, SF_OBPNN yielded the highest accuracy (92.3%) and precision (90.9%) when compared with the other three models (AF_BPNN, AF_OBPNN, SF_BPNN). For Breast cancer also, SF_OBPNN yielded the highest accuracy (98.5%) and precision (97.0%) when compared with the other three models (AF_BPNN, AF_OBPNN, SF_BPNN). Similarly, the SF_OBPNN model yielded the highest accuracy (98.1%) and precision (95.8%) for PD. This proves that applying GA for feature selection and optimization improves the model's performance. For all three diseases (Diabetes, Breast Cancer, and PD), Selected Features with Optimized Neural Network (SF_OBPNN) yielded the highest accuracy. This application can be further used in other diseases also. Future work will explore datasets from different sources to improve accuracy and experiment with new techniques for better results.

References

Ahmad F et al (2013) Intelligent breast cancer diagnosis using hybrid GA-ANN. 2013 Fifth international conference on computational intelligence, communication systems and networks, IEEE, pp 9–12

Alharbi A, Alghahtani M (2019) Using genetic algorithm and ELM neural networks for feature extraction and classification of type 2-diabetes mellitus. Appl Artif Intell 33(4):311–328

Amma NG (2012) Bhuvaneswari. Cardiovascular disease prediction system using genetic algorithm and neural network. 2012 International conference on computing, communication and applications, pp 1–5

Arabasadi Z et al (2017) Computer aided decision making for heart disease detection using hybrid neural network-genetic algorithm. Comput Methods Programs Biomed 141:19–26

Arabgol S, Ko HS (2013) Application of artificial neural network and genetic algorithm to healthcarewaste prediction. J Artif Intell Soft Comput Res 3(4)

Asmae O et al (2020) Parkinson's disease identification using KNN and ANN algorithms based on voice disorder. 2020 1st international conference on innovative research in applied science, engineering and technology (IRASET), IEEE, pp 1–6

Breast Cancer. https://www.who.int/news-room/fact-sheets/detail/breast-cancer. Accessed 17 July 2023.

Choi H et al (2017) Refining diagnosis of parkinson's disease with deep learning-based interpretation of dopamine transporter imaging. NeuroImage: Clinical 16:586–94

Diabetes—India. https://www.who.int/india/health-topics/mobile-technology-for-preventing-ncds. Accessed 13 May 2023.

Farran B et al (2013) Predictive models to assess risk of type 2 diabetes, hypertension and comorbidity: machine-learning algorithms and validation using national health data from Kuwait—a cohort study. BMJ Open 3(5):e002457

Grover S et al (2018) Predicting severity of Parkinson's disease using deep learning. Procedia Comput Sci 132:1788–94

Hosseinpoor M (2022) Predicting gestational diabetes using an intelligent algorithm based on artificial neural network. J Modern Med Inform Sci 8(2):126–139

Hota HS (2013) Diagnosis of breast cancer using intelligent techniques. In J Emerg Sci Eng (IJESE) 1(3):9

Hota HS (2014) Identification of breast cancer using ensemble of support vector machine and decision tree with reduced feature subset. Blue eyes intelligence engineering & sciences Publication Pvt. Ltd. Int J Innov Technol Exploring Eng (IJITEE) 3(9):4

Huang C-Y et al (2009) Evaluating the process of a genetic algorithm to improve the back-propagation network: a monte carlo study. Expert Syst Appl 36(2), Part 1:1459–65

Karegowda A et al (2011) Application of genetic algorithm optimized neural network connection weights for medical diagnosis of PIMA indians diabetes. Int J Soft Comput (IJSC) 2

Lin C-H et al (2022) Early detection of Parkinson's disease by neural network models. IEEE Access 10:19033–44

Malakar S et al (2022) Computer based diagnosis of some chronic diseases: a medical journey of the last two decades. Arch Comput Methods Eng 29(7):5525–67

Mall PK et al (2022) Early warning signs of parkinson's disease prediction using machine learning technique. J Pharmaceutical Negative Results, 4784–92

Mansour RF (2018) Deep-learning-based automatic computer-aided diagnosis system for diabetic retinopathy. Biomed Eng Lett 8(1):41–57

Mitchell TM (1997) Machine learning. McGraw-Hill Science/Engineering/Math

Monirujjaman Khan M et al (2022) Machine learning based comparative analysis for breast cancer prediction. J Healthcare Eng, e4365855

Nilashi M et al (2022) Predicting Parkinson's disease progression: evaluation of ensemble methods in machine learning. J Healthcare Eng, e2793361

Nancy NRS (2022) "Diagnosis of Alzheimer's and Parkinson's disease using artificial neural network. Int J Scientific Technol Res 9:3659

Parkinson Disease. https://www.who.int/news-room/fact-sheets/detail/parkinson-disease. Accessed 13 May 2023.

Rajasekaran S, Pai GAV (2003) Neural networks. Fuzzy logic and genetic algorithms (synthesis and Applications), PHI

Sexton RS et al (1998) Toward global optimization of neural networks: a comparison of the genetic algorithm and backpropagation. Decision Support Syst 22(2):171–85

Templeton JM et al (2022) Classification of Parkinson's disease and its stages using machine learning. Scientific Reports 12(1):14036

Temurtas H et al (2009) A comparative study on diabetes disease diagnosis using neural networks. Expert Syst Appl 36(4):8610–15

Übeyli ED (2007) Implementing automated diagnostic systems for breast cancer detection. Expert Syst Appl 33(4):1054–1062

Veena VV, Anjali C (2015) Prediction and diagnosis of diabetes mellitus—a machine learning approach. 2015 IEEE recent advances in intelligent computational systems (RAICS), IEEE, pp 122–27

Yahyaoui A et al (2019) A decision support system for diabetes prediction using machine learning and deep learning techniques. 2019 1st international informatics and software engineering conference (UBMYK), IEEE, pp 1–4

Chapter 15
A Machine Learning Approach for Forecasting Energy Use in the Transportation Sector of the USA

Rabin K. Jana and Bidushi Chakraborty

Abstract Transportation is an important sector for any country and its economy. The sector uses energy from different available sources. Due to the extremely high volatility in energy demand, predicting energy use patterns from various sources is difficult. This study considers three major sources of energy used in the transportation sector of the USA. They are Natural Gas (NG), Petroleum (PT), and Fossil Fuel (FF). We perform a predictive modeling exercise based on monthly data collected from January 1973 to August 2023. We use Extreme Gradient Boosting (XGBoost), an ensemble Machine Learning (ML) approach. The results indicate that the consumption of NG is much less compared to FF. The consumption of NG started to increase during the last four years. Among the three sources, energy used from FF is the most predictable, followed by PT and NG. The findings of this study will be useful for practitioners and investors in the field of energy used in the transportation sector of the USA.

Keywords Machine learning · XGBoost · Predictive modeling · Energy

15.1 Introduction

Energy consumption in any sector is directly linked to greenhouse gas emissions, sustainability, and economic growth (Gorus and Aydin 2019). Understanding the consumption patterns and the evolution dynamics of various energy sources is crucial for all sectors (Elavarasan et al. 2023). Demands for such sources of energy are highly volatile. Policymakers' role is pivotal in regulating consumption patterns. These factors immensely impact the evolutionary dynamics and the future movements of each energy source (Su et al. 2023a). The importance and magnitude of this topic,

R. K. Jana (✉)
Indian Institute of Management Raipur, Atal Nagar, Raipur, CG 493661, India
e-mail: rkjana@iimraipur.ac.in

B. Chakraborty
Institute of Engineering and Management, Kolkata, WB 70091, India

D. K. Sharma et al. (eds.), *Machine Learning for Real World Applications*, Transactions on Computer Systems and Networks, https://doi.org/10.1007/978-981-97-1900-6_15

owing to environmental protection and sustainability, has attracted researchers and practitioners around the globe (Jana et al. 2022; Ma et al. 2023; Su et al. 2023b). Studying the present state of use and examining the governing patterns of various energy sources is essential for framing guidelines for sustainable uses. Therefore, exploring behavioral dynamics and predictive energy consumption modeling is truly important and needed.

All concerns mentioned above are valid for energy used from various sources related to the transportation sector (Sharma and Jana 2009; Chukwu et al. 2015). Transportation is a huge sector that has a relationship with almost all other sectors (Saboori et al. 2014; Raza et al. 2019). Researchers studied this field from diverse angles (Zhu et al. 2020). They reported a few primary energy sources used in the transportation sector (Ortmeyer and Pillay 2001; Ong et al. 2012). The primary sources are Petroleum (PT), Fossil Fuel (FF), and Natural Gas (NG). The US economy is the largest in the world. Therefore, in this research, we selected the transportation sector of the USA to study the temporal patterns and the predictive modeling. These two aspects of the transportation sector of the USA are not explored much.

Econometric techniques-based predictive modeling approaches are prevalent. These techniques have been criticized for several rigid assumptions and satisfaction of different conditions (Suganthi and Samuel 2012). Over the years, these techniques have been put aside to the other side and gradually become a lesser choice option. Contrary to this, Machine learning (ML)-driven predictive modeling techniques were found to replace the econometric-based approaches due to their capability of handling large and complex datasets, having fewer or no stringent assumptions, and performing fast and efficient computation. The literature reports diverse applications of ML-driven predictive modeling approaches (Jana et al. 2023, 2020; Ghosh and Jana 2023; Ghosh et al. 2023, 2020; Jana and Ghosh 2022). We utilized Boruta and Extreme Gradient Boosting (XGBoost) algorithms to achieve our research endeavors (Kursa and Rudnicki 2010; Chen and Carlos 2016).

Through this chapter, we mainly contribute in two ways. First, we study the evolutionary patterns of energy sources used in the transportation sector of the USA. In doing so, we found that the NG must be used more for the transportation sector to reduce the adverse environmental impacts. There is a significant gap in the use of NG compared to FF and PT. A lack of infrastructural support may be one of the causes of less use in this sector. An appropriate policy may be framed to intensify the use of NG in the said sector. Second, we explored the predictability of the leading energy sources for the transportation sector through an ensemble ML approach. We found that the time series corresponding to FF and PT are more predictable than NG. This essentially indicates the instability of the NG supply. This study will encourage researchers to apply ML algorithms in domains that have not been explored yet.

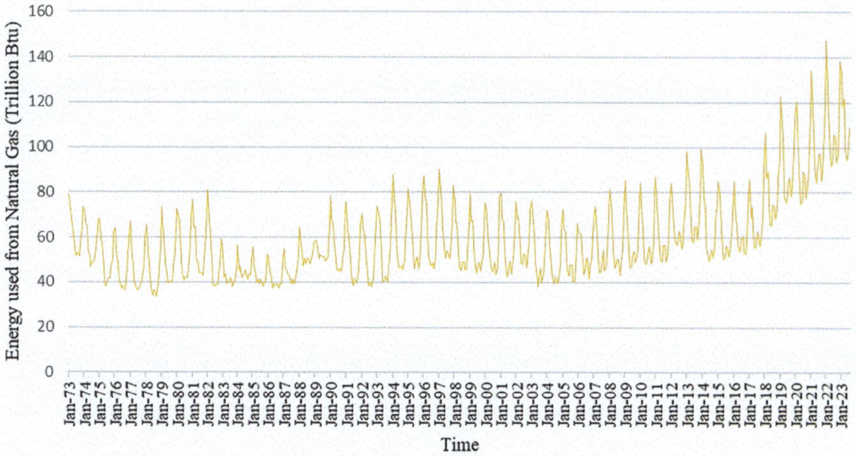

Fig. 15.1 Energy used from NG in the transportation sector (Trillion Btu)

15.2 Data and Descriptive Statistics

We collected data from the official website of the US Energy Information Administration (https://www.eia.gov/[1]). Monthly data are available from the repository. The data spans over more than half of a century. The start and end months are January 1973 and August 2023, respectively. Figures 15.1–15.3 portray the temporal patterns of NG, PT, and FF.

From Figs. 15.1, 15.2, 15.3, we found a highly non-linear behavior in the temporal patterns of all three time series representing the energy used in the transportation sector of the USA. Together with the non-linear patterns, NG also showed a cyclical behavior. It is interesting to note that there are similarities in the temporal movements of PT and FF. The only difference is that the volume of energy consumed from FF is more than PT. Consumption of NG in the transportation sector has started to increase from 2018–2019. On the contrary, NG consumption was the lowest from 1983 to 1990. Consumption of PT and FF reached their peak during 2008–2009. After that, PT and FF consumption started to decrease. Consumption was the least during the COVID-19 pandemic. The transportation sector witnessed severe restrictions during the major span of the COVID-19 pandemic.

We obtained the descriptive statistics of all the time series to gain further insights. Table 15.1 presents the descriptive statistics values.

The descriptive statistics values in Table 15.1 validate the information depicted by the temporal patterns. The transportation sector chose FF as the primary energy source, followed by PT. NG consumption was the least. The time series data represented by NG, PT, and FF are significantly non-normal. Both the normality test results strongly reject the null hypothesis that assumes that the time series under

[1] Accessed on October 15, 2023.

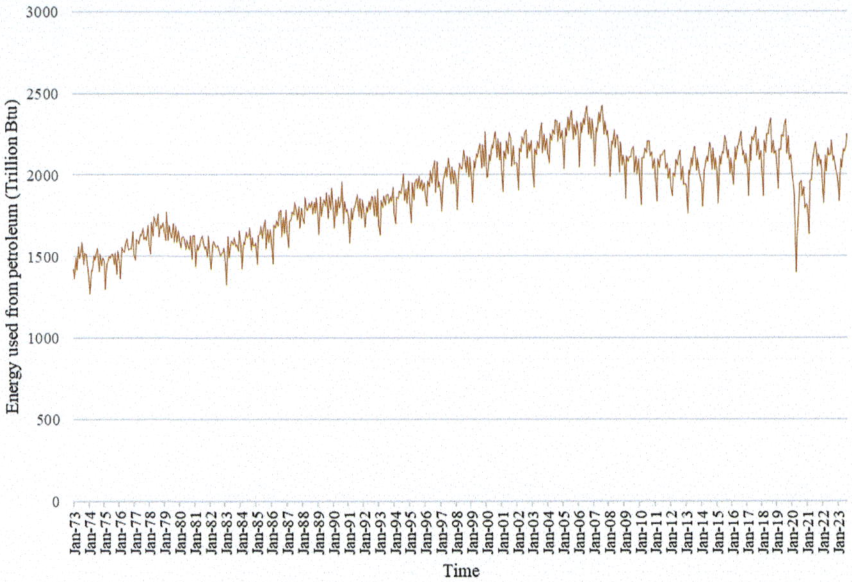

Fig. 15.2 Energy used from PT in the transportation sector (Trillion Btu)

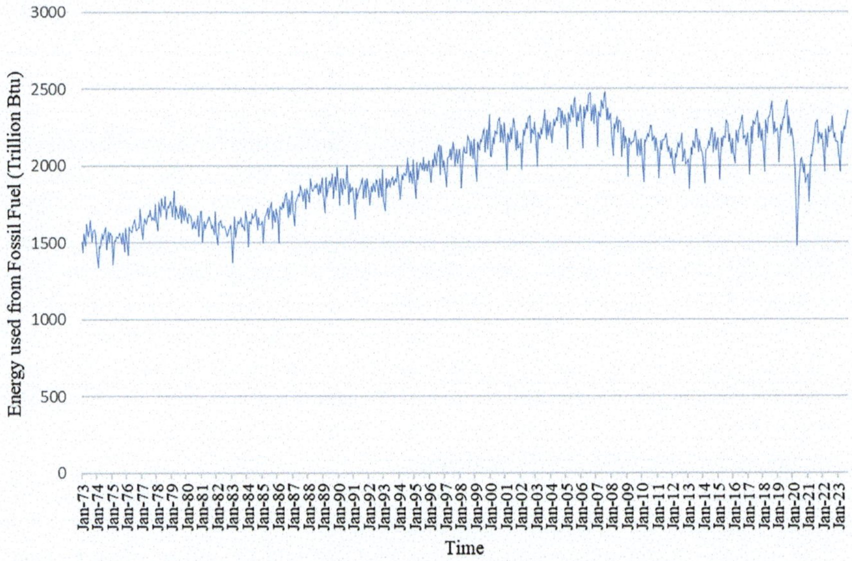

Fig. 15.3 Energy used from FF in the transportation sector (Trillion Btu)

Table 15.1 Descriptive statistics

Measure	NG	PT	FF
Mean	59.71	1904.96	1964.68
Std.Dev	19.06	262.36	266.88
Min	33.91	1272.63	1338.91
Q1	46.07	1688.00	1736.00
Median	54.21	1937.54	2008.00
Q3	68.83	2124.00	2197.00
Max	148.05	2422.55	2476.99
MAD	15.72	312.75	323.68
Range	114.14	1149.92	1138.08
SE	0.77	10.64	10.82
Skewness	1.45	-0.20	-0.23
Kurtosis	2.37	-1.08	-1.13
Jarque–Bera Test	344.12***	31.31***	35.72***
Shapiro–Wilk Test	0.88***	0.96***	0.95***
Teraesvirta NN Test	2.56#	12.36***	18.13***

NN: Neural Network, ***: significant at 1% level, #: Not significant.

consideration are normal. We observed an interesting fact through the Teraesvirta NN Test. It offers a significant test statistic value for PT and FF, implying that these two time series are highly non-linear. Unfortunately, the NN test result contradicts the findings of the temporal behavioral patterns for the NG time series. The test statistic value turned out to be insignificant for NG.

15.3 Methodology

Machine learning algorithms are powerful computational techniques. The literature reported several applications of such techniques (Jana et al. 2023; Ghosh and Jana 2023). Ensemble ML algorithms have proven more potent in forecasting, classification, and other tasks (Ghosh et al. 2023, 2020; Jana and Ghosh 2022). This research utilizes the XGBoost ML algorithm to predict the future movements of the NG, PT, and FF.

XGBoost stands for Extreme Gradient Boosting. It is a robust ensemble ML algorithm belonging to the gradient-boosting family (Chen and Carlos 2016). The algorithm prevents the over-fitting problem of other boosting algorithms by performing row and column samplings. By doing this, XGBoost incorporates a regularization quotient with the objective function. It is designed based on the concept of distributed and parallel computing, making the algorithm computationally efficient and powerful. The XGBoost algorithm can briefly be introduced as follows:

For a dataset having m features and n examples, let us denote.

$P = \{(x_i, y_i)\}$ such that $|P| = n, x_i \in \mathcal{R}^m, y_i \in \mathcal{R}$.

If K is the number of additive functions used to forecast the output of an ensemble tree model, then

$$\overline{y}_i = \int_{k=1}^{K} f_k(x_i), f_k \in F \tag{15.1}$$

where F is the space for the regression tree. Then,

$$F = f(x) = w_q(x)(q : R^m \to T, w_q \in R^T), \tag{15.2}$$

where q denotes the tree structure with T number of leaves in the tree, $f_k(.)$ is a function representing the tree structure q with leaf weight w. The objective function to reduce the errors in the XGBoost is formed as follows:

$$F^t = \sum_{i=1}^{n} l\{\left(y_i, \overline{y}_j^{(t-1)}\right) + f_t(x_i)\} + \Phi(f_k) \tag{15.3}$$

where l helps to calculate the error between the observed and predicted values, Φ denotes the complexity penalized with the regression trees. Now,

$$\Phi(f_k) = \mu T + \frac{1}{2}\lambda||w||^2 \tag{15.4}$$

where w is the score vector, μ is the required minimum loss, and the regularization function is λ. The Taylor series expansion leads to the following:

$$F = \sum_{i=1}^{n} \left[\frac{1}{2}(f_t(x_i))^2 h_i + f_t(x_i) g_i\right] + \mu T + \frac{\mu}{2}\sum_{j=1}^{T} w_j^2 \tag{15.5}$$

where h_i and g_i

Along with the XGBoost, we also utilized the Boruta feature selection algorithm (Kursa and Rudnicki 2010) to identify the contributions of the technical indicators-based features set. These features are now popularly used to extract hidden information from the time series under consideration (Ghosh et al. 2019). The features used are summarized in Table 15.2. The existing literature forms the basis for selecting these features.

15.4 Results and Discussion

The main results of this study are presented in this section. We used 18 features for the predictive modeling exercise. The importance of features for individual time series NG, PT, and FF are presented next. Figures 15.4, 15.5, 15.6 present the findings of

Table 15.2 List of features

Sl. No	Feature
1	One-month apart value (LAG1)
2	Two-months apart value (LAG2)
3	Three-months apart value (LAG3)
4	Four-months apart value (LAG4)
5	Five-months apart value (LAG5)
6	Moving average of cycle 5 (MA5)
7	Moving average of cycle 10 (MA10)
8	Five-month bias (B5)
9	Ten-month bias (B10)
10	Five-month momentum (MTM5)
11	Ten-month momentum (MTM10)
12	Five-month rate of change (ROC5)
13	Ten-month rate of change (ROC10)
14	Exponential MA of five-month (EMA5)
15	Exponential MA of ten-month (EMA10)
16	Five-month relative strength index (RSI5)
17	Relative strength index of ten-month (RSI10)
18	William indicator (WMS)

the Boruta feature selection algorithm. The figures containing box plots with green color indicate that the corresponding features are significant in determining future values.

Figure 15.4 shows that the box plots of all 18 features are green. Therefore, they all are significant in determining the future movements of NG. LAG1, RSI10, and B5 are the top three features for NG. RSI5, EMA10, and MA10 are the top three features in terms of their importance scores for PT. RSI5, MA10, and MA5 appeared as the top three features for FF. As all 18 features are significant in determining the future movements of NG, PT, and FF, we carried them all and performed the predictive modeling exercise.

Quality of prediction is the most important concern in predictive modeling. We considered three traditional measures, MAE, RMSE, and MAPE, and two advanced measures, Nash–Sutcliffe Efficiency (NSE) and Index of Agreement (IA). For better prediction accuracy, the first three measures should preferably have values close to zero. On the other hand, NSE and IA values should be tending to 1 for the superior prediction. We performed the predictive modeling exercise using the 'XGBoost' package of '*R*' with the default parameter setting. 85% of the data have been used to train the model, and the rest 15% have been used to test it. We separately reported the predictive performance for test, train, and the full sample datasets in Tables 15.3–15.5.

Table 15.3 shows that the first three measures are much larger than desired. So, the predictive accuracy may appear very low. This may not always be the case. The

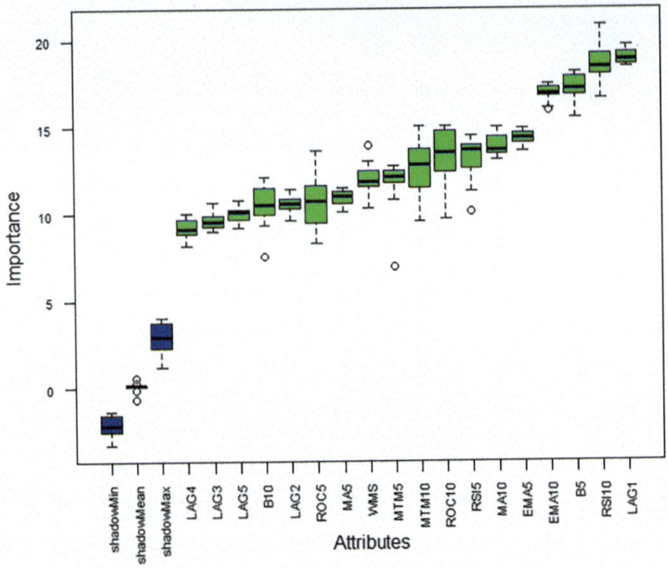

Fig. 15.4 Feature importance for NG

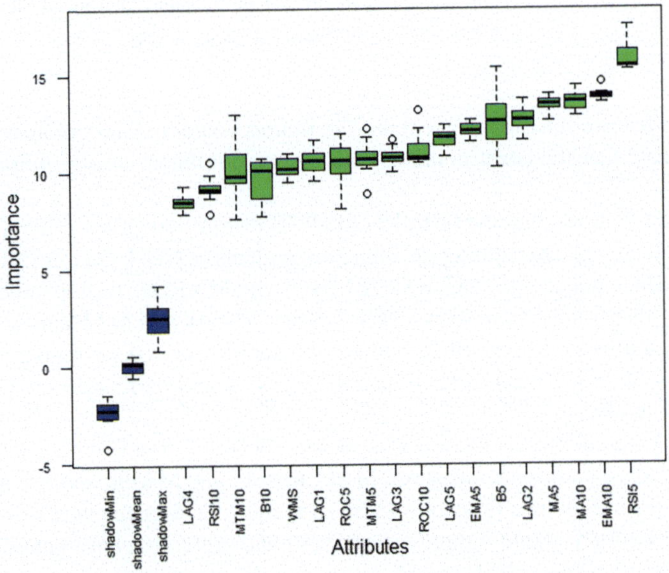

Fig. 15.5 Feature importance for PT

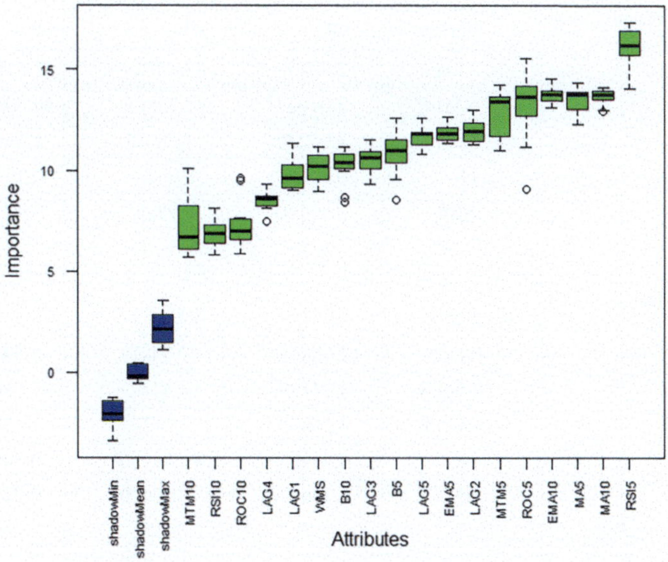

Fig. 15.6 Feature importance for FF

Table 15.3 Results for the test dataset

Series	NG	PT	FF
MAE	4.32668	98.19942	55.92735
RMES	5.91940	120.55696	74.00053
MAPE	7.15895	5.05088	2.97253
IA	0.99766	0.99897	0.99965
NSE	0.99093	0.99606	0.99861

volume of use of FF and PT in the transportation sector is much higher than NG. Therefore, RMSE, MAPE, and MAE values are higher in PT and FF than in NG. NSE and IA can address this limitation and provide a better picture. Both indexes tend to be 1 for all the time series. In particular, the maximum NSE (0.99965) and IA (0.99861) values correspond to FF. They occupied the second spot for PT, and the values are the lowest for NG. Therefore, we conclude that the future movements of FF are better predictable compared to PT and NG, as its predictive accuracy is the highest. The predictive accuracy of NG is the lowest compared to the other two. However, it may also be noted that its corresponding NSE and IA values are 0.99093 and 0.99766, respectively. This implies that although the predictive performance of NG is the lowest among the three, it is also highly predictable. So, in summary, we may say that the future movements of all three time series representing the energy used for the transportation sector of the USA are highly predictable based on the test dataset.

Table 15.4 Results for the training dataset

Series	NG	PT	FF
MAE	2.30853	33.13748	31.84803
RMES	3.07627	42.75967	40.90991
MAPE	3.95230	1.78924	1.65795
IA	0.99940	0.99988	0.99989
NSE	0.99760	0.99951	0.99958

Table 15.5 Results for the full sample

Series	NG	PT	FF
MAE	2.68345	39.52819	35.39741
RMES	3.72159	54.44323	47.26644
MAPE	4.60858	2.15162	1.85172
IA	0.99912	0.99980	0.99986
NSE	0.996476	0.99920	0.99944

We further investigated the results presented in Tables 15.4 and 15.5. Table 15.4 shows the results based on the training dataset. Table 15.5 presents the results based on the full dataset. We found a very similar trend to that of the test dataset. Like the test dataset, the predictability of FF is the best for the training dataset, followed by PT and NG. The NSE and IA values are obtained as 0.99861 and 0.99989 for FF. We arrived at the same conclusion for the full dataset. For the series FF, the NSE and IA values are obtained as 0.99944 and 0.99986, respectively. Therefore, the future values of all three time series related to the transportation sector of the USA are highly predictable based on the test, train, and full datasets. Finally, we displayed the actual and predicted NG, PT, and FF values in Figs. 15.7, 15.8, 15.9.

Figures 15.7–15.9 present the observed and forecasted values of all three energy sources for the US transportation network. These figures support the results obtained from the predictive modeling.

15.5 Conclusions

This study explored the behavioral dynamics and predictability of energy used for the transportation sector of the USA. We selected three primary sources, NG, PT, and FF, and analyzed the dataset spanning over 50 years. Our analysis revealed that the time series corresponding to all the energy sources exhibits a non-normal and highly non-linear pattern. We observed that FF is still the most popular and number one energy source for the transportation sector. There is a significant gap between the use of NG and FF. Recently, after the COVID-19 pandemic, NG consumption started to gain momentum compared to many previous decades. More NG must be used in the transportation sector to reduce carbon emissions.

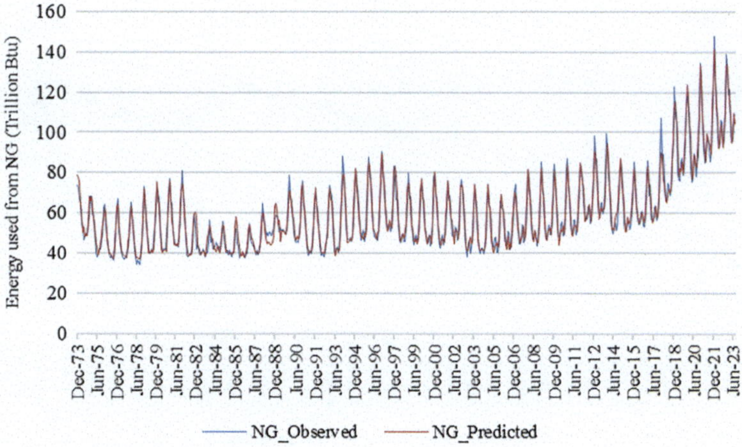

Fig. 15.7 NG: Observed vs. forecasted

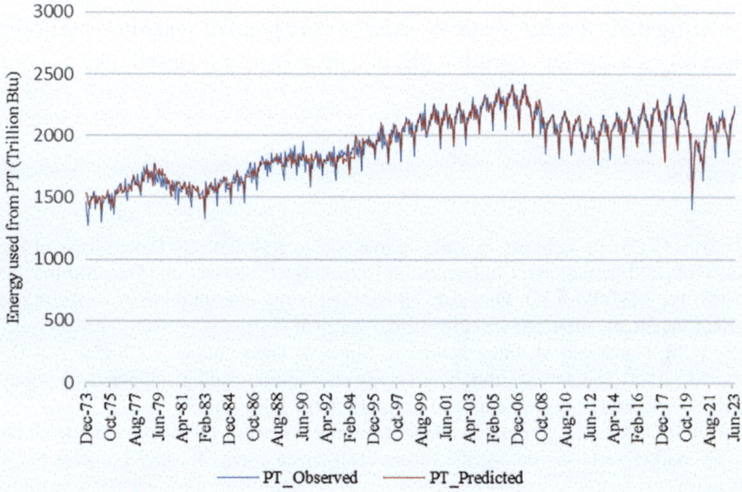

Fig. 15.8 PT: Observed vs. forecasted

We also examined the future movements of these energy sources using XGBoost, an ensemble ML algorithm. In the predictive modeling exercise, we used 18 technical indicators as features. All the features appeared to be significant. We found that the time series corresponding to FF is easier to predict than PT and NG. This finding was supported by test, train, and full datasets. The predictability of the PT and NG time series is also very high. One of the significant challenges of this work was the availability of data. Only monthly consumptions are available. That leads to a small dataset for the analysis using an ML algorithm. The present work may be extended

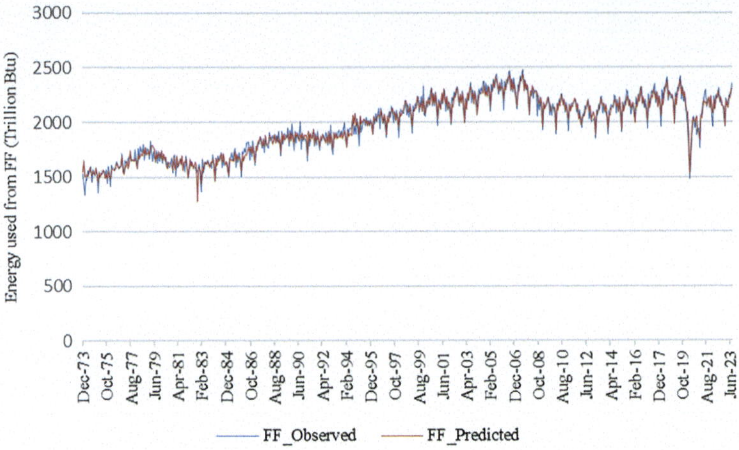

Fig. 15.9 FF: Observed vs. forecasted

by considering daily data, if available. Also, a comparative study may be performed by considering a series of ensemble ML and deep learning algorithms.

References

Chen T, Carlos G (2016) Xgboost: A scalable tree boosting system. In: Proceedings of the 22nd ACM SIGKDD International Conference on Knowledge Discovery and Data Mining 785–794

Chukwu PU, Isa AH, Ojosu JO, Olayande JS (2015) Energy consumption in transport sector in Nigeria: Current situation and ways forward. Energy 5(1)

Elavarasan R M, Nadarajah M, Pugazhendhi R, Sinha A, Gangatharan S, Chiaramonti D, Abou Houran M (2023) The untold subtlety of energy consumption and its influence on policy drive towards Sustainable Development Goal 7. Appl Energy 334, 120698

Ghosh I, Jana RK (2023) A granular machine learning framework for forecasting high-frequency financial market variables during the recent black swan event. Technol Forecast Soc Chang 194:122719

Ghosh I, Jana RK, Sanyal MK (2019) Analysis of temporal pattern, causal interaction and predictive modeling of financial markets using non-linear dynamics, econometric models and machine learning algorithms. Appl Soft Comput 82:105553

Ghosh I, Jana RK, Abedin MZ (2023) An ensemble machine learning framework for Airbnb rental price modeling without using amenity-driven features. Int J Contemp Hosp Manag 35(10):3592–3611

Ghosh I, Sanyal MK, Jana RK (2020) An ensemble of ensembles framework for predictive analytics of commodity market. In: 2020 4th International Conference on Computational Intelligence and Networks (CINE) (pp. 1–6). IEEE, Singapore

Gorus MS, Aydin M (2019) The relationship between energy consumption, economic growth, and CO2 emission in MENA countries: Causality analysis in the frequency domain. Energy 168:815–822

Jana RK, Ghosh I, Sanyal MK (2020) A granular deep learning approach for predicting energy consumption. Appl Soft Comput 89:106091

Jana RK, Ghosh I, Wallin MW (2022) Taming energy and electronic waste generation in bitcoin mining: Insights from Facebook prophet and deep neural network. Technol Forecast Soc Chang 178:121584

Jana RK, Ghosh I (2022) A residual driven ensemble machine learning approach for forecasting natural gas prices: Analyses for pre-and during-COVID-19 phases. Ann Oper Res 1–22

Jana RK, Ghosh I, Kumar PR (2023) An explainable AI-enabled granular ensemble machine learning framework to demystify fertilizer price movements. J Oper Res Soc 1–18

Kursa MB, Rudnicki W (2010) Feature Selection with the Boruta Package. J Stat Softw 36:1–13

Ma B, Sharif A, Bashir M, Bashir MF (2023) The dynamic influence of energy consumption, fiscal policy and green innovation on environmental degradation in BRICST economies. Energy Policy 183:113823

Ong HC, Mahlia TMI, Masjuki HH (2012) A review on energy pattern and policy for transportation sector in Malaysia. Renew Sustain Energy Rev 16(1):532–542

Ortmeyer TH, Pillay P (2001) Trends in transportation sector technology energy use and greenhouse gas emissions. Proc IEEE 89(12):1837–1847

Raza SA, Shah N, Sharif A (2019) Time frequency relationship between energy consumption, economic growth and environmental degradation in the United States: Evidence from transportation sector. Energy 173:706–720

Saboori B, Sapri M, bin Baba M (2014) Economic growth, energy consumption and CO2 emissions in OECD (Organization for Economic Co-operation and Development)'s transport sector: A fully modified bi-directional relationship approach. Energy 66, 150–161

Sharma DK, Jana RK (2009) A hybrid genetic algorithm model for transshipment management decisions. Int J Prod Econ 122(2):703–713

Su F, Chang J, Li X, Fahad S, Ozturk I (2023a) Assessment of diverse energy consumption structure and social capital: a case of southern Shaanxi province China. Energy 262:125506

Su S, Qamruzzaman M, Karim S (2023b) Charting a sustainable future: the impact of economic policy, environmental taxation, innovation, and natural resources on clean energy consumption. Sustainability 15(18):13585

Suganthi L, Samuel AA (2012) Energy models for demand forecasting—A review. Renew Sustain Energy Rev 16(2):1223–1240

Zhu Q, Li X, Li F, Wu J, Zhou D (2020) Energy and environmental efficiency of China's transportation sectors under the constraints of energy consumption and environmental pollutions. Energy Econ 89:104817